AF358838

In the Arms of Morpheus: Recent Advances in Dreaming and in Other Sleep-Related Metacognitions

In the Arms of Morpheus: Recent Advances in Dreaming and in Other Sleep-Related Metacognitions

Guest Editors

Sergio A. Mota-Rolim
Brigitte Holzinger
Michael Nadorff
Luigi De Gennaro

Basel • Beijing • Wuhan • Barcelona • Belgrade • Novi Sad • Cluj • Manchester

Guest Editors

Sergio A. Mota-Rolim
Brain Institute
Federal University of Rio
Grande do Norte
Natal
Brazil

Brigitte Holzinger
Sleep Coaching
Medical University Vienna
Vienna
Austria

Michael Nadorff
Department of Psychology
Mississippi State University
Starkville
United States

Luigi De Gennaro
Department of Psychology
Sapienza University of Roma
Rome
Italy

Editorial Office
MDPI AG
Grosspeteranlage 5
4052 Basel, Switzerland

This is a reprint of the Special Issue, published open access by the journal *Brain Sciences* (ISSN 2076-3425), freely accessible at: www.mdpi.com/journal/brainsci/special_issues/S9B171AEY1.

For citation purposes, cite each article independently as indicated on the article page online and using the guide below:

Lastname, A.A.; Lastname, B.B. Article Title. *Journal Name* **Year**, *Volume Number*, Page Range.

ISBN 978-3-7258-3150-0 (Hbk)
ISBN 978-3-7258-3149-4 (PDF)
https://doi.org/10.3390/books978-3-7258-3149-4

Contents

Sérgio Mota-Rolim, Brigitte Holzinger, Michael R. Nadorff and Luigi De Gennaro
In the Arms of Morpheus: Recent Advances in Dreaming and in Other Sleep-Related Metacognitions
Reprinted from: *Brain Sci.* **2024**, *14*, 1017, https://doi.org/10.3390/brainsci14101017 1

J. F. Pagel
The Persistent Paradox of Rapid Eye Movement Sleep (REMS): Brain Waves and Dreaming
Reprinted from: *Brain Sci.* **2024**, *14*, 622, https://doi.org/10.3390/brainsci14070622 5

Anna C. van der Heijden, Jade Thevis, Jill Verhaegen and Lucia M. Talamini
Sensational Dreams: The Prevalence of Sensory Experiences in Dreaming
Reprinted from: *Brain Sci.* **2024**, *14*, 533, https://doi.org/10.3390/brainsci14060533 22

Gianluca Ficca, Oreste De Rosa, Davide Giangrande, Tommaso Mazzei, Salvatore Marzolo and Benedetta Albinni et al.
Quantitative–Qualitative Assessment of Dream Reports in Schizophrenia and Their Correlations with Illness Severity
Reprinted from: *Brain Sci.* **2024**, *14*, 568, https://doi.org/10.3390/brainsci14060568 34

Simon Kempe, Werner Köpp and Lutz Wittmann
Personality Functioning Improvement during Psychotherapy Is Associated with an Enhanced Capacity for Affect Regulation in Dreams: A Preliminary Study
Reprinted from: *Brain Sci.* **2024**, *14*, 489, https://doi.org/10.3390/brainsci14050489 49

Emma Desjardins, Lina Gaber, Emily Larkin, Antoine Benoit, Addo Boafo and Joseph De Koninck
The Dream Experience and Its Relationship with Morning Mood in Adolescents Hospitalized after a Suicide Attempt
Reprinted from: *Brain Sci.* **2024**, *14*, 804, https://doi.org/10.3390/brainsci14080804 62

Milena Camaioni, Serena Scarpelli, Valentina Alfonsi, Maurizio Gorgoni, Rossana Calzolari and Mina De Bartolo et al.
How COVID-19 Affected Sleep Talking Episodes, Sleep and Dreams?
Reprinted from: *Brain Sci.* **2024**, *14*, 486, https://doi.org/10.3390/brainsci14050486 83

Elena Gerhardt and Benjamin Baird
Frequent Lucid Dreaming Is Associated with Meditation Practice Styles, Meta-Awareness, and Trait Mindfulness
Reprinted from: *Brain Sci.* **2024**, *14*, 496, https://doi.org/10.3390/brainsci14050496 96

Abel A. Oldoni, André D. Bacchi, Fúlvio R. Mendes, Paula A. Tiba and Sérgio Mota-Rolim
Neuropsychopharmacological Induction of (Lucid) Dreams: A Narrative Review
Reprinted from: *Brain Sci.* **2024**, *14*, 426, https://doi.org/10.3390/brainsci14050426 117

Don Kuiken
The Epistemic Limits of Impactful Dreams: Metacognition, Metaphoricity, and Sublime Feeling
Reprinted from: *Brain Sci.* **2024**, *14*, 528, https://doi.org/10.3390/brainsci14060528 139

Ajar Diushekeeva, Santiago Hidalgo and Antonio Zadra
Impact of Pre-Sleep Visual Media Exposure on Dreams: A Scoping Review
Reprinted from: *Brain Sci.* **2024**, *14*, 662, https://doi.org/10.3390/brainsci14070662 159

brain
sciences

MDPI

Editorial

In the Arms of Morpheus: Recent Advances in Dreaming and in Other Sleep-Related Metacognitions

Sérgio Mota-Rolim [1,*], **Brigitte Holzinger** [2], **Michael R. Nadorff** [3] **and Luigi De Gennaro** [4]

[1] Brain Institute, Federal University of Rio Grande do Norte, Natal 59078-970, Brazil
[2] Institute for Consciousness and Dream Research, 1180 Vienna, Austria
[3] Department of Psychology, Mississippi State University, Starkville, MS 39762, USA
[4] Department of Psychology, Sapienza University of Roma, 00185 Roma, Italy
[*] Correspondence: sergioarthuro@neuro.ufrn.br

Citation: Mota-Rolim, S.; Holzinger, B.; Nadorff, M.R.; De Gennaro, L. In the Arms of Morpheus: Recent Advances in Dreaming and in Other Sleep-Related Metacognitions. *Brain Sci.* **2024**, *14*, 1017. https://doi.org/10.3390/brainsci14101017

Received: 1 October 2024
Accepted: 8 October 2024
Published: 14 October 2024

Dreams have always fascinated humans. The first written references to dreaming come from Greek mythology: Nyx, the goddess of night, gave birth to Hypnos, who represented sleep. Hypnos had a twin brother called Thanatos, the god of death, and fathered Morpheus, the god of dreams. In addition, most religions, such as the Abrahamic monotheisms—Judaism, Christianity, and Islam—recognize dreams as a way to communicate with God to understand the present and predict the future. Thus, ancient civilizations tend to interpret dreams as supernatural phenomena [1].

This began to change with Freud. In 1900, his famous book *"The Interpretation of Dreams"* was the first attempt to understand dreams as a natural process of the brain [2]. Nowadays, dreams are currently defined as any mental activity that occurs during sleep. More specifically, dreams are sense–perceptual, emotional, and cognitive experiences that are more frequently recalled after awakening from rapid eye movement (REM) sleep [3–7]. However, dreams can also occur during non-REM (NREM) sleep stages, with specific characteristics. During sleep onset (N1), dreams tend to be fast, and appear as unspecific images and sounds, or strange ideas. Dreams during light sleep (N2) are more related to the thoughts and memories of the previous day(s). During deep sleep (N3), dreams are remembered less often [8–14].

Despite all the significant scientific advances in dream research in the last century, we still do not know why we dream, and why some people remember dreams every day, while others only rarely. Modern theories such as the continuity hypothesis [15], the threat-simulation theory [16], the activation hypothesis [17,18], and neuropsychoanalysis [5] further our understanding of dreaming, but they cannot explain all oneiric features.

For this Special Issue, we invited experts in dream research to submit original research articles and review papers. We aimed to provoke and articulate ideas to foster a broad discussion on dream research, and to convey the challenges and misunderstandings in this complex research field. In addition to stimulating the submission of neuroscientific studies, we also encouraged philosophical works, because we believe that this interaction between biomedical and social sciences is fundamental to fostering our understanding of dreams and human consciousness. In all, ten papers—including six original research articles and four review works—were published in this Special Issue.

As previously stated, typical dreams are more related to REM sleep. However, according to the comprehensive review of Pagel (contribution 1), there are still many controversies about this sleep state. REM dreams are longer, more bizarre, transformative, and remembered, but none of these characteristics are exclusive to REM sleep. Moreover, REM sleep is also known as paradoxical sleep, because it presents the intracranial electrical theta rhythm (5-8Hz), but this oscillation is the marker of REM sleep in non-human animals. Thus, REM sleep and dreaming continue as a paradox, and more work directly addressing the electrophysiology of REM sleep and the phenomenological features of REM dreams is necessary.

Dreams include perceptual experiences, but what are the main sensory modalities that appear in dreams? To clarify this issue, van der Heijden et al. (contribution 2) developed a dream diary with questions about sensory experiences during dreaming, and found that visual modality was the most common, followed by auditory and tactile ones. Olfactory and gustatory modalities had similar lowest rates. These results, at least in terms of sensorial and perceptual modalities, are in accordance with the continuity hypothesis [15], which postulates that dreaming is similar to the waking life.

Dreams are also relevant for clinical issues. In historical terms, the German philosophers Kant and Schopenhauer suggested that "a lunatic is a wakeful dreamer" and that "a dream is a short-lasting psychosis, and a psychosis is a long-lasting dream", respectively. Wundt, who was also one of the founders of experimental psychology, affirmed that "we can experience in dreams all the phenomena we find in the hospice". All of these authors influenced Freud, who postulated that psychosis is an abnormal intrusion of a dreaming activity into waking life. In this way, dreams can be seen as a model of psychosis, since both include internally generated perceptions and lack of rational judgment, probably due to the decreased activity of the frontal regions during REM sleep [19]. Ficca et al. (contribution 3) assessed the dream reports of subjects diagnosed with schizophrenia using both quantitative and qualitative methods and found correlations with illness severity.

Kempe et al. (contribution 4) observed, preliminarily, that the improvement in personality functioning during psychotherapy is associated with an increased ability to regulate the affects one experiences during dreams. Additionally, Desjardins et al. (contribution 5) investigated the relation between dreams and morning mood in adolescents hospitalized after a suicide attempt. Consistent with the previous literature [20], the authors found that suicidal adolescents had nightmares more frequently, and that their dreams had more negative moods, destructive themes, failures, and aggressions.

In the last few years, an increase in dreams' and nightmares' recall frequency was also observed due to the COVID-19 outbreak [21–23]. Camaioni et al. (contribution 6) investigated how the COVID-19 pandemic affected sleep talking and dreams, and found a higher frequency of sleep talking episodes during the pandemic compared to before the pandemic. In addition, the authors also observed that sleep talking episodes were associated with the emotional intensity of dreams, independent of the pandemic.

An interesting kind of dreaming that has been gaining more attention recently is lucid dreaming, when subjects know that they are dreaming during the dream, and may also have some degree of control over the oneiric plot [24]. However, the main challenge in lucid dream research is developing a reliable way to induce them, because they are rare for most people. Since lucid dreaming has similarities with the notion of mindfulness and meta-awareness, Gerhardt and Baird (contribution 7) investigated its association with meditation, and found that daily frequent meditators experience more lucid dreams than non-frequent meditators, and that meta-awareness is higher for meditators and weekly lucid dreamers. In addition to practicing meditation, another way to induce lucid dreams is trough exogenous substances. Oldoni et al. (contribution 8) reviewed natural plants and artificial drugs that increase metacognition, REM sleep, and/or dream recall. The authors found that the main candidates are substances that increase cholinergic and/or dopaminergic transmission, such as galantamine.

We also received an important philosophical work by Kuiken (contribution 9), who reviewed the epistemic significance of impactful dreams, dividing them in nightmares, existential dreams, and transcendent dreams. The author also defines a neo-Kantian account of "sublime feeling" as the cumulative effect of successive metaphoric/literal categorical transformations that produces a higher-level form of metacognition. Sublime feeling occurs as either sublime disquietude (existential dreams) or as sublime enthrallment (transcendent dreams).

Finally, one of the main features of contemporary urban societies is that everybody has a smartphone. In a scoping review, Diushekeeva et al. (contribution 10) investigated the impact of pre-sleep visual media exposure on dreams, and found that the range of stimulus-related incorporation was 3–43% for REM dream reports, and 4–30% for NREM dream reports. In addition to entering dreams, smartphone use is forcing people to sleep later, but subjects still need to go to school, university, or work early in the morning. Since REM sleep episodes tend to be longer in the second half of the night, modern humans are partly deprived of this sleep stage [25]. Recent works have shown that REM sleep and dreaming are associated with emotional regulation [26]; thus, we hypothesize that the contemporary increase in depression diagnostics may be related to this REM sleep deprivation. We also hypothesize that this emotional regulation failure caused by dream deprivation may also contribute to our insensibility in destroying the natural resources that constitute our planet, and that are fundamental for all living beings, including us humans. Future works are necessary to clarify this crucial issue.

Author Contributions: All authors contributed to writing this work. All authors have read and agreed to the published version of the manuscript.

Acknowledgments: We are grateful to the authors and reviewers who contributed to this Special Issue.

Conflicts of Interest: The authors declare no conflicts of interest.

List of Contributions

1. Pagel, J.F. The Persistent Paradox of Rapid Eye Movement Sleep (REMS): Brain Waves and Dreaming. *Brain Sci.* **2024**, *14*, 622. https://doi.org/10.3390/brainsci14070622.
2. van der Heijden, A.C.; Thevis, J.; Verhaegen, J.; Talamini, L.M. Sensational Dreams: The Prevalence of Sensory Experiences in Dreaming. *Brain Sci.* **2024**, *14*, 533. https://doi.org/10.3390/brainsci14060533.
3. Ficca, G.; De Rosa, O.; Giangrande, D.; Mazzei, T.; Marzolo, S.; Albinni, B.; Coppola, A.; Lustro, A.; Conte, F. Quantitative–Qualitative Assessment of Dream Reports in Schizophrenia and Their Correlations with Illness Severity. *Brain Sci.* **2024**, *14*, 568. https://doi.org/10.3390/brainsci14060568.
4. Kempe, S.; Köpp, W.; Wittmann, L. Personality Functioning Improvement during Psychotherapy Is Associated with an Enhanced Capacity for Affect Regulation in Dreams: A Preliminary Study. *Brain Sci.* **2024**, *14*, 489. https://doi.org/10.3390/brainsci14050489.
5. Desjardins, E.; Gaber, L.; Larkin, E.; Benoit, A.; Boafo, A.; De Koninck, J. The Dream Experience and Its Relationship with Morning Mood in Adolescents Hospitalized after a Suicide Attempt. *Brain Sci.* **2024**, *14*, 804. https://doi.org/10.3390/brainsci14080804.
6. Camaioni, M.; Scarpelli, S.; Alfonsi, V.; Gorgoni, M.; Calzolari, R.; De Bartolo, M.; Mangiaruga, A.; Couyoumdjian, A.; De Gennaro, L. How COVID-19 Affected Sleep Talking Episodes, Sleep and Dreams? *Brain Sci.* **2024**, *14*, 486. https://doi.org/10.3390/brainsci14050486.
7. Gerhardt, E.; Baird, B. Frequent Lucid Dreaming Is Associated with Meditation Practice Styles, Meta-Awareness, and Trait Mindfulness. *Brain Sci.* **2024**, *14*, 496. https://doi.org/10.3390/brainsci14050496.
8. Oldoni, A.A.; Bacchi, A.D.; Mendes, F.R.; Tiba, P.A.; Mota-Rolim, S. Neuropsychopharmacological Induction of (Lucid) Dreams: A Narrative Review. *Brain Sci.* **2024**, *14*, 426. https://doi.org/10.3390/brainsci14050426.
9. Kuiken, D. The Epistemic Limits of Impactful Dreams: Metacognition, Metaphoricity, and Sublime Feeling. *Brain Sci.* **2024**, *14*, 528. https://doi.org/10.3390/brainsci14060528.
10. Diushekeeva, A.; Hidalgo, S.; Zadra, A. Impact of Pre-Sleep Visual Media Exposure on Dreams: A Scoping Review. *Brain Sci.* **2024**, *14*, 662. https://doi.org/10.3390/brainsci14070662.

References

1. Mota-Rolim, S.A.; Bulkeley, K.; Campanelli, S.; Lobão-Soares, B.; de Araujo, D.B.; Ribeiro, S. The Dream of God: How Do Religion and Science See Lucid Dreaming and Other Conscious States during Sleep? *Front. Psychol.* **2020**, *11*, 555731. [CrossRef] [PubMed]
2. Freud, S. *The Interpretation of Dreams*; Encyclopedia Britannica: London, UK, 1900.
3. Hobson, J.A. REM sleep and dreaming: Towards a theory of protoconsciousness. *Nat. Rev. Neurosci.* **2009**, *10*, 803–813. [CrossRef] [PubMed]
4. Rados, R.; Cartwright, R.D. Where do dreams come from? A comparison of presleep and REM sleep thematic content. *J. Abnorm. Psychol.* **1982**, *91*, 433–436. [CrossRef] [PubMed]
5. Solms, M. Dreaming and REM sleep are controlled by different brain mechanisms. *Behav. Brain Sci.* **2000**, *23*, 843–850. [CrossRef] [PubMed]
6. Foulkes, D. Dreaming and REM sleep. *J. Sleep Res.* **1993**, *2*, 199–202. [CrossRef] [PubMed]
7. Hobson, J.A.; Stickgold, R.; Pace-Schott, E.F. The neuropsychology of REM sleep dreaming. *NeuroReport* **1998**, *9*, R1–R14. [CrossRef]
8. Hobson, J.A.; Pace-Schott, E.F.; Stickgold, R. Dreaming and the brain: Toward a cognitive neuroscience of conscious states. *Behav. Brain Sci.* **2000**, *23*, 793–842. [CrossRef]
9. Blagrove, M.; Edwards, C.; van Rijn, E.; Reid, A.; Malinowski, J.; Bennett, P.; Carr, M.; Eichenlaub, J.-B.; McGee, S.; Evans, K.; et al. Insight from the Consideration of REM Dreams, Non-REM Dreams, and Daydreams. Psychology of Consciousness: Theory, Research, and Practice. *Psychol. Conscious. Theory Res. Pract.* **2019**, *6*, 138–162. [CrossRef]
10. McNamara, P.; Johnson, P.; McLauren, D.; Harris, E.; Beauharnais, C.; Auerbach, S. REM and NREM sleep mentation. *Int. Rev. Neurobiol.* **2010**, *92*, 69–86. [CrossRef]
11. McNamara, P.; McLaren, D.; Durso, K. Representation of the self in REM and NREM dreams. *Dreaming* **2007**, *17*, 113–126. [CrossRef]
12. Schredl, M. Dreams in patients with sleep disorders. *Sleep Med. Rev.* **2009**, *13*, 215–221. [CrossRef] [PubMed]
13. Takeuchi, T.; Miyasita, A.; Inugami, M.; Yamamoto, Y. Intrinsic dreams are not produced without REM sleep mechanisms: Evidence through elicitation of sleep onset REM periods. *J. Sleep Res.* **2001**, *10*, 43–52. [CrossRef] [PubMed]
14. Stumbrys, T.; Erlacher, D. Lucid dreaming during NREM sleep: Two case reports. *Int. J. Dream Res.* **2012**, *5*, 151–155. [CrossRef]
15. Schredl, M.; Hofmann, F. Continuity between waking activities and dream activities. *Conscious. Cogn.* **2003**, *12*, 298–308. [CrossRef] [PubMed]
16. Revonsuo, A. The reinterpretation of dreams: An evolutionary hypothesis of the function of dreaming. *Behav. Brain Sci.* **2000**, *23*, 877–901. [CrossRef]
17. Antrobus, J. Dreaming: Cognitive processes during cortical activation and high afferent thresholds. *Psychol. Rev.* **1991**, *98*, 96–121. [CrossRef]
18. Koulack, D.; Goodenough, D.R. Dream recall and dream recall failure: An arousal-retrieval model. *Psychol. Bull.* **1976**, *83*, 975–984. [CrossRef]
19. Mota-Rolim, S.A.; Araujo, J.F. Neurobiology and clinical implications of lucid dreaming. *Med. Hypotheses* **2013**, *81*, 751–756. [CrossRef] [PubMed]
20. Bolstad, C.J.; Holzinger, B.; Scarpelli, S.; De Gennaro, L.; Yordanova, J.; Koumanova, S.; Mota-Rolim, S.; Benedict, C.; Bjorvatn, B.; Chan, N.Y.; et al. Nightmare frequency is a risk factor for suicidal ideation during the COVID-19 pandemic. *J. Sleep Res.* **2024**, *33*, e14165. [CrossRef]
21. Parrello, S.; Sommantico, M.; Lacatena, M.; Iorio, I. Adolescents' dreams under COVID-19 isolation. *Int. J. Dream Res.* **2021**, *14*, 10–20.
22. Scarpelli, S.; Nadorff, M.R.; Bjorvatn, B.; Chung, F.; Dauvilliers, Y.; Espie, C.A.; Inoue, Y.; Matsui, K.; Merikanto, I.; Morin, C.M.; et al. Nightmares in people with COVID-19: Did coronavirus infect our dreams? *Nat. Sci. Sleep* **2022**, *14*, 93–108. [CrossRef] [PubMed]
23. Pesonen, A.-K.; Lipsanen, J.; Halonen, R.; Elovainio, M.; Sandman, N.; Mäkelä, J.-M.; Antila, M.; Béchard, D.; Ollila, H.M.; Kuula, L. Pandemic dreams: Network analysis of dream content during the COVID-19 lockdown. *Front. Psychol.* **2020**, *11*, 573961. [CrossRef] [PubMed]
24. La Berge, S.P.; Nagel, L.E.; Dement, W.C.; Zarcone, V.P. Lucid dreaming verified by volitional communication during REM sleep. *Percept. Mot. Ski.* **1981**, *52*, 727–732. [CrossRef]
25. Naiman, R. Dreamless: The silent epidemic of REM sleep loss. *Ann. N. Y. Acad. Sci.* **2017**, *1406*, 77–85. [CrossRef]
26. Palmer, C.A.; Alfano, C.A. Sleep and emotion regulation: An organizing, integrative review. *Sleep Med. Rev.* **2017**, *31*, 6–16. [CrossRef] [PubMed]

Review

The Persistent Paradox of Rapid Eye Movement Sleep (REMS): Brain Waves and Dreaming

J. F. Pagel [1,2]

[1] Family Medicine Department, University of Colorado Medical School System, P.O. Box 6, Arroyo Seco, NM 87514, USA; pueo34@juno.com; Tel.: +1-719251707

[2] Psychology Department, Cape Breton University, 38 Gull Cove Rd., Glace Bay, NS B1K 3S6, Canada

Abstract: The original conceptualization of REM sleep as paradoxical sleep was based on its EEG resembling wakefulness and its association with dreaming. Over time, the concept of paradox was expanded to include various associations with REM sleep, such as dream exclusivity, high recall, and pathophysiology. However, none of these associations are unique to REM sleep; they can also occur in other sleep states. Today, after more than fifty years of focused research, two aspects of REMS clearly retain paradoxical exclusivity. Despite the persistent contention that the EEG of human REMS consists of wake-like, low-voltage, non-synchronous electrical discharges, REMS is based on and defined by the intracranial electrical presence of 5–8 Hz. theta, which has always been the marker of REMS in other animals. The wake-like EEG used to define REMS on human polysomnography is secondary to a generalized absence of electrophysiological waveforms because the strong waves of intracranial theta do not propagate to scalp electrodes placed outside the skull. It is a persistent paradox that the theta frequency is restricted to a cyclical intracranial dynamic that does not extend beyond the lining of the brain. REMS has a persistent association with narratively long and salient dream reports. However, the extension of this finding to equate REMS with dreaming led to a foundational error in neuroscientific logic. Major theories and clinical approaches were built upon this belief despite clear evidence that dreaming is reported throughout sleep in definingly different physiologic and phenomenological forms. Few studies have addressed the differences between the dreams reported from the different stages of sleep so that today, the most paradoxical aspect of REMS dreaming may be how little the state has actually been studied. An assessment of the differences in dreaming between sleep stages could provide valuable insights into how dreaming relates to the underlying brain activity and physiological processes occurring during each stage. The brain waves and dreams of REMS persist as being paradoxically unique and different from waking and the other states of sleep consciousness.

Keywords: rapid eye movement sleep (REMS); dream; sleep; theta; electrophysiology; parasomnias; nightmares; resonance; brain waves; neuroelectrophysiology; bizarreness; gamma; dream recall; consciousness; neuroconsciousness; paralysis; lucidity; paradoxical; dream content

Citation: Pagel, J.F. The Persistent Paradox of Rapid Eye Movement Sleep (REMS): Brain Waves and Dreaming. *Brain Sci.* **2024**, *14*, 622. https://doi.org/10.3390/brainsci14070622

Academic Editors: Sergio A. Mota-Rolim, Brigitte Holzinger, Michael Nadorff and Luigi De Gennaro

Received: 28 May 2024
Revised: 18 June 2024
Accepted: 20 June 2024
Published: 21 June 2024

1. Introduction

The brain waves and dreams of REMS persist as being paradoxically unique and different from waking and the other states of sleep consciousness. Today, the most paradoxical aspect of REMS dreaming may be how little the state has actually been studied. Few studies have addressed the differences between the dreams reported from different stages of sleep. Addressing the similarities and differences between electrophysiology and dreaming of different forms of sleep-associated consciousness provides valuable insight into how rapid eye movement sleep (REMS) dreaming relates to underlying brain activity and physiological processes.

Rapid eye movement sleep (REMS) was identified in the 1950s after Stages 1–4 had already been discovered. At that time, there was considerable debate as to what to call this physiologically unique pattern of sleep. Based on its association with both intense episodes of dreaming and activated non-synchronous EEG activity that looked much like wakefulness, the state was initially designated "paradoxical" sleep [1,2]. REMS is a state of sleep consciousness with complex anatomic, chemical, and electrophysiological physiology that is conceptually far larger than its limited attribute of rapid eye movements. This sleep/dream state became a seminal construct for the developing fields of psychoanalysis, psychology, dream science, and sleep medicine. Persistently contradictory REMS elements continue to inspire equivocation, cognitive dissonance, and research, producing logically counter-intuitive results. Today, primary aspects of this state can still be considered paradoxical. Recent research addressing its neuroelectrophysiology and associated dreaming continues to produce major changes in both applied neuroscientific theory and our comprehension of this synonymous state of paradox and REMS [3].

Early work with REMS emphasized its associated neuroanatomy. In decorticate cats, the eye movements, muscle atonia, and spatial orientation associated with REMS were maintained even when the brain was resected above the brainstem [1]. Since REMS was equated with dreaming, this brainstem-based neuroanatomic perspective was integrated with Freudian psychodynamics to produce neuroconsciousness theories of brain functioning, suggesting that dreams were emanations of the primitive Id-associated brainstem [4]. Current understandings of REMS neuroanatomy based on electrophysiologic scanning and animal models indicate that extensive areas of the cortex, hypothalamus, midbrain, and medulla are involved in the control and/or suppression of the globally projected neuroprocessing [5]. Some animal neuropsychologists have interpreted the wide spectrum of neurologic activation sites during REMS to suggest that this wake-like neuronal activity is proof that dogs dream of their waking experience [6,7]. Yet, neither REMS nor dreaming are associated with clearly described neuroanatomic markers or consistent patterns of neuronal activation [8]. Sleep and dreaming are global states defined and best described by their associated neuroelectrophysiology with neuroanatomic control and effector sites poorly described and loosely reflective of state electrophysiology [9].

2. REMS Paradox Part 1: The Brain Waves

Soon after its discovery, neuroscientists begin to refer to paradoxical REMS as desynchronized sleep. There were no oscillatory waveforms seen on the scalp EEG like those seen in the other stages of sleep, but rather a disordered pattern of electrical activity, a "wracked storm-tossed ocean" of intersecting spikes and poorly formed waves, much like that seen during waking. Polysomnographic monitoring using multiple sensors to detect the presence of a low-voltage, wake-like electroencephalogram (EEG), conjugate eye movements recorded by electrooculogram (EOG), and diminished chin electromyographic (EMG) activity was developed as a technique that could be used to record REM sleep in humans [9]. Today, some neuroscientists and many sleep clinicians persist in their description of human REMS electrophysiologic activity as disordered, low-voltage, and wake-like [10,11].

REMS is more evident in animal models, where it is defined by long, well-synchronized runs of recorded theta (5–8 Hz). This frequency is only minimally apparent on recordings from the scalp EEG leads used in human studies. It has even been suggested that this "paradoxical" lack of REMS synchronicity might describe a basic physiologic difference between humans and other animals [12].

However, the wake-like human EEG seen during REMS has turned out to describe a false dichotomy. The low-voltage asynchronous EEG associated with the human EEG is an artifact related to recording technique. Since the EEG frequencies associated with the other sleep stages easily propagate to scalp electrodes, it seemed rational to expect that the brain wave system associated with REMS, the well-described coherent frequency of theta, would also propagate beyond the brain to the scalp leads of the EEG. However, REMS theta differs from the other frequencies of sleep. It is difficult to observe and record outside the constraints of the brain container (skull, dura, and scalp). The major difference noted in the REMS EEG between animals and humans has turned out to be the electrode placement (Figure 1). Recording electrodes are implanted intracranially in animal models, and after recordings are completed, the animals are sacrificed. In human EEG recordings, the electrodes are non-invasively attached to the scalp. When electrodes were eventually placed within the human brain during intracranial surgeries for recalcitrant seizure disorders, the tightly coupled REMS theta rhythm that had been seen in other animals became readily apparent [13].

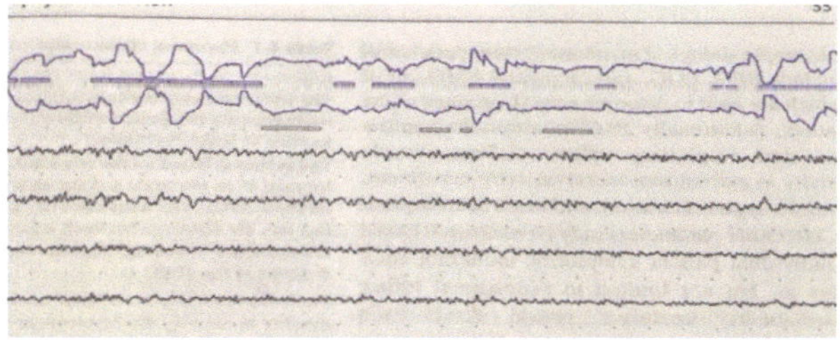

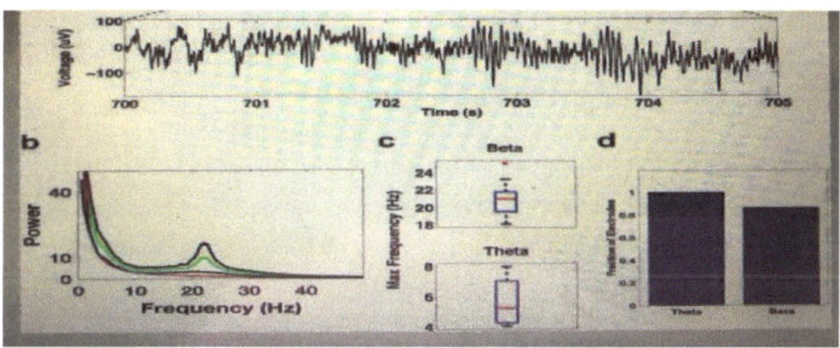

Figure 1. Comparative REMS monitoring. Adapted from Vijayan et al. [13].

3. Ponto-Geniculo-Occipital (PGO) Waves

Ponto-geniculo-occipital (PGO) waves are low-voltage discrete spike/waves that can only be seen using intracranial electrodes. PGO spikes begin as electrical pulses in the pons and extend to the lateral geniculate nucleus of the thalamus, before extending to the primary visual cortex of the occipital lobe. These small signals have an amplitude of 100 microvolts (μV), last only 350 milliseconds (ms), and often occur in clusters of more than 25 spikes per minute. PGO waves often precede the onset of REM sleep, occurring 30–90 s before the first conjugate eye movements, EMG dropout, and the onset of synchronous runs of theta. This close association with REMS has led some neuroscientists to suggest that PGO spikes are the neurological trigger for REMS [14]. Based on the belief that REMS is dream sleep, it has been suggested that PGO waves are an on/off trigger for dreaming [15]. More recently, the relationship with REMS has been reconfigured so that PGO spikes are now theorized to have significant roles with hippocampal theta in visual field localization and marking [15].

4. QEEG and MEG

The electrical fields associated with REMS can also be recorded indirectly in humans using 256-lead quantitative QEEG systems. The QEEG signals are processed digitally, transformed, and analyzed using complex mathematical algorithms so that the otherwise hidden, low-voltage scalp signals of REMS-associated theta and gamma can be revealed [16]. QEEG recordings of the theta and gamma are obtained from short, clean windows of data that require computer-adjudicated filtering and comparative analysis to produce usable recordings [17,18]. These systems have limitations, are prone to multiple lead dysfunctions and artifacts, and produce a large data set confounded by sometimes uncontrollable variables. Gamma (35–60 Hz) recorded during REMS is also difficult to record using scalp electrodes [19]. REMS gamma is low-voltage and in the same frequency range as more powerful muscle EMG activity and the 60 Hz gamma of the environmental electrical grid. Physiologic gamma is often obscured by the 40–60 Hz block filters required for removing these artifactual frequencies from the recording. One small QEEG study is purported to describe a posterior cortical "dream center" of the brain that can be monitored to predict whether an individual will report dreams or the absence of dreaming [9]. This study is, however, minimally consistent with other work, with the proposed neural correlate of dreaming low-frequency (1–4 Hz) delta, a frequency that one of the study's author's had previously proposed as a marker for dreamless sleep [20]. It is also possible that low-voltage wake-like EEG activity, as in the case of lucidity, may be involved in the incorporation of sleep mentation into waking consciousness [21,22].

Attempts have also been made to use magnetoelectroencephalography (MEG) to record sleep. The earliest use of the sleep MEG was to detect gamma, a frequency theoretically tied to cellular assembly, cognitive binding, and the development of consciousness [23]. Rudolfo Llineas and Urs Ribery, among the most famous neuroscientists of this era, initially used MEG to describe 40 Hz activity during REMS. Since, at the time, REMS was viewed as dreaming, 40 Hz gamma was proposed to be a marker for dream consciousness [24]. Despite multiple attempts, other researchers have not been able to repeat these results. Neural magnetic field changes occurring in sleep are very small, and strict positional controls are required for MEG studies (few, if any, research subjects are able to attain REM sleep in the MEG laboratory) [25,26].

5. REMS Theta

Theta is the electrophysiologic marker for REMS. Some neuroscientists have suggested that theta does not project outside the skull because of its origin in the deep "primitive" brain structures of the hippocampus and brainstem [27]. This supposition, however, is inconsistent with the extra-cranial propagation of sigma (also a hippocampal rhythm), thetas' lack of consistent association with QEEG vectors, and the strong theta signals observed just inside the skull on the surface of the brain. Beyond its lack of propagation, theta is also unlike the other sleep state-associated rhythms in that it has no clearly described cellular ionic origin. At the neuronal membrane, alpha is formed by the time frequency of the potassium ionic flux; sigma is tied to calcium, and delta to calcium modulated potassium channel cycling [28]. Theta seems to first appear in the rhythm modulation of hippocampal pyramidal cells [29]. It is possible that theta originates in the nano-vibrations of cellular microtubules [30]. However, it is also possible that theta has no direct intracellular origin. Global resonance theory (GRT) suggests that theta is part of a nested hierarchy of electromagnetic (EM) fields that spans the entire human physiology, encompassing both the physiologic brain rhythms and an extracerebral network of endogenous neural rhythms. These electrical frequencies can be nested in a hierarchy of resonate frequencies built upon the 5 Hz frequency of slow theta [31]. It is possible that theta forms as a resonant frequency able to entrain and form harmonic interactions with gamma and other higher-frequency fields, including ripples and gamma bursts [32,33].

6. Theta Waveforms—Operation and Functions

Animal research indicates that waking theta has concrete roles in waking cognitive functioning. In the hippocampus, theta functions as a timekeeper, a parser of memory, and an imagery marker for orientation in space. In mice and rats, theta is prominent during exploratory movements, such as when navigating through a maze looking for food [27]. During spatial navigation, the spike timing of hippocampal neurons systematically shifts to an earlier phase of the theta cycle as the rat moves across the field, a phenomenon called "phase precession" [34]. During sleep immediately after spatial tasks in which rats repeatedly run through a sequence of place fields, hippocampal neurons fire in the same sequential order as during waking [35]. Theta can set the timing of neuronal firing, playing an important role in the formation and retrieval of episodic and spatial memory [36].

Cognitive reaction time is comparatively better in animals with synchronization when compared to those with a desynchronization at the theta frequency [37]. Studies have implicated theta phase modulation in memory formation and retrieval [32].

7. The Paradox of REMS Neuroelectrophysiology

The electrophysiology of REMS continues to be a paradox but for other reasons than as initially described. The EEG of REMS has only an artifactual resemblance to the EEG of waking. REMS is defined by the presence of a synchronous electrical field: 5–8 Hz theta. The low-voltage wake-like EEG used to define REMS on human polysomnography is a null state of generalized wave absence existing secondary to the non-propagation of intracranial theta. This is the most paradoxical and unexplained persisting aspect of REMS electrophysiology. The other sleep/dream frequencies of consciousness (alpha, sigma, and delta) propagate outside the skull. REMS theta is restricted in time and space to a homunculus-like dynamic of cyclical intracranial electromagnetic fields that do not extend beyond the brain.

8. REMS Paradox Part 2: Dreaming

Sleep is comprised of a group of very active central nervous system (CNS) processing states. Except for millisecond interludes of neuronal quiet during deep sleep (Stage 3), biomental activity takes place throughout sleep. This neural processing is often associated with some form of reported dreaming.

REMS dreaming is the sleep consciousness associated with the state. For much of the last era, all dreams were presumed to have occurred during REMS. Much of the research and literature addressed dreaming as a unitary state based on the belief that REMS was the physiologic marker for dream consciousness. However, even just after the state's discovery, it was clear that dreams were being reported from all of the stages of sleep [38]. Early studies indicated that 80% of dreaming took place in REM sleep, with 6.9% of dreams reported from the other sleep stages [39]. With improved methodology and the expansion of the definition of dreaming to include other types of mentation, dream report rates from the non-REMS stages have risen to levels greater than 70% [40]. Some individuals will report sleep-associated mentation (dreaming) whenever they are awakened [41]. REMS and dreaming are doubly dissociable [42], with REM sleep occurring without reported dreaming and dreaming reported outside REMS [43] (Table 1). The belief that REMS is dreaming is a meme and bias that taints a large portion of the research and literature addressing dreaming.

Table 1. Summary of primary evidence approaches that indicate that REMS is not equivalent to dreaming [8,32,40,41,43–47].

- Dreaming occurs and is reported throughout sleep [Dreaming is reported from all sleep stages].
- REMS occurs without associated dreaming.
- Each sleep stage is associated with state-specific forms of dreaming.
- Sleep state-specific parasomnias are associated with profoundly different forms of frightening dreaming.
- Medications and illnesses that affect REMS do not necessarily affect reported dream recall or content.
- Scanning systems indicate that a wide spectrum of different CNS activity is associated with dreaming during all stages of sleep.
- Damage to brainstem CNS areas involved in controlling REMS does not affect dreaming if the patient can report. Damage to basal-frontal areas of the brain that affect dream recall does not affect REMS.

9. Activation-Synthesis

Activation-synthesis theory postulates that neuronal activation is the marker for brain functioning [48]. The association of intense dreaming with the "activated" neuro-electrophysiology of REMS was presented as the primary proof of activation-synthesis theory. Over the next fifty years, various versions of this theory, including AIM (activation, input–output gating, and modulation), covert dreaming, and protoconsciousness, have dominated the fields of sleep and dream neuroscience [11,48,49]. All neuroconsciousness theories have been built upon the founding construct of REMS as being equivalent to dream sleep. Despite the loss of this primary experimental proof, neuroconsciousness theories continue to delimit neuroscientific and neuroanatomic understandings of brain functioning (mind). REMS = dreaming continues to be presented as fact, incorporated into the theories of neuroconsciousness built upon activation-synthesis, basic psychology texts, neuro-medical literature addressing sleep/ dream associations, and in many of the popular sleep and dreaming texts [20,49–51]. REMS equals dreaming has become an apparently irrefutable meme.

10. Psychodynamics

At the time of the discovery of REMS, almost all of the literature addressing sleep and dreaming was based on psychoanalytic theory. REMS was construed as a "smoking gun," proof that dreaming was a state of activation of the Freudian "Id", a primitive neuronal system throbbing in the brainstem, activating our brains and behaviors [52]. Freud had proposed dreams as a "royal" road to the unconscious, an interpretive path that could be used to explore the psychodynamics of the human psyche [53]. Extensive literature has documented the use of dream interpretation in the diagnosis and treatment of psychiatric illness. Unfortunately, beyond anecdote, little evidence indicated that dream-based psychoanalysis actually worked to treat psychiatric illness [54]. Dream content, while reflective of an individual's waking experience of psychiatric disturbance, was not particularly useful in diagnosis. While dream interpretation can be used to expand memory associations and explore the significance of dream stories, today, the psychoanalytic use of dream content exists primarily in approaches to self-actualization, self-understanding, creativity, and art [55]. A general perspective persists, however, that dreams are strange, bizarre experiences that function in intrapersonal psychodynamics and cognitive CNS operations.

11. REMS and Dreaming: The Special Relationship

Over the last seventy years, the physiology of REMS has been the focus of extensive study. There is a strong record of failed attempts to maintain the REMS–dreaming correlate, but far fewer studies comparing REMS dreaming to the dreams of other sleep stages. REMS dreams continue to be construed as "special;" the classic dreams most likely to be remembered and written down, characterized by vivid imagery, bizarre or implausible storylines, potential lucidity, increased and often negative emotionality, and high levels of motivational significance [56]. They are often described as the most bizarre, most emotional, most lucid, and most significant of dreams. More recent literature has questioned definitions, the actual sleep stage association of the reported dreams, and the bias of the subject and researcher [57]. In studies strictly controlling for transference effects and bias, it has been difficult to demonstrate significant differences in REMS dream content from reports obtained from the other stages of sleep [58]. Today, it is unclear as to whether any special relationship exists between REM sleep and dreaming. While it seems likely that most of the reports of associated dream content included in the psychoanalytic literature are from REMS, dreams and dream-associated parasomnias are reported from all sleep stages. It has been argued that the typical REM sleep dream has higher recall, greater length and narrative development, greater negativity and association with nightmares, increased emotional content, and higher levels of bizarreness and salience than the dreams that are typically reported from the other stages of sleep. There is contrary evidence, however, for all these contentions, except for report length [41].

So, what remains of the "special" relationship between REMS and dreaming? Many of the accepted foundational correlates were incorrect, insubstantial, or tenuous at best so that REMS = dreaming has become an exemplary example of a foundational reference failure. What follows is a short review of recent research and theory addressing REMS and its association with dreaming.

12. REMS Dream Recall

REMS is a state of high dream recall, with a dream report on awakening frequency of 70–80%, in the same range as sleep onset—Stage 1. If recall assessment is restricted to the last REMS period of the night using a sensitive QEEG system, the retrieval rate of REMS dreaming can be increased to above 90% [59]. Both REMS and dreaming are experienced into extreme old age, continuing to be reported after major levels of CNS damage [60]. When a subject is still able to consciously communicate, the loss of dreaming is not typically associated with damage to brainstem areas controlling REMS; the loss of dreaming is most often reported after significant basilar bi-frontal CNS damage such as that created by frontal lobotomy psychosurgeries [44].

13. REMS Dream Content

REMS dreams have been described as the classic, long, "special" dreams most likely to be remembered and written down, replete with vivid imagery, a bizarre or implausible storyline, potential lucidity, increased and often negative emotion, and high levels of motivational significance [61]. Freud proposed that such dreams marked a royal road into the unconscious, an interpretive path that could be used to expand memory associations, the significance of narrative stories, and explore the psychodynamics of the human psyche [53]. This psychoanalytic perspective has contributed to the belief that dreams are evolutionarily "primitive" compared to more advanced brain functioning, a point-of-view that persists in the terminology of neuroanatomy contrasting neocortex, executive, and higher brain functions, with the conceptual perspective of dreaming as an evolutionarily inferior brain stem process [62].

Dream content has proven difficult to study. An amazing number of variables are known to affect reported dream content. Dream content is altered by the gender of both subject and technician; the type, site and time of reporting; dream definition, bias, and expectations of both subject and tech; and socioeconomic variables, including economic, language, and educational levels. Transference is particularly difficult to control. The strongest consistent variable known to affect dream content is the individual's waking experience. The "continuity hypothesis" describes the only variations in content that clearly persist when methodology, bias, and transference variables are eliminated [63]. In most studies of dream content, confounding variables have been ignored, so that today it is unclear whether dream content is significantly affected by age, personality, culture, psychiatric illness, creativity, psychodynamic style, and/or trauma. Recent studies suggest that there may be few differences in dream content between REMS and the dreams collected from other stages [45]. Today, explorations of dream content are utilized primarily in approaches to self-actualization, self-understanding, creativity, and art [57].

14. Report Length

The dreams and the nightmares that occur during REMS-associated theta are longer, narratively bizarre, and include more wake-like cognitive thought than the dreams of the other sleep states [64]. Dream content has been described as the thinking of the body, an example of the birth of the literary process. Dream plots, like those of literary fiction, are a continually evolving pattern of imagery and events [65]. Dream reports rarely form full narratives but more often incomplete fragments of stories that simplify past happenings by generating tight fictional narratives around associated schemas [46]. REM sleep dreams are often subjectively perceived as story-like and autobiographically meaningful [21]. The longer REMS dream reports lend themselves to literature-based narrative analysis, an approach that has been used in the attempt to prove that REMS dreams differ from non-REMS dreams in having content that is more improbable and bizarre in the domains of dream plot, cognition, and affect.

14.1. Bizarreness

Bizarreness, the quality of being strange and unusual, is a matter of contention in dream science. In 1987, researchers at Harvard designed a bizarreness scale that used the storyline narrative content of a dream to rate bizarreness. Dreams that included incongruous, uncertain, and discontinuous statements were rated as the most bizarre. Since REMS dreams are longer and have greater narrative structure, studies using this scale consistently demonstrate that REMS dreams are the most bizarre [61]. Based on this rating scale, the shorter dreams of the other stages rate as less bizarre despite apparently "bizarre" characteristics such as hallucinations, extreme emotional distress, and intense disassociation from reality. An alternative bizarreness scale taking such dream-associated behaviors into account indicates that dreams reported from sleep onset (Stage 1) and deep sleep (Stage 3) may be more bizarre than the dreams of REMS [66]. The rating of bizarreness is an excellent example of how researchers can manipulate experimental methodology.

14.2. Lucidity

Lucid dreaming is the experience of achieving conscious awareness of dreaming while asleep. Most dreams have lucid characteristics in that they are described from the dreamer's point of view with the dreamer present and, at some level, aware during the mental activity of every dream. Much of the research into this state has focused on identifying and attempting to prove that lucid dreaming occurs during REMS [67]. Yet, even in early descriptions of lucidity, it was apparent that lucid dreaming took place during sigma and alpha sleep (with 18% of lucid dreams occurring at sleep onset) [68]. In the lab, some lucid dreamers can push buttons, move, or fix their gaze while asleep and use this change in gaze to signal an external observer [69]. This volitional signaling capability has allowed researchers to analyze lucid dreaming using real-time scanning systems. The initial hope was that scanning would display specific areas of the CNS active during REMS. However, the results of such studies have been confusing. Signaling is almost always associated with arousal from sleep, in which the dreamer consciously reinstitutes the perceptual and motor controls associated with wakefulness to signal researchers. Lucidity with signaling may be best viewed as a form of conscious sleep offset—the opposite of the turning away from waking consciousness required for sleep onset [22]. During lucid dreaming, there are bursts of alpha and gamma, and activation of CNS sites involved in working memory and the analysis of visual perception—brain areas normally de-activated during REMS [70]. Lucid dreaming is a transitional state between different stages of sleep and waking, a type of dreaming that has been reported from all stages of sleep except deep sleep delta.

15. REMS Physiology and Dreaming

The REMS state differs from the other forms of sleep/dream consciousnesses in its association with conjugate eye movements, the suppression of skeletal muscle activity (EMG), sporadic muscle twitches, heat loss, accelerated gastric transit, hypoventilation, autonomic nervous system activation, genital arousal, increased heart rate variability, and oscillations in body temperature [71,72]. These physiologic characteristics, however, cannot be considered REMS state-specific. REMS can occur without these attributes, which also occur, albeit less often and in associated forms, in the other stages of sleep [73] (Table 2). The dreaming-associated REMS parasomnias of sleep paralysis, nightmares, dream hallucinations, and even the acting out of dreams associated with REMS behavior disorder (RBD) have all been described in association with the other stages of sleep.

Table 2. Characteristic REMS-associated phenomenology and physiology.

	REMS Association	Other Sleep State Associations	Associated Pathophysiology
Eye movements	Defining (conjugate movements)	Rolling eye movements	Eye movement pathologies
High dream recall (>70% awakenings)	Increases through night	S1 and S3 parasomnias	
Long report length	Exclusive to REMS		
Extreme emotions	Nightmares and sleep paralysis	S3 night terrors and confusional arousals	S1 hallucinations, PTSD; S2 panic attacks
Use in creativity	Incubation	S1—induced images; S3—parasomnias	
Salience	Transcendent, persistent, and life-changing	S1 hallucinations and S3 night terrors	
Bizarreness	Narrative	S2, S3 behavioral, S1 hyynogognic hallucinations	Psychiatric hallucinations
Lucidity	Predominate state	S1, S2, and sleep offset	Confusion, psychosis
Genital arousal	Males and females	Near awakening	Drug-induced sexsomnia
Autonomic system destabilization	Predominate state	Can occur throughout sleep	Sleep position, reflux, SIDS
EMG suppression	Defining	Sleep onset S1	RBD: EMG suppression lost
Sleep paralysis	Most often	Common—S1	Narcolepsy
Dream acting-out	Most often	S3 parasonias	RBD, PTSD, Parkinson's disease
Nightmares	Exclusive to REMS		Throughout sleep in PTSD
Cardiac irregularity	Characteristic		Associated with apneas
Hypoventilation	Increased in some	Throughout sleep	Central apneas, altitude, Cardiopulmonary disease
EEG spike waves	Suppressed		Seizure disorders
Periodic limb movements	Suppressed	Throughout sleep	Restless leg syndrome
Gastric transit	Accelerated		Reflux, irritable bowel
Body temperature changes	Characteristic	Throughout sleep	Systemic illness

Key: sleep onset Stage 1 (S1); light sleep Stage 2 (S2); deep sleep (S3); rapid eye movement sleep (REMS); REMS behavior disorder (RBD); post-traumatic stress disorder (PTSD); sudden infant death syndrome (SIDS); [9,67,68,70,72–77].

While there is good evidence for the continuity of waking experience with dream content, the relationship between sleep-associated physiologic events and dream content is less clear. The muscle twitches, disordered breathing, apneas, temperature changes, and gastric reflux associated with REMS are rarely incorporated into dream content [74]. REMS-associated genital erections and muscular paralysis, however, have an extended history and diagnostic association with both REMS dreaming and dream-associated parasomnias.

16. REMS Associated Genital Arousal

Psychoanalytic dream theory posits that sexual "wet" dreams occur during REMS as a reflection of brainstem activation and the influence of the primitive "Id." Yet, beyond anecdotal reports, there is little evidence supporting the association between REMS genital erections and sexually thematic dreams [78]. For many years, sleep studies were used to prove or disprove the presumed psychogenic origin of male impotence using strain gauges that could detect penile tumescence during REMS. With time, it has become obvious that impotence is better viewed as a neurophysiologic dysfunction independent of etiology and best treated with phosphodiesterase type 5 inhibitors, such as Viagra. Today, nocturnal penile tumescence (NPT) studies are rarely performed and reserved for selected cases where the exact determination of the underlying case for impotence is deemed necessary. Sleep-associated genital arousals and wet dreams, like lucidity, signally, and dream report hot zones, might be better viewed as near-awakening transition experiences [79].

17. Electro-Muscular Suppression and Sleep Paralysis

An abrupt drop in mentalis EMG innervation is a primary polysomnographic criterion for human REMS [8]. The motor paralysis associated with REMS sleep paralysis has been tied to multiple well-described chemical and motor trigger sites in the pontomedullary reticular formation interconnected synaptically and utilizing the neurotransmitters glycine and Gamma aminobutyric acid (GABA) (+) and the neuromodulators norepinephrine and serotonin (−) [80,81] (Figure 2). Sleep paralysis presents as a dissociated state in which the REMS atonia of major skeletal muscle systems persists into wakefulness. The Parasomnia is well titled—the experience of awakening from sleep with an inability to move. Although sleep paralysis can be reported during Stage 1 (sleep onset alpha), it occurs most often on arousal from REM sleep [75]. Episodes of sleep paralysis often include negative and frightening content, developed in detail, and associated with distress that extends into awakening. Predisposing factors include sleep deprivation, irregular sleep–wake schedules, and PTSD [82].

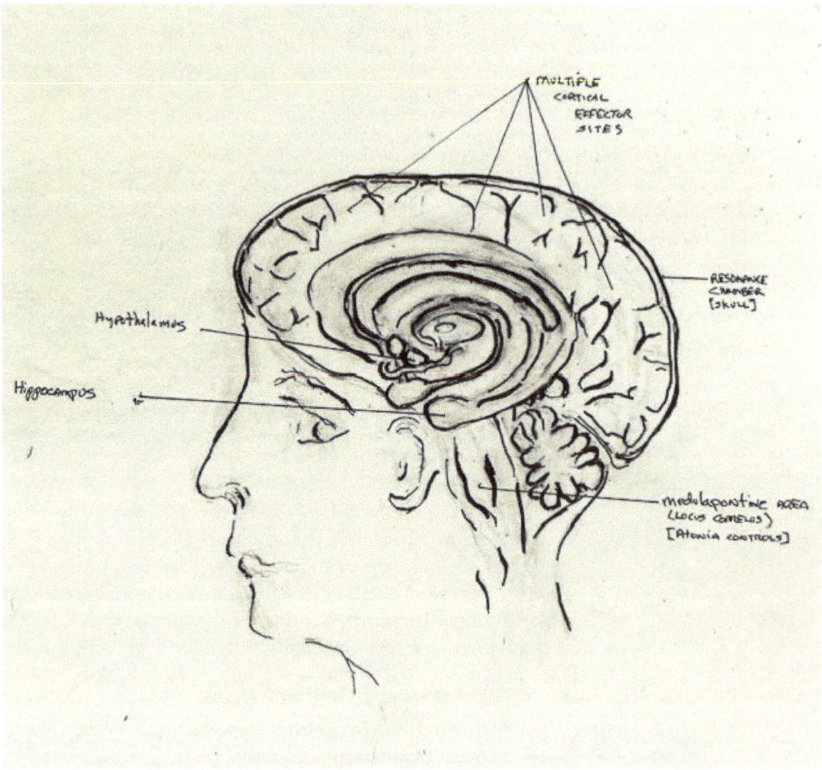

Figure 2. REMS dreaming: associated neuroanatomy [art work—Pagel & Broyles 2024].

During episodes of REM sleep behavior disorder (RBD), individuals sometimes physically act out their dreams, behaviors that can be dangerous to both the dreamer and their sleep-mate. As we currently understand RBD, the neural mechanisms that serve to disable motor activity during REMS dreaming do not function normally, resulting in the physical acting out of the dream experience. RBD, most common in older males, is closely associated with progressive neurologic illnesses, such as Parkinson's disease [83]. Some studies have failed to find a clear correlation between RBD behaviors and described dream content [84].

18. Nightmares

The most experienced parasomnia is the nightmare, a REMS dream of utter and incomprehensible dread, beginning as a seemingly real and coherent dream sequence and becoming increasingly more disturbing as it unfolds [76]. The emotions of nightmares include anxiety and fear, as well as anger, rage, embarrassment, and disgust, with a nightmare's story most often focusing on imminent physical danger [85]. Other negative dreams rarely have the length and terrifyingly real detail of the nightmare [77]. Nightmares are common experiences, with up to 70% of us having more than one per month [86]. Many nightmares induce awakening, sometimes as an escape from the storyline of the dream. Nightmares can be secondary to stress, anxiety, irregular sleep, and medication. Surprisingly, the medications inducing nightmares at the highest frequency increase CNS levels of serotonin, a REMS-suppressant neurochemical [47]. The most common exogenous cause of nightmares is unreconciled trauma. Nightmares are the most common symptom of post-traumatic stress disorder (PTSD), a diagnosis in which up to 50% include re-experience of the trauma. Except for the notable exception of those individuals with the diagnosis of PTSD, nightmares occur only during REMS [87].

19. SUMMARY: Brain Waves and Dreaming, The Paradoxes of REMS

The original conception of REMS as paradoxical sleep was based on EEG electro-physiology resembling the "activated" EEG of waking and the association of REMS with non-perceptual sleep-associated consciousness (dreaming). Through the years, the concept of the REMS paradox expanded to include concepts as diverse as sleep-associated dream consciousness, neuroconsciousness, parasomnias, lucidity, sleep paralysis, wet dreams, and emotional neuroprocessing. While each of these constructs has at some point been considered exclusive to REMS, in every case, evidence indicates that equivalent cognitions and behaviors take place in other states of sleep and waking consciousness (Table 2). The REMS state is associated with a loose pattern of non-exclusive correlates, so that attempts to apply scientific metaphors to REMS often produce unclear, contradictory findings and paradoxes [66].

19.1. Neuroelectrophysiology

While the association of REMS with the electrophysiology of waking consciousness is provably incorrect, philosophically, this construct persists in neuroanatomic semiotics of executive and higher brain functioning, and in the cognitive and electrophysiologic concept of activation. There is a clear bias to equate higher more-activated frequencies, such as gamma, with functions of consciousness. Consistently, the 'on' state of activated neurons is viewed as the operative phase of neuronal functioning. Yet, there is good evidence that off and periodically disconnected states, such as deep sleep dreaming and the default waking state of mind-wandering, are states with important roles in cognitive processing and functioning [77,88]. Despite such evidence, such off and disconnected default states have received minimal study.

CNS electrophysiology is potentially as complex and important for physiologic func-tioning as cerebral neuroanatomy [41,89,90]. Theta functions in the hippocampus as a time-keeping marker for orientation in space and memory [36,91]. Theta is likely to have other functions as it interacts as a base frequency in resonance with short higher-frequency ripples and bursts, as well as with the slower frequencies of respiration, circulation, sym-pathetic and parasympathetic discharge, and the circadian and ultra-circadian endocrine conveyed patterns of light [33,89]. REMS theta is formed in intracerebral isolation independent of neuronal ionic time sequences and protected from environmental contamination by the same system that confines it within the protective structure of the skull and the dura. An electromagnetic homunculus forms in the central brain during episodes of REMS. This fluxing area of hippocampal theta induces the neuronal net to fire in concert with the polarizing upstroke of the waveform so that the firing of each affected neuron augments the field. There are billions of neurons available, and the wave frequency builds on the summation of neural spikes. Synchronous extracellular field potentials develop, go through feedback loops, affect neurons, and propagate throughout the CNS [28]. The complex series of neuronal spikes utilized digitally in creating each wave can become information incorporated into the traveling wave, sometimes referred to as phase coding. At distant CNS sites, that pattern of neuronal firing can be reproduced, reflected, and reconstituted into a complementary series of the same neural firing pattern that was initially utilized in producing the wave [92]. Some recent studies suggest that neuroelectrophysiology is of even greater complexity, with multiple forms of sleep consciousness present in the CNS at the same time, competing for available resources and affecting the functioning of different neural network systems in the CNS [93].

19.2. REMS Dreaming

Equating REMS with dreaming was a foundational error of logic. Almost all studies of dream content addressed dreaming as a unified state, and major theories of neuroscience were built upon this loose and provably false correlation. Research into dreaming became focused on REMS as the physiologic marker for dreaming, and thus, the differences in dreaming between the different sleep stages were rarely addressed. The profound differences in the dreams reported from the different sleep stages are perhaps most easily described by focusing on the differences between sleep-state-associated types of frightening dreams. The visual, high recall, creatively frightening sleep onset hallucinations of sleep onset Stage 1 differ markedly from the brief moments of terror we call panic attacks that occur in light sleep (Stage 2). The intense, bizarre, confused arousals of Stage 3 deep sleep are profoundly different from the other states of sleep and waking. And these intensely frightening and sometimes extraordinarily bizarre dreams have little in common with the reality-like, narrative, and complex storylines of REMS nightmares. Few studies have compared the differences between the non-frightening dreams of the various sleep stages, but they are also associated with profoundly different EEG activity, physiology, content, and patterns of neural activation.

REMS is a state that became defined by the belief that REMS was equivalent to dreaming, a construct adapted to support the theoretical understandings of sleep, dreams, and neuroconsciousness. As noted in this brief presentation, almost all REMS-exclusive beliefs were conflated. Lucidity, bizarre content, intense emotion, paralysis, genital erections, autonomic disarray, dispersed brain activation, and eye movements are associated with but not restricted to the state of REMS.

20. Conclusions: REMS the Persistent Paradox

Early sleep neuroscientists were among the first to step outside the presumed role of dreaming in psychodynamics proposed by Freud, suggesting that the electrophysiologic aspects of REMS might induce phenomenological aspects of the associated dreaming. Allen Rechtschaffen published his now-famous paper on the single-mindedness and isolation of dreams [94]. David Foulks emphasized the intrapersonal processing of associative memory systems during dreaming [39]. However, their work was deemphasized by neuroscientists who fully equated REMS with dreaming as a unitary state. Today, after more than fifty years of confusing REMS with dreaming and the therapeutic fallacies of dream interpretation, their insights can seem prescient. The two aspects of the state that have retained their exclusivity to REMS are the state's unique electrophysiology and the long dreams with intense intrapersonal content originally associated with the term paradoxical. REMS—the paradoxical state—persists as a path that can be utilized to understand both dreaming and consciousness.

21. Future Explorations

As originally conceived, the wake-like EEG of REMS was an artifact of the technique of recording. Yet, the reason for this slewed perspective, the non-propagation of the REMS theta, retains its paradoxical fascination. The initial Berger EEG recordings from the 1920s were recordings of 10 Hz. Alpha [95]. Alpha and the other physiologic brain wave systems (sigma and delta) are easily detected on electrodes placed outside the brain. These frequencies have a clear intracellular origin based on neuronal ionic transport systems. All can be affected by applied external electromagnetic fields. REMS theta differs. It is not harmonic with the other physiologic CNS electrical fields, with the rhythms varying on a logarithmic base so that they minimally affect one another [89]. Theta exists independently of the other frequencies of sleep consciousness. Yet, theta interacts within a nested spectrum of other physiologic and consciousness-associated frequencies [33,90]. The theta of REMS is clearly very different, existing in a very different electrophysiologic environment from the other frequencies used to define the conscious states of sleep.

22. REMS Research Limitations

Since REMS theta does not propagate outside the brain, it is difficult to observe using scalp electrodes. This limitation has restricted human REMS studies to those that define REMS based on the absence of associated EEG frequencies on scalp EEG leads. Intracranial electrodes are required to directly observe REMS theta. QEEG can be utilized to retrospectively detect REMS-associated theta and gamma that is otherwise visually hidden in the recording. These systems are complex and produce a huge amount of filtered data that are difficult to assess using current statistical techniques. MEG has been used to detect REMS theta; however, methodologic limitations make it difficult for individuals to sleep when attached to the equipment. Without ready alternatives, in the near future, low-voltage, theta-absent EEG, recorded from scalp electrodes coupled with the presence of lowered EMG and conjugate eye movements, will continue to be used to describe the presence of REMS in humans. This derivative approach can be used to identify the occurrence of REMS but provides little information about the vagaries of its important and unique system of brain waves.

Up to this point, almost all studies have addressed dreaming as a ubiquitous, undifferentiated form of consciousness; therefore, little work has considered how dream-associated neuroprocessing, physiology, and content might differ between sleep stages. It is unclear how previous work from the era of REMS = dreaming might best be applied to understanding the dream forms and thought content collected from the physiologically and phenomenologically different states of sleep.

23. Conclusions

Perhaps the most paradoxical aspect of REMS dreaming is how little we actually know about the state. REMS dreams are likely to be those that we remember most often, those incorporated into our literature, and those transformative dreams more likely to be analyzed and interpreted. REMS dreams are among those most likely to be recalled, to be associated with sleep paralysis, classic nightmares, lucidity, genital erections, and conjugate eye movements. The content of REMS dreams may be the most bizarre, single-minded, and the most salient, associated with long reports that potentially reflect the psychodynamics of our functioning consciousness. Yet, none of these characteristics are exclusive to REMS.

Dreaming occurs and is reported throughout sleep. Clear evidence for the uniqueness of REMS dreaming is limited to longer report length (a variable that contributes to the narrative aspects of REMS dream reports), and the association of REMS with classic nightmares (in individuals without the diagnosis of PTSD). There have been few studies addressing the non-unitary nature of dreaming—the differences between the dreams reported from the different stages of sleep. This has produced a corpus of literature in which the mental activity associated with very different states of consciousness is conflated into something loosely referred to as dreaming. Today, dreaming is used to describe the metaphors of our existence, the poetics of literature, and the essence of our hopes and beliefs. REMS science has lagged far behind, as have the theories and constructs built upon the defined attributes and content of its associated dreaming. There has been little work directly addressing the electrophysiology of human REMS or the phenomenological characteristics of REMS dreams. The dreams of the different states of sleep consciousness are profoundly different. An exploration of those differences provides a path that can be used to understand the forms of human consciousness.

Funding: This research received no external funding.

Conflicts of Interest: The author declares no conflicts of interest.

References

1. Jouvet, M.; Michel, F.; Courjon, J. Sur un stade d'activité électrique cérébrale rapide au cours du sommeil physiologique. *C R. Soc. Biol.* **1959**, *153*, 1024–1028.
2. Jouvet, M. Recherches sur les structures nerveuses et les mécanismes responsables des differentes phases du sommeil physiologique. *Arch. Ital. Biol.* **1962**, *100*, 125–206. [PubMed]
3. Rasch, B.; Pommer, J.; Diekelmann, S.; Born, J. Pharmacological REM sleep suppression paradoxically improves rather than impairs skill memory. *Nat. Neurosci.* **2009**, *12*, 396–397. [CrossRef] [PubMed]
4. Hobson, J.A. *The Dreaming Brain: How the Brain Creates Both the Sense and Nonsense of Dreams*; Basic Books: New York, NY, USA, 1989.
5. Park, S.H.; Weber, F. Neural and Homeostatic Regulation of REM Sleep. *Front. Psychol.* **2020**, *11*, 531747. [CrossRef] [PubMed]
6. Corhen, S. Do Dogs dream? Dogs dream like humans and about similar things. In *Psychology Today*; Sussex Publishers: New York, NY, USA, 2019.
7. Google. Now, When You Want to Know If Dogs Dream, or Anything Else, the Google App Is There for You. Available online: https://www.ispot.tv/ad/7Mal/google-app-dreams (accessed on 12 June 2024).
8. McKenna, J.T.; Chen, L.; McCarley, R.W. Neuronal Models of REM-Sleep Control: Evolving Concepts. In *Rapid Eye Movement Sleep—Regulation and Function*; Mallick, B.N., Pandi-Perumal, S.R., McCarley, R.W., Morrison, A.R., Eds.; Cambridge University Press: Cambridge, UK, 2011; pp. 285–299.
9. Iber, C.; Ancoli-Israel, S.; Chesson, A.; Quan, S.F. American Academy of Sleep Medicine. In *The AASM Manual for the Scoring of Sleep and Associated Events: Rules, Terminology and Technical Specifications*; American Academy of Sleep Medicine: Westchester, NY, USA, 2007.
10. Siclari, F.; Baird, B.; Perogamvros, L.; Bernardi, G.; LaRocque, J.J.; Riedner, B.; Boly, M.; Postle, B.R.; Tononi, G. The neural correlates of dreaming. *Nat. Neurosci.* **2017**, *20*, 872–878. [CrossRef]
11. Lendner, J.D.; Helfrich, R.F.; Mander, B.A.; Romundstad, L.; Lin, J.J.; Walker, M.P.; Larsson, P.G.; Knight, R.T. An electrophysiological marker of arousal level in humans. *Elife* **2020**, *9*, e55092. [CrossRef]
12. Hobson, J.A. REM sleep and dreaming: Towards a theory of protoconsciousness. *Nat. Rev. Neurosci.* **2009**, *10*, 803–862. [CrossRef]
13. Vijayan, S.; Lepage, K.Q.; Kopell, N.J.; Cash, S.S. Frontal beta-theta network during REM sleep. *Elife* **2017**, *6*, e18894. [CrossRef] [PubMed]
14. Antrobus, J.; Kondo, T.; Reinsel, R.; Fein, G. Dreaming in the late morning: Summation of rem and diurnal cortical activation. *Conscious. Cogn.* **1995**, *4*, 275–299. [CrossRef] [PubMed]
15. McCarley, R.W.; Nelson, J.P.; Hobson, J.A. Ponto-Geniculo-Occipital (PGO) burst neurons: Correlative evidence. *Science* **1978**, *4352*, 269–272. [CrossRef]
16. Gott, J.A.; Liley, D.T.J.; Hobson, J.A. Towards a functional understanding of PGO waves. *Front. Hum. Neurosci.* **2017**, *11*, 00089. [CrossRef] [PubMed]
17. Rochart, R.; Liu, Q.; Fonteh, A.N.; Harrington, M.G.; Arakaki, X. Compromised Behavior and Gamma Power During Working Memory in Cognitively Healthy Individuals With Abnormal CSF Amyloid/Tau. *Front. Aging Neurosci.* **2020**, *12*, 574214. [CrossRef] [PubMed]
18. Nuwer, M. Assessment of digital EEG, quantitative EEG and EEG brain mapping: Report of the American Academy of Neurology and the American Clinical Neurophysiology Society. *Neurology* **1997**, *49*, 277–292. [CrossRef]
19. Popa Dragos, H.; Pantelemon, C.; Verisezan, R.O.; Strilciuc, S. The Role of Quantitative EEG in the Diagnosis of Neuropsychiatric Disorders. *J. Med. Life* **2020**, *13*, 8–15. [CrossRef] [PubMed]
20. Bergel, A.; Deffieux, T.; Demené, C.; Tanter, M.; Cohen, I. Local hippocampal fast gamma rhythms precede brain-wide hyperemic patterns during spontaneous rodent REM sleep. *Nat. Commun.* **2018**, *9*, 5364. [CrossRef] [PubMed]
21. Edelman, G.M.; Tononi, G. *A Universe of Consciousness: How Matter Becomes Imagination*; Basic Books: New York, NY, USA, 2000.

22. Voss, U.; Klimke, A. Dreaming during REM sleep: Autobiographically meaningful or a simple reflection of a Hebbian-based memory consolidation process? *Arch. Ital. Biol.* **2018**, *156*, 99–111. [CrossRef] [PubMed]
23. Pagel, J.F. *Lucid Dreaming as Sleep Offset Waking—Sleep 34—Abstract Supplement*; Oxford Academic Press: New York, NY, USA, 2013.
24. Crick, F.; Koch, C. Towards a neurobiological theory of consciousness. *Semin. Neurosci.* **1990**, *2*, 263–275.
25. Llinás, R.; Ribary, U. Coherent 40-Hz Oscillation Characterizes Dream State in Humans. *Proc. Natl. Acad. Sci. USA* **1993**, *90*, 2078–2081. [CrossRef]
26. Braeutigam, S. Magnetoencephalography: Fundamentals and established and emerging clinical applications in radiology. *ISRN Radiol.* **2013**, *2013*, 529463. [CrossRef]
27. Brancaccio, A.; Tabarelli, D.; Bigica, M.; Baldauf, D. Cortical source localization of sleep-stage specific oscillatory activity. *Sci. Rep.* **2020**, *10*, 6976. [CrossRef] [PubMed]
28. Buzsáki, G. Theta oscillations in the hippocampus. *Neuron* **2002**, *33*, 325–340. [CrossRef] [PubMed]
29. Steriade, M.; McCarley, R.W. *Brain Control of Wakefulness and Sleep*; Springer: New York, NY, USA, 2005.
30. Noguchi, A.; Yamashiro, K.; Matsumoto, N.; Ikegaya, Y. Theta oscillations represent collective dynamics of multineuronal membrane potentials of murine hippocampal pyramidal cells. *Commun. Biol.* **2023**, *6*, 398. [CrossRef] [PubMed]
31. Cantero, M.; Cantiello, H. Microtubule Electrical Oscillations and Hippocampal Function. *J. Neurol. Neuromed.* **2020**, *5*, 1–5. [CrossRef]
32. Groppe, D.M.; Bickel, S.; Keller, C.J.; Jain, S.K.; Hwang, S.T.; Harden, C.; Mehta, A.D. Dominant frequencies of resting human brain activity as measured by the electrocorticogram. *NeuroImage* **2013**, *202*, 116066. [CrossRef] [PubMed]
33. Klimesch, W. The frequency architecture of brain and brain body oscillations: An analysis. *Eur. J. Neurosci.* **2018**, *48*, 2431–2453. [CrossRef] [PubMed]
34. Simor, P.; van Der Wijk, G.; Gombos, F.; Kovács, I. The paradox of rapid eye movement sleep in the light of oscillatory activity and cortical synchronization during phasic and tonic microstates. *NeuroImage* **2019**, *79*, 223–233. [CrossRef]
35. O'Keefe, J.; Recce, M.L. Phase relationship between hippocampal place units and the EEG theta rhythm. *Hippocampus* **1993**, *3*, 317–330. [CrossRef] [PubMed]
36. Lee, A.K.; Wilson, M.A. Memory of sequential experience in the hippocampus during slow wave sleep. *Neuron* **2002**, *36*, 1183–1194. [CrossRef] [PubMed]
37. Hasselmo, M.E. What is the function of hippocampal theta rhythm?—Linking bahavioral data to phasic properties of field potential and unit recording data. *Hippocampus* **2005**, *15*, 936–949. [CrossRef]
38. Polanía, R.; Nitsche, M.A.; Korman, C.; Batsikadze, G.; Paulus, W. The importance of timing in segregated theta phase-coupling for cognitive performance. *Curr. Biol.* **2012**, *22*, 1314–1318. [CrossRef]
39. Foulks, D. *Dreaming: A Cognitive-Psychological Analysis*; Hilsdale, N.J., Ed.; Erlbaum: Hillsdale, IL, USA, 1985.
40. Dement, W.; Kleitman, N. The relation of eye movements during sleep to dream activity: An objective method for the study of dreaming. *J. Exp. Psychol.* **1957**, *53*, 339. [CrossRef]
41. Nielsen, T. Variations in dream recall frequency and dream theme diversity by age and sex. *Front. Neurol.* **2012**, *3*, 106. [CrossRef]
42. Solms, M. Dreaming and Rem sleep are controlled by different brain mechanisms. *Behav. Brain Sci.* **2000**, *23*, 843–850. [CrossRef]
43. Pagel, J.F. The neuro-electric alignment of dreaming—The sleep/dream state frequencies of consciousness. *Dreaming* **2023**, *33*, 252–263. [CrossRef]
44. Kaplan-Solms, K.; Solms, M. *Clinical Studies in Neuro-Psychoanalysis: Introduction to a Depth Neuropsychology*; Karnac Books: London, UK, 2000.
45. Yu, C.K.C. We dream about typical themes in both REM and non-REM sleep. *Dreaming* **2020**, *30*, 317–328. [CrossRef]
46. Montangero, J. Dreams are Narrative Simulations of Autobiographical Episodes, Not Stories or Scripts: A Review. *Dreaming* **2012**, *22*, 157–172. [CrossRef]
47. Pagel, J.F.; Helfter, P. Drug induced nightmares—An etiology based review. *Hum. Psychopahrmacol Clin. Exp.* **2003**, *18*, 59–67. [CrossRef] [PubMed]
48. McCarley, R.W.; Hobson, J.A. Neuronal excitability modulation over the sleep cycle: A structural and mathematical model. *Science* **1975**, *189*, 58–60. [CrossRef] [PubMed]
49. Nielsen, T. *A Review of Mentation in REM NREM Sleep: "Covert" REM Sleep as a Possible Reconciliation of Two Opposing Models in Sleep Dreaming: Scientific Advances Reconsiderations*; Pace-Schott, E., Solms, M., Blagrove, M., Harand, S., Eds.; Cambridge University Press: Cambridge, UK, 2003; pp. 59–74.
50. Hobson, J.A.; Pace-Schott, E.F.; Stickgold, R. Dreaming and the brain: Toward a cognitive neuroscience of conscious states. In *Sleep and Dreaming: Scientific Advances and Reconsiderations*; Pace-Schott, E.F., Solms, M., Blagrove, M., Harnad, S., Eds.; Cambridge University Press: Cambridge, UK, 2003; pp. 1–50. [CrossRef]
51. Koch, C.; Massimini, M.; Boly, M.; Tononi, G. Neural correlates of consciousness: Progress and problems. *Nat. Rev. Neurosci.* **2016**, *17*, 307–321. [CrossRef]
52. Walker, M. *Why We Sleep*; Scribner: New York, NY, USA, 2017.
53. Pfaff, D. *Brain Arousal and Information Theory*; Harvard University Press: Cambridge, MA, USA, 2006. [CrossRef]
54. Freud, S. *The Interpretation of Dreams, Standard ed.*; Hogarth Press: London, UK, 1900; Volume 4.
55. Grunbum, A. *The Foundations of Psychoanalysis: A Philosophical Critique*; University of California Press: Berkeley, CA, USA, 1984.

56. Pagel, J.F. The Sleep-Dream State: Historic Philosophic Perspectives. In *Rapid Eye Movement Sleep—Regulation and Function*; Mallick, B.N., Pandi-Perumal, S.R., McCarley, R.W., Morrison, A.R., Eds.; Cambridge University Press: Cambridge, UK, 2011; pp. 1–7.

57. Pagel, J.F. REM Sleep Dreaming: An Elegy for the Special Relationship. In *Dreams and Dreaming: Analysis, Interpretation and Meaning*; Nebeolisa, O., Ed.; Nova: New York, NY, USA, 2018; pp. 27–54. ISBN 978-1-53613-017-1.

58. Domhoff, G.W.; Schneider, A. Much Ado about Very Little: The Small Effect Sizes When Home and Laboratory Collected Dreams Are Compared. *Dreaming* **1999**, *9*, 139–151. [CrossRef]

59. Yu, C.K.C. Toward 100% Dream Retrieval by Rapid-Eye-Movement Sleep Awakening: A High-Density Electroencephalographic Study. *Dreaming* **2014**, *24*, 1–17. [CrossRef]

60. McCormick, L.; Nielsen, T.; Ptito, M.; Hassainia, F.; Ptito, A.; Villemure, J.G.; Vera, C.; Montplaisir, J. REM sleep dream mentation in right hemispherectomized patients. *Neuropsychologia* **1997**, *35*, 695–701. [CrossRef]

61. Hobson, J.A.; Hoffman, S.A.; Helfand, R.; Kostner, D. Dream bizarreness and the activation-synthesis hypothesis. *Hum. Neurobiol.* **1987**, *6*, 157–164.

62. Solms, M. The conscious id. *Neuropsychoanalysis* **2013**, *15*, 5–85. [CrossRef]

63. Domhoff, G. *The Neurocognitive Theory of Dreaming: The Where, How, When, What, and Why of Dreams*; M.I.T. Press: Cambridge, MA, USA, 2022. [CrossRef]

64. Martin, J.M.; Andriano, D.W.; Mota, N.B.; Mota-Rolim, S.A.; Araújo, J.F.; Solms, M.; Ribeiro, S. Structural differences between REM and non-REM dream reports assessed by graph analysis. *PLoS ONE* **2020**, *15*, 7. [CrossRef]

65. States, B. *Seeing in the Dark—Reflections on Dreams and Dreaming*; Yale University Press: New Haven, CT, USA, 1993.

66. Hunt, H. *The Multiplicity of Dreams: Memory, Imagination and Consciousness*; Yale University Press: New Haven, CT, USA, 1989.

67. Voss, U.; Holzmann, R.; Tuin, I.; Hobson, J.A. Lucid dreaming: A state of consciousness with features of both waking and non-lucid dreaming. *Sleep* **2009**, *32*, 1191–1200. [CrossRef]

68. LaBerge, S. *Lucid Dreaming*; JP Tarcher: Los Angeles, CA, USA, 1985.

69. Holzinger, B.; LaBerge, S.; Levitan, L. Psychological correlates of lucid dreaming. *Dreaming* **2006**, *16*, 88–96. [CrossRef]

70. Dresler, M.; Wehrle, R.; Spoormaker, V.I.; Koch, S.P.; Holsboer, F.; Steiger, A.; Obrig, H.; Sämann, P.G.; Czisch, M. Neural correlates of dream lucidity obtained from contrasting lucid versus non-lucid REM sleep: A combined EEG/fMRI case study. *Sleep* **2012**, *35*, 1017–1020. [CrossRef] [PubMed]

71. White, J.; Drinnan, M.; Smithson, A.; Griffiths, C.; Gibson, G. Respiratory muscle activity and oxygenation during sleep in patients with muscle weakness. *Eur. Respir. J.* **1995**, *8*, 807–814. [CrossRef]

72. Peng, A.; Ji, S.; Li, W.; Lai, W.; Qiu, X.; He, S.; Dong, B.; Huang, C.; Chen, L. Gastric Electrical Dysarrhythmia in Probable Rapid Eye Movement Sleep Behavior Disorder. *Front. Neurol.* **2021**, *12*, 687215. [CrossRef]

73. Vetrugno, R.; Montagna, P. From REM sleep behaviour disorder to status dissociatus: Insights into the maze of states of being. *Sleep. Med.* **2011**, *12*, S68–S71. [CrossRef] [PubMed]

74. MacFarlane, J.; Wilson, T. A Relationship Between Nightmare Content and Somatic Stimuli in a Sleep-Disordered Population: A Preliminary Study. *Dreaming* **2006**, *16*, 53–59. [CrossRef]

75. Mainieri, G.; Maranci, J.B.; Champetier, P.; Leu-Semenescu, S.; Gales, A.; Dodet, P.; Arnulf, I. Are sleep paralysis and false awakenings different from REM sleep and from lucid REM sleep? A spectral EEG analysis. *J. Clin. Sleep Med.* **2021**, *17*, 719–727. [CrossRef]

76. Levin, R.; Nielsen, T. Nightmares, bad dreams, and emotion dysregulation: A review and new neurocognitive model of dreaming. *Curr. Dir. Psychol. Sci.* **2009**, *18*, 84–88. [CrossRef]

77. Pagel, J.F. *Parasomnia Dreaming—Exploring Other forms of Sleep Consciousness*; Pagel, J.F., Ed.; Nova Science Publishers Inc.: New York, NY, USA, 2020; pp. i–ix. ISBN 978-3-030-55908-3.

78. Wasserman, M.D. Psychoanalytic dream theory and recent neurobiological findings about rem sleep. *J. Am. Psychoanal. Assoc.* **1984**, *32*, 831–846. [CrossRef]

79. Mann, K.; Sohn, M. Spontaneous Nocturnal Erections—Physiology and Clinical Applications. *Somnologie* **2005**, *9*, 119–126. [CrossRef]

80. Brooks, P.L.; Peever, J.H. Identification of the transmitter and receptor mechanisms responsible for REM sleep paralysis. *J. Neurosci.* **2012**, *32*, 9785–9795. [CrossRef]

81. Lai, Y.Y.; Seigal, J.M. Pontomedullary mediated REM-sleep atonia. In *Rapid Eye Movement Sleep—Regulation and Function*; Mallick, B.N., Pandi-Perumal, S.R., McCarley, R.W., Morrison, A.R., Eds.; Cambridge University Press: Cambridge, UK, 2012; pp. 121–129.

82. Stefani, A.; Högl, B. Nightmare Disorder and Isolated Sleep Paralysis. *Neurotherapeutics* **2021**, *18*, 100–106. [CrossRef]

83. Schenck, C.H.; Mahowald, M.W. Rapid eye movement sleep parasomnias. *Neurol. Clin.* **2005**, *23*, 1107–1126. [CrossRef]

84. Valli, M.; Uribe, C.; Mihaescu, A.; Strafella, A.P. Neuroimaging of rapid eye movement sleep behavior disorder and its relation to Parkinson's disease. *J. Neurosci. Res.* **2022**, *100*, 1815–1833. [CrossRef] [PubMed]

85. Pagel, J.F.; Nielsen, T. Parasomnias: Recurrent Nightmares. In *The International Classification of Sleep Disorders—Diagnostic and Coding Manual (ICD-10)*; American Academy of Sleep Medicine: Westchester, NY, USA, 2005.

86. Pagel, J.F.; Shocknasse, S. Dreaming and insomnia: Polysomnographic correlates of reported dream recall frequency. *Dreaming* **2007**, *17*, 140–151. [CrossRef]

87. Phelps, A.J.; Kanaan, R.A.A.; Worsnop, C.; Redston, S.; Ralph, N.; Forbes, D. An Ambulatory Polysomnography Study of the Post-traumatic Nightmares of Post-traumatic Stress Disorder. *Sleep* **2018**, *41*, zsx188. [CrossRef]
88. Domhoff, G.W. Fox KC. Dreaming and the default network: A review, synthesis, and counterintuitive research proposal. *Conscious. Cogn.* **2015**, *33*, 342–353. [CrossRef]
89. Buzsaki, G. *The Brain from Inside Out*; Oxford University Press: Oxford, UK, 2019.
90. Hunt, T.; Schooler, J.W. The Easy Part of the Hard Problem: A Resonance Theory of Consciousness. *Front. Hum. Neurosci.* **2019**, *13*, 00378. [CrossRef]
91. Mehta, M.R.; Quirk, M.C.; Wilson, M.A. Experience-dependent asymmetric shape of hippocampal receptive fields. *Neuron* **2000**, *25*, 707–715. [CrossRef]
92. Guo, W.; Fouda, M.E.; Eltawil, A.M.; Salama, K.N. Neural Coding in Spiking Neural Networks: A Comparative Study for Robust Neuromorphic Systems. *Front Neurosci.* **2021**, *15*, 638474. [CrossRef]
93. Antonioni, A.; Raho, E.M.; Sensi, M.; Di Lorenzo, F.; Fadiga, L.; Koch, G. A new perspective on positive symptoms: Expression of damage or self-defence mechanism of the brain? *Neurol. Sci.* **2024**, *4*, 2347–2351. [CrossRef]
94. Rechtschaffen, A. The Single-Mindedness and Isolation of Dreams. *Sleep* **1978**, *1*, 97–109. [CrossRef] [PubMed]
95. Tudor, M.; Tudor, L.; Tudor, K.I. Hans Berger (1873–1941)—The history of electroencephalography. *Acta Medica Croat.* **2005**, *59*, 307–313. [PubMed]

Article

Sensational Dreams: The Prevalence of Sensory Experiences in Dreaming

Anna C. van der Heijden [1], Jade Thevis [1,†], Jill Verhaegen [1,†] and Lucia M. Talamini [1,2,*]

1   Department of Psychology, Brain and Cognition, University of Amsterdam, 1018 WT Amsterdam, The Netherlands; a.c.vanderheijden@uva.nl (A.C.v.d.H.); jill.kirchhof@student.uva.nl (J.V.)
2   Amsterdam Brain and Cognition, University of Amsterdam, 1001 NK Amsterdam, The Netherlands
*   Correspondence: l.m.talamini@uva.nl
†   These authors contributed equally to this work.

Abstract: Dreaming, a widely researched aspect of sleep, often mirrors waking-life experiences. Despite the prevalence of sensory perception during wakefulness, sensory experiences in dreams remain relatively unexplored. Free recall dream reports, where individuals describe their dreams freely, may not fully capture sensory dream experiences. In this study, we developed a dream diary with direct questions about sensory dream experiences. Participants reported sensory experiences in their dreams upon awakening, over multiple days, in a home-based setting (n = 3476 diaries). Our findings show that vision was the most common sensory dream experience, followed by audition and touch. Olfaction and gustation were reported at equally low rates. Multisensory dreams were far more prevalent than unisensory dreams. Additionally, the prevalence of sensory dream experiences varied across emotionally positive and negative dreams. A positive relationship was found between on the one hand sensory richness and, on the other emotional intensity of dreams and clarity of dream recall, for both positive and negative dreams. These results underscore the variety of dream experiences and suggest a link between sensory richness, emotional content and dream recall clarity. Systematic registration of sensory dream experiences offers valuable insights into dream manifestation, aiding the understanding of sleep-related memory consolidation and other aspects of sleep-related information processing.

Keywords: dreams; dream diary; sensory dream experiences

Citation: van der Heijden, A.C.; Thevis, J.; Verhaegen, J.; Talamini, L.M. Sensational Dreams: The Prevalence of Sensory Experiences in Dreaming. *Brain Sci.* **2024**, *14*, 533. https://doi.org/10.3390/brainsci14060533

Academic Editors: Luigi De Gennaro, Sergio A. Mota-Rolim, Brigitte Holzinger and Michael Nadorff

Received: 23 April 2024
Revised: 15 May 2024
Accepted: 19 May 2024
Published: 24 May 2024

1. Introduction

Dreaming, which may be defined as a form of mental activity during sleep [1], is a long-studied phenomenon. Previous research has shown that daytime activities can shape dream content [2–6]. Conversely, dreaming affects daytime functioning [5,6] by enhancing memory consolidation [7,8], contributing to problem-solving [9,10] and facilitating emotional processing [11]. Thus, there appears to be a bidirectional relationship between wakefulness and dreaming.

While waking life provides a sensory rich experience, the prevalence of sensory experiences in dreams remains understudied. A few studies investigated sensory dream experiences using free recall dream reports, in which participants were instructed to freely describe their dream content [1,12,13]. These dream reports were subsequently scored for sensory experiences by an independent rater. A limitation of this approach is that participants might omit sensory dream experiences in the free recall report, obscuring the prevalence of sensory dream experiences [1,12,13].

Another study on sensory dream experiences used a single, retrospective questionnaire to assess participants' dream experiences over the previous year [14–16]. This approach included direct questions on sensory dream experiences. While direct questions may help to capture the prevalence of sensory dream experiences, the long period from which dreams needed to be recalled may have led to the forgetting of dream content. Consequently, dream recall over long retention periods may again lead to distortion in the assessment of the prevalence of sensory dream experiences [14–16].

In this study, we employed a seven-day dream diary on sensory experiences, collected at final morning awakenings, in the home environment. By explicitly asking about sensory experiences in a dream diary, no subjective interpretation of dream content was required by an independent rater. Specifically, participants were asked to quantify the number of dreams in which they experienced vision, audition, touch, olfaction, and gustation during the previous night. By assessing dream reports upon final morning awakening, we reduced the time between the occurrence of a dream and the report, resulting in a shorter recall interval than the delayed questionnaire methods that require recalling dream details overthe past year.

We will compare the prevalence of sensory dream experiences across all sensory modalities and quantify the prevalence of combined experience of different senses in multisensory dream experiences. Prevalence will be determined as a percentage of both the total number of reports and participants; the latter to evaluate potential interindividual differences in the ability to experience each sensory modality. Additionally, the prevalence of sensory modalities will be compared between emotionally positive and negative dreams. Finally, we will explore the relationship between sensory richness and dream emotionality, as well as the relationship between sensory richness and clarity of dream recall, for both positive and negative dreams.

2. Materials and Methods

2.1. Sample and Procedure

In total, 611 participants were recruited for this study. All participants were asked to fill in the dream diary for seven consecutive days upon final morning awakening. Incomplete diary reports for any day resulted in the exclusion of only the corresponding diary entry. Participants who completed the diary on fewer than two days were excluded, resulting in a final sample of 533 participants (78.6% female, 21.4% male) with a mean age of 22.0 (SD ± 7.24, ranging from 18 to 70 years old; see Supplemental Figure S1). Participants were recruited through the participant pool of the Department of Psychology from the University of Amsterdam, various online social media platforms, and flyers. Inclusion criteria were a minimum age of 18 years old, normal or corrected-to-normal hearing and vision and no other sensory impairments. Participation was voluntarily; no financial compensation was provided. All procedures were approved by the ethical committee of the University of Amsterdam.

Participation occurred online via the questionnaire application Qualtrics (Qualtrics, Provo, UT, USA). Participants completed an entry survey to provide their contact details, age, biological sex, and country of residence. Next, participants digitally received the sensory dream diary for seven consecutive days. To aid participants, email reminders containing a link to the diary were sent each morning.

2.2. Sensory Dream Diary

A customized dream diary, comprising twenty-two items, was developed for this study. Participants reported the total number of dreams experienced during the previous night. If participants reported having had one or more dreams, they were next asked to indicate the presence of specific sensory experiences, such as vision, audition, touch, olfaction, and gustation, cumulatively across all reported dreams. Follow-up, open-ended questions were presented for each indicated sensory modality, asking about the number of dreams in which the sense occurred.

In addition, participants were asked to rate the emotionality of the dream eliciting the strongest positive emotion and the one evoking the strongest negative emotion using a seven-point Likert scale (0 = "none", 6 = "extreme" [17]). Subsequently, sensory dream experiences were assessed for the most positive and negative dream separately. For all questions, an "I do not remember" or "None of the above" option was present to avoid the reporting of false positive sensory experiences.

The diary included basic questions on sleep quality and duration [18], as well as questions on the consumption of drugs and medication. It allowed multiple dreams per night to be reported. Reporting the actual dream content was optional. In case the participant reported to not recall any dreams, questions on dream content were automatically skipped. The expected completion time ranged from five to ten minutes. The complete diary has been provided in Supplemental Material S1.

2.3. Statistical Analyses

Jamovi 2.4.8 was used for all statistical analyses. Two-tailed tests were performed for all analyses.

2.3.1. Sensory Dream Experiences

To compare the prevalence of sensory dream experiences among all sensory modalities, a generalized linear mixed model (GLMM) was employed, executed using the GAMLj3 package of Jamovi. The fixed factor was the sensory modalities (vision, audition, touch, gustation, and olfaction), and the dependent variable was the dream report count per sensory modality. The multiday diary reports per participant were included as a random effect to take into account the interindividual variability in the number of dream experiences. The following settings were applied: Poisson distribution, logit link function, the Restricted Maximum Likelihood (REML) estimation method, the random effects test Likelihood Ratio test (LRT), and a Bonferroni post hoc correction. All participants were included in this analysis, comprising 3476 dream diaries.

In addition, to obtain a measure of intersubject variability in the prevalence of sensory dream experiences, we calculated a percentage of all participants who reported this sensory experience at least once during the seven-day study period. This percentage was calculated for all sensory modalities independently.

2.3.2. Multisensory Dream Experiences

To assess how different sensory components co-occur within a single dream, we selected diaries that included only one dream per night to prevent overlap across multiple dreams. This resulted in a subsample of 1221 diaries, obtained from 498 participants. For all possible (two- to five-way) combinations of the five sensory modalities, a percentage of the total dream reports was calculated.

2.3.3. Link between Sensory Richness, Emotionality, and Clarity

To study a potential link between sensory richness and emotionality in dreams, reports were selected based on the presence of emotion. Exclusions were only made when emotion was indicated as not remembered; otherwise, reports were included even if emotion was reported to be absent. This resulted in a subsample of 1650 diaries from 497 participants for positive dreams and 1567 from 494 participants for negative dreams. The relationship between sensory richness and emotionality was examined separately for positive and negative emotions. Sensory richness was quantified as the percentage of sensory modalities present (e.g., presence of all five senses yielded a 100% score). Dream emotionality was assessed using a seven-point Likert scale (0 = None, 6 = Extreme). Spearman's correlation tests were conducted including sensory richness and emotional intensity for positive and negative emotions.

In addition, a potential relationship between sensory richness and dream clarity was studied. Dream clarity denotes the extent to which dream content could be recalled upon waking and was assessed by a rating on a scale from 1 to 10 (1 = extremely unclear, 10 = extremely clear). Spearman's correlation was applied. Additionally, a generalized linear mixed model was applied to compare dream clarity between recent and remote dreams, with dream clarity used as dependent variable, dream remoteness as fixed factor, and the participant as random factor. Recent dreams were defined as occurring within the same hour of final awakening, whereas remote dreams occurred longer than one hour before final awakening. To be able to link the dream occurrence to an individual dream, reports were selected with a total of one dream per night, in which the dream occurred before final awakening and diary entry after final awakening, resulting in a subsample of 607 reports.

Finally, to assess whether the prevalence of sensory modalities differed between positive and negative dreams, we utilized a 2 (positive and negative emotion) $\times$ 5 (sensory modalities: vision, audition, touch, gustation, olfaction) chi-square test. Here, a subsample of dreams that included emotion was selected (based on the 7-point Likert scale, emotion was ≥ 1). To address imbalances in the counts of different emotionality types, we employed undersampling. This involved matching the count of the most frequent emotionality type to the least frequent, resulting in a subsample of 1117 reports per emotionality type. Post hoc comparisons were conducted in R 4.3.1 [19] using the chisq.posthoc.test package, including a Bonferroni correction.

3. Results

3.1. Demographics

In total, 533 participants (78.6% female, 21.4% male) with a mean age of 22.0 (SD $\pm$ 7.24) were included, accounting for a total of 3476 reported dreams. Across all diaries, participants slept for 7 h and 38 min on average and reported a moderate-to-good sleep quality (score 3.5 out of 5). The average number of dreams per night was determined by averaging the number of dreams over all days per participant and subsequently taking the group average. Participants reported an average of 0.80 ($\pm$0.74) dreams per night. Zero dreams per night were reported most frequently, while one dream per night was the most common quantity if a dream occurred (Figure 1A). On average, participants completed the diary for 6.52 ($\pm$1.02) out of 7 days (Supplemental Figure S2). Most dreams were reported to occur within the same hour as the final awakening (Figure 1B). The average clarity of the dream recall was 5.16 ($\pm$2.32, on a scale from 1 (extremely unclear) to 10 (extremely clear). See Table 1 for all descriptives.

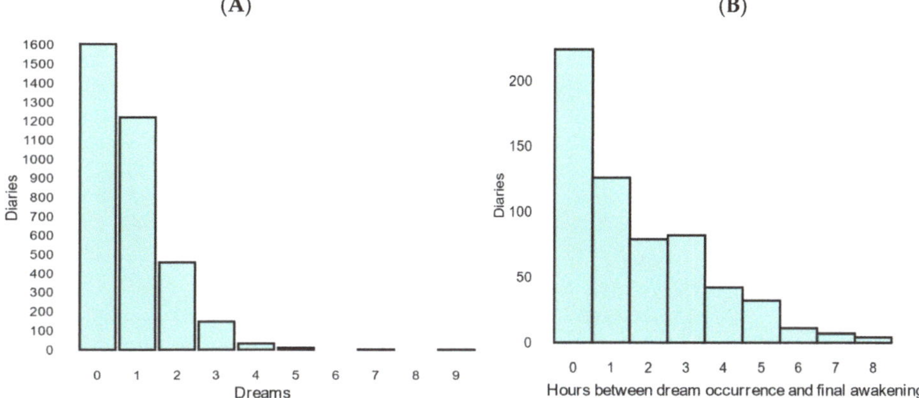

Figure 1. (**A**) Number of dreams per night and (**B**) hours between dream occurrence and final awakening. (**A**) The distribution of the number of dreams per night has been illustrated. Most diaries reported zero dreams, while one dream per night was the most common quantity if a dream occurred. The range of reported dreams per night spanned from zero to nine, displaying a descending pattern where higher numbers of dreams tended to occur less frequently. (**B**) The histogram displays the distribution of dream occurrences, expressed in hours before final awakening. In order to link the time of dream occurrence to a single dream, only reports with one dream per night were selected.

Table 1. Descriptives of dream reports for overall dream reports and sleep quality.

	Mean	SD
Dream diary entries	6.52	1.02
Sleep quality	3.50	0.93
Total sleep time	7:38 h	1.28 h
Bedtime	00:45 h	01:39 h
Wake-up time	08:41h	01:42 h
Number of dreams	0.80	0.74
Dream clarity	5.16	2.32
Dream occurrence (prior to final awakening)	01:39 h	1:52 h
Dream retention interval (between dream occurrence and diary entry)	02:43 h	01:39 h
Diary entry interval (between waking up and diary entry)	1:13 h	2:16 h
Number of nightmares	0.152	0.446

3.2. Sensory Dream Experiences

The prevalence of sensory dream experiences was compared across sensory modalities using a generalized linear mixed model (GLMM), using sensory modality as a fixed factor, report count as a dependent variable, and participant as a random factor. The GLMM revealed a significant difference in the prevalence of different sensory modalities in dreams ($\chi^2 = 2621$, df = 4.00, $p < 0.001$). In the subsequent post hoc analyses with Bonferroni correction, significant differences in prevalence emerged across all senses, except for olfactory and gustatory experiences. More specific, vision was the most prevalent sensory dream experience (51.7%), followed by audition (39.4%) and touch (18.2%) ($p < 0.001$). Olfaction (2.6%) and gustation (2.6%) occurred at equally low rates ($p > 0.05$) (Figure 2).

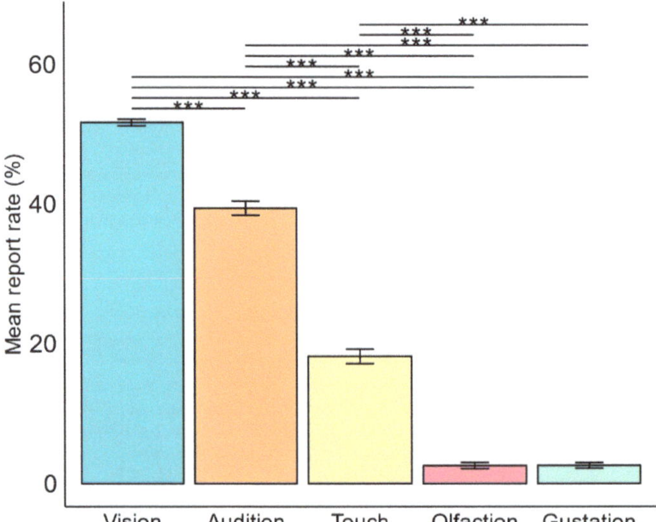

Figure 2. Comparison of the prevalence of sensory dream experiences. In analyzing the prevalence of sensory dream experiences, vision emerged as the most frequent sensation compared to all other senses ($p < 0.001$). Audition followed as the second most prevalent sensation, surpassing touch, olfaction, and gustation ($p < 0.001$). Touch ranked third in prevalence, exceeding olfaction, and gustation ($p < 0.001$). Olfaction and gustation exhibited equally low rates of occurrence ($p > 0.05$). The mean report rate percentage was calculated in two steps: first, the percentage of dreams containing the sensory dream experience was calculated over all reports per participant, and then this percentage was averaged across all participants. Error bars represent standard deviation. *** = $p < 0.001$.

To assess potential interindividual differences in the occurrence of sensory dream experiences, we computed, per sensory modality, the percentage of participants who reported this modality at least once during the seven-day diary period. We found that nearly all participants experienced vision (95.9%). In addition, both audition (85.6%) and touch (62.1%) were experienced by most participants. In contrast, gustation (16.5%) and olfaction (14.6%) were reported by a minority. While gustation and olfaction occurred at equal rates across all dream reports (2.6%), gustation was experienced by a numerically higher number of participants than olfaction. Moreover, amongst participants who reported dreams, 0.2% reported to not have had any sensory experiences, whereas 1.4% reported to have dreamt but to be unsure whether sensory experiences were present. The latter supports the idea that sensory experiences are paired with dream recall in most, but not all, cases. These findings demonstrate a notable intersubject variability in sensory dream experiences, highlighting that no sensory modality was experienced by all participants.

3.3. Multisensory Dream Experiences

Next, we evaluated the prevalence of combinations of sensory dream experiences within a single dream. A combination of audition and vision was most frequently present (40.0%), across all unique two- to five-way combinations of sensations (Figure 3). The second most prevalent sensory dream experience was a combination of vision, audition and touch (23.6%). These combinations exceeded the independent occurrence of vision (21%), audition (1.4%), touch (0.5%), olfaction (0%), and gustation (0%), demonstrating that multisensory dream experiences are more prevalent than unisensory ones. However, not all potential sensory combinations were represented, with only 58.1% of the potential multisensory combinations being reported (see Figure 3). The occurrence of dreams involving all five senses was relatively uncommon, being present in only 0.9% of reports. The least frequent, but still reported, sensory combination was composed of vision, olfaction, and gustation (0.1%), as well as vision, olfaction, and touch (0.1%). Overall, these findings indicate that multisensory dream experiences are more prevalent than unisensory dreams, with high variability in the sensory modalities involved.

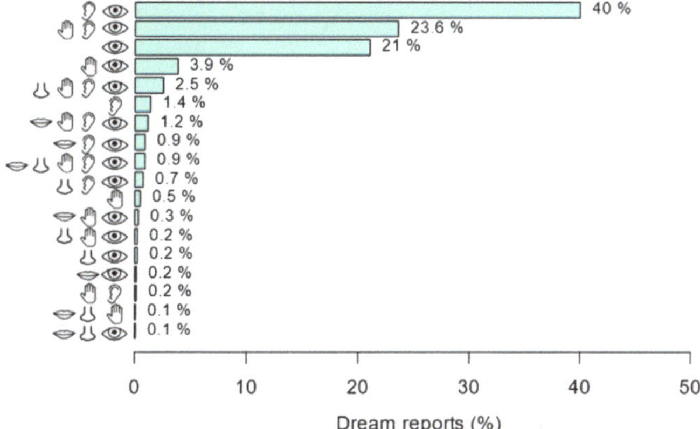

Figure 3. Prevalence of combinations of sensory modalities in dream experiences. This chart represents the composition and prevalence of multisensory dream experiences (prevalence is expressed as a percentage of total dream reports that contained a single dream). Sensory combinations not represented in the chart were not reported by any participant (report rate of 0%). These include the less frequently reported senses, gustation, and olfaction, occurring in isolation or in 2-, 3- and 4-way combinations with other senses. Moreover, 2.1% of dreams did not include any sensory experiences. Icons represent the following sensory dream experiences: eye = vision, ear = audition, hand = touch, nose = olfaction, mouth = gustation.

3.4. Sensory Rich Dreams Are Associated with Higher Emotional Intensity and Clarity

To study the relationship between dream sensory richness and emotional intensity, Spearman's rank test was performed. Sensory richness was defined as the percentage of senses present in the dream experience. A positive relationship between sensory richness and emotional intensity was found both for dreams with positive (r = 0.367) and negative dream emotion (r = 0.465) ($p < 0.001$). This indicates that dreams with richer sensory experiences tend to have more intense emotional content.

In addition, sensory richness was positively correlated with dream clarity for both positive (r = 0.247) and negative dreams (r = 0.245) ($p < 0.001$). This suggests that dreams characterized by richer sensory experiences are reported to be more clearly remembered. Of note, the strength of this correlation was less (r = 0.247) than the correlation between emotional intensity and sensory richness (r = 0.367) ($p < 0.001$). Additionally, we found that recent dreams, which were reported to occur within one hour before final awakening, tended to be more clearly remembered than remote dreams (trend-level significance; $\chi^2 = 3.78$, df = 1, $p = 0.052$). While dreams were most clearly remembered when occurring within one hour before final awakening, clarity did not decrease linearly with the time elapsed since the dream (see Figure S3).

Finally, we evaluated whether the prevalence of individual sensory modalities differed between dreams featuring positive and negative emotions. A chi-square test revealed that the prevalence of sensory modalities differed significantly between positive and negative dreams ($\chi^2 = 10.1$, $p = 0.038$). Positive dreams appeared to have a higher frequency of gustatory, olfactory, and visual sensations relative to negative dreams, whereas negative dreams seemed to exhibit more frequent auditory and tactile sensations (Figure 4). An uncorrected post hoc test demonstrated more frequent gustatory experiences in positive than negative dreams ($p < 0.05$) and a trend towards more frequent auditory experiences in negative than positive dreams ($p = 0.05$); however, these post hoc differences did not survive correction for multiple comparisons ($p > 0.01$).

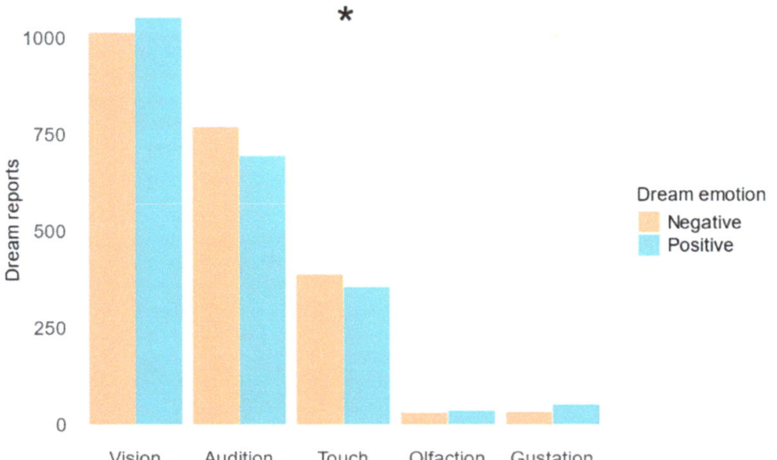

Figure 4. Prevalence of sensory dream experiences in emotionally positive versus negative dreams. The number of dream reports are shown for each sensory dream experience for emotionally negative (orange) and positive (blue) dreams. A chi-square test confirmed that the prevalence of sensory modalities differed amongst positive and negative dreams ($\chi^2 = 10.1$, $p = 0.038$). Post hoc tests comparing the prevalences of positive and negative dream reports per sensory modality did not survive correction for multiple comparisons ($p > 0.01$). * = p<0.05 of chi-square test.

In summary, sensory richness was positively associated with the intensity of dream emotion and clarity of dream recall. This relationship was present for both emotionally positive and negative dream content. In addition, sensory experiences occurred at different rates during emotionally positive and negative dreams.

4. Discussion

In this study, we investigated sensory dream experiences using a multiday dream diary, administered upon final morning awakenings, in a home-based setting. Our findings revealed that vision was the most prevalent sensory dream experience, followed by audition and, subsequently, touch. In contrast, olfaction and gustation occurred at similar, low rates. We observed large interindividual differences in the prevalence of sensory dream experiences. Nearly all participants reported vision (95.9%); a majority reported audition (85.6%) and touch (62.1%). Gustation (16.5%) and olfaction (14.6%) were reported by a minority. A small percentage of participants (2.1%) reported dreams in the absence of any sensory experiences. This means that during these dreams, these participants reported not experiencing any vision, sound, touch, smell, or taste. Furthermore, our results indicate that multisensory dreams were far more prevalent than unisensory dreams and that the prevalence of sensory modalities differed for emotionally positive and negative dreams. Additionally, a positive relationship was found between sensory richness and emotional intensity of dreams, both for positive and negative emotion. Similarly, sensory richness was positively associated with the clarity of dream recall, again, for both positive and negative dream emotions. These findings highlight the complexity and variability of sensory dream experiences and suggest links between sensory richness, emotional content, and clarity of dream recall.

Previous studies on this topic described a 100% report rate for the presence of visual dream experiences [1,12,20]. While vision was indeed the most prevalent sensory dream experience in our study, vision was not reported for all dreams, and some participants did not report it at all. This may indicate participants having thought like dreams without any visual imagery. It may not be possible for an independent rater to distil such dream experiences from dream narratives, resulting in a potential overestimation of visual dream experiences in previous studies [1,12,20]. Direct questions about experiencing vision in dreams, as adopted in the current study, likely improve the accuracy of classifying visual dream experiences. This approach may be particularly useful for evaluating non-rapid eye movement (NREM) dreams, which have been described as thought-like [21–23]. In conclusion, while sensory experiences are frequently involved in dream content, they appear not be a standard feature, even in the context of morning dreams.

While previous research overestimated visual experiences, somatosensory experiences appear to have been somewhat underestimated. Previous studies assessing sensory dream experiences in the home environment did not evaluate somatosensory experiences [12,15,16]. Studies conducted in laboratory settings, based on narrative dream reports, only showed low rates (1%) of tactile experiences [13,20]. However, our study revealed that tactile dream sensations were present in 18.2% of dream reports and experienced by a majority of participants (62.1%). This suggests a substantially higher prevalence than previously reported, again demonstrating a methodological benefit of including direct questions on sensory experiences in dream diaries. To specify the diversity of somatosensory dream experiences, follow-up studies may consider subtyping cutaneous sensations (e.g., touch, temperature, and pain) and include proprioception and kinesthesia.

In dream and memory research, a link between waking experiences and dreams has been widely acknowledged [6,8,24]. However, there seems to be a discrepancy between the prevalence of daytime and dream-related sensory experiences. While chemosensory sensations like olfaction and gustation are common during wake experience, they seem to be less frequently reported during dreaming than other sensations, like audition and vision. This applies to both our study and previous studies [2,12,25–27]. The putative mechanism underlying this phenomenon is uncertain. Speculatively, contributing factors may include the following: (1) less (conscious) daytime stimulation of gustation and olfaction relative to vision and audition, which reflects in sensory dream experiences; (2) infrequent replay of gustatory and olfactory memories during sleep; (3) neurobiological mechanisms in the sleeping brain that somehow inhibit chemosensory dream experiences; (4) alternatively, such dream experiences do occur but are (preferentially) forgotten in the transition to wakefulness.

Several limitations should be considered. Firstly, our study primarily captured morning dreams, as dreams occurred most frequently within the same hour as final morning awakening. An average of less than one dream per night was reported. The late dream occurrence and low report rate may indicate a lack of dream recall from earlier parts of the night [28,29]. Speculatively, given that REM (rapid eye movement) sleep is most prevalent in the second half of the night [30,31] (and dream occurrence was close to final morning awakening in our sample, this study might predominantly represent REM dreams. However, the lack of polysomnography (PSG) precluded the ability to reliably differentiate between NREM and REM dreams. Recent advancements in wearable electroencephalography (EEG) technology may facilitate sleep-monitored dream studies in the home environment and enable home-based, serial awakening paradigms [32] to distinguish NREM and REM dream content.

While self-reporting dream experiences minimizes interpretation biases by independent raters, it remains reliant on the dreamer's evaluation, introducing potential biases. For instance, anecdotal evidence suggests instances where dreamers inferred auditory dream sensations based on visually perceiving conversations. However, upon explicit inquiry by the researcher, the participant could not confirm whether auditory dream experiences had actually been present. Addressing such nuances effectively may involve employing semi-structured interviews or providing explicit instructions in sensory dream diaries.

A final limitation concerns the composition of the sample of participants in this study, which predominantly consisted of young, female adults. Sensory dream experiences may alter with age, as sleep patterns change [33–36] and daytime sensory perception deteriorates [37–42]. Given the young sample, the findings of this study may not be generalizable to older age groups. Additionally, the sample was predominantly represented by females (78.6%). Females tend to report their dreams more frequently, as indicated by a meta-analysis based on 175 studies [43], suggesting that the dream-related prevalences reported in our study may be lower in a male population.

Future research could explore the potential link between interindividual differences in daytime sensory processing [44] and sensory experiences in dreams. Although vision dominates human perception during both waking [45] and dreaming [1,12,20], there may be subpopulations who experience certain sensory modalities more frequently than others. For instance, a previous study showed that individuals who were particularly aware of odors reported higher rates of olfactory dreams than individuals with low odor awareness [25]. Interindividual differences in sensory dream experiences that link to daytime sensory processing are further supported by evidence from studies on individuals with sensory impairments. For instance, dream experiences of the blind contain a higher prevalence of auditory and tactile sensations than those of the non-blind [46]. In deaf individuals, hearing was less frequently present compared to non-deaf dreamers, whereas gustatory, olfactory, and somatosensory dream experiences were increased compared to non-deaf dreamers [47]. Additionally, pain is more prevalent in dreams of chronic pain patients compared to healthy individuals [48]. Further investigation into sensory dream experiences involving other sensory-impaired populations may provide valuable insights into the relationship between daytime sensory experiences and dreams. Examples of such populations may consist of individuals with olfactory dysfunction, which is prevalent among long-COVID patients [49–51], or sensory paralysis.

5. Conclusions

By shifting from examining individual dream content to systematically assessing dreams through targeted inquiries about experiences (such as sensory experiences, color perception, and language), we may gain more insight into how dreams manifest. This approach could yield valuable insights into sleep-related neural processing and how this affects memory consolidation over time, further elucidating sleep's role in strengthening and weakening memories [52–55].

Supplementary Materials: The following supporting information can be downloaded at: https://www.mdpi.com/article/10.3390/brainsci14060533/s1, Figure S1: Age distribution of participants in the sample; Figure S2: Quantity of diary reports per participant; Figure S3: (A) Intervals between dream occurrence and diary entry, (B) hours between waking up and diary completion; Figure S4: Distribution of dream clarity and time interval between dream occurrence and final awakening.

Author Contributions: Conceptualization, A.C.v.d.H.; Data curation, J.T. and J.V. Formal analysis, J.T. and J.V., Investigation, J.T. and J.V., Methodology, A.C.v.d.H. and L.M.T.; Project administration, A.C.v.d.H. Supervision, A.C.v.d.H. and L.M.T.; Visualization, A.C.v.d.H.; Writing—original draft, A.C.v.d.H.; Writing—review and editing, L.M.T. All authors have read and agreed to the published version of the manuscript.

Funding: This research received no external funding.

Institutional Review Board Statement: All procedures were approved by the ethical committee of the University of Amsterdam (Approval code: FMG-4352; Approval date: 4 October 2023).

Informed Consent Statement: Informed consent was obtained from all subjects involved in the study.

Data Availability Statement: The raw data supporting the conclusions of this article will be made available by the authors on request due to privacy reasons.

Acknowledgments: We thank L. van Wechem for contributing to the data collection and being involved in studying sensory dream experiences prior to this project. We thank S.M.D.D. Fitzsimmons for proofreading the manuscript.

Conflicts of Interest: The authors declare no conflicts of interest.

References

1. Schredl, M.; Wittmann, L. Dreaming: A Psychological View. *Schweiz. Arch. Neurol. Psychiatr.* **2005**, *156*, 484–492.
2. Schredl, M. Characteristics and Contents of Dreams. *Int. Rev. Neurobiol.* **2010**, *92*, 135–154. [CrossRef] [PubMed]
3. Klepel, F.; Schredl, M.; Göritz, A.S. Dreams Stimulate Waking-Life Creativity and Problem Solving: Effects of Personality Traits. *Int. J. Dream Res.* **2019**, *12*, 95–102. [CrossRef]
4. Wamsley, E.J.; Tucker, M.; Payne, J.D.; Benavides, J.A.; Stickgold, R. Dreaming of a Learning Task Is Associated with Enhanced Sleep-Dependent Memory Consolidation. *Curr. Biol.* **2010**, *20*, 850–855. [CrossRef] [PubMed]
5. Van den Bulck, J.; Çetin, Y.; Terzi, Ö.; Bushman, B.J. Violence, Sex, and Dreams: Violent and Sexual Media Content Infiltrate Our Dreams at Night. *Dreaming* **2016**, *26*, 271–279. [CrossRef]
6. Eichenlaub, J.-B.; Cash, S.S.; Blagrove, M. Daily Life Experiences in Dreams and Sleep-Dependent Memory Consolidation. In *Cognitive Neuroscience of Memory Consolidation*; Springer: Cham, Switzerland, 2017; pp. 161–172. [CrossRef]
7. Wamsley, E.J.; Stickgold, R. Dreaming of a Learning Task Is Associated with Enhanced Memory Consolidation: Replication in an Overnight Sleep Study. *J. Sleep. Res.* **2019**, *28*, e12749. [CrossRef] [PubMed]
8. Wamsley, E.J.; Stickgold, R. Memory, Sleep, and Dreaming: Experiencing Consolidation. *Sleep Med. Clin.* **2011**, *6*, 97–108. [CrossRef]
9. White, G.L.; Taytroe, L. Personal Problem-Solving Using Dream Incubation: Dreaming, Relaxation, or Waking Cognition? *Dreaming* **2003**, *13*, 193–209. [CrossRef]
10. Barrett, D. Dreams and Creative Problem-Solving. *Ann. N. Y. Acad. Sci.* **2017**, *1406*, 64–67. [CrossRef]
11. Scarpelli, S.; Bartolacci, C.; D'Atri, A.; Gorgoni, M.; De Gennaro, L. The Functional Role of Dreaming in Emotional Processes. *Front. Psychol.* **2019**, *10*, 442641. [CrossRef]
12. Zadra, A.L.; Nielsen, T.A.; Donderi, D.C. Prevalence of Auditory, Olfactory, and Gustatory Experiences in Home Dreams. *Percept. Mot. Ski.* **1998**, *87*, 819–826. [CrossRef] [PubMed]
13. Snyder, F. The Phenomenology of Dreaming. In *The Psychodynamic Implications of the Physiological Studies on Dreams*; Charles C Thomas Publisher, Limited: Springfield, IL, USA, 1970; pp. 124–151.
14. Yu, C.K.C. Dream Intensity Inventory and Chinese People's Dream Experience Frequencies. *Dreaming* **2008**, *18*, 94–111. [CrossRef]
15. Yu, C.K.C. Testing the Factorial Structure of the Dream Intensity Scale. *Dreaming* **2012**, *22*, 284–309. [CrossRef]
16. Yu, C.K.C. Dream Intensity Scale: Factors in the Phenomenological Analysis of Dreams. *Dreaming* **2010**, *20*, 107–129. [CrossRef]
17. Harmon-Jones, C.; Bastian, B.; Harmon-Jones, E. The Discrete Emotions Questionnaire: A New Tool for Measuring State Self-Reported Emotions. *PLoS ONE* **2016**, *11*, e0159915. [CrossRef]

18. Carney, C.E.; Buysse, D.J.; Ancoli-Israel, S.; Edinger, J.D.; Krystal, A.D.; Lichstein, K.L.; Morin, C.M. The Consensus Sleep Diary: Standardizing Prospective Sleep Self-Monitoring. *Sleep* **2012**, *35*, 287. [CrossRef]
19. R Core Team. *R: A Language and Environment for Statistical Computing*; R Foundation for Statistical Computing: Vienna, Austria, 2021.
20. Mccarley, R.W.; Hoffman, E. REM Sleep Dreams and the Activation-Synthesis Hypothesis. *Am. J. Psychiatry* **1981**, *138*, 904–912.
21. Fosse, M.J.; Fosse, R.; Hobson, J.A.; Stickgold, R.J. Dreaming and Episodic Memory: A Functional Dissociation? *J. Cogn. Neurosci.* **2003**, *15*, 1–9. [CrossRef] [PubMed]
22. Foulkes, D.; Rechtschaffen, A. Presleep determinants of dream content: Effects o f two films. *Percept. Mot. Ski.* **1964**, *19*, 5–19. [CrossRef]
23. Marti, J.M.; Andriano, D.W.; Mota, N.B.; Mota-Rolim, S.A.; Araujo, J.F.; Solms, M.; Ribeiro, S. Structural Differences between REM and Non-REM Dream Reports Assessed by Graph Analysis. *PLoS ONE* **2020**, *15*, e0228903. [CrossRef]
24. Perogamvros, L.; Dang-Vu, T.T.; Desseilles, M.; Schwartz, S. Sleep and Dreaming Are for Important Matters. *Front. Psychol.* **2013**, *4*, 51581. [CrossRef]
25. Martinec Nováková, L.; Kliková, M.; Miletínová, E.; Bušková, J. Olfaction-Related Factors Affecting Chemosensory Dream Content in a Sleep Laboratory. *Brain Sci.* **2021**, *11*, 1225. [CrossRef] [PubMed]
26. Stevenson, R.J.; Case, T.I. Olfactory dreams: Phenomenology, relationship to volitional imagery and odor identification. *Cogn. Personal.* **2004**, *24*, 69–90. [CrossRef]
27. Zhong, Y.; Jiang, R.; Zou, L. Do You Remember If You Have Olfactory Dreams? A Content Analysis of LOFTER and a Questionnaire Survey Conducted in China. *Physiol. Behav.* **2022**, *252*, 113849. [CrossRef] [PubMed]
28. Noreika, V.; Valli, K.; Lahtela, H.; Revonsuo, A. Early-Night Serial Awakenings as a New Paradigm for Studies on NREM Dreaming. *Int. J. Psychophysiol.* **2009**, *74*, 14–18. [CrossRef]
29. Picard-Deland, C.; Konkoly, K.; Raider, R.; Paller, K.A.; Nielsen, T.; Pigeon, W.R.; Carr, M. The Memory Sources of Dreams: Serial Awakenings across Sleep Stages and Time of Night. *Sleep.* **2023**, *46*, zsac292. [CrossRef] [PubMed]
30. Carskadon, M.A.; Dement, W.C. Chapter 2-Normal Human Sleep: An Overview. *Princ. Pract. Sleep Med.* **2011**, *5*, 16–26.
31. Feriante, J.; Araujo, J.F. Physiology, REM Sleep. In *StatPearls [Internet]*; StatPearls Publishing LLC: St. Petersburg, FL, USA, 2018; pp. 1–6.
32. Siclari, F.; LaRocque, J.J.; Postle, B.R.; Tononi, G. Assessing Sleep Consciousness within Subjects Using a Serial Awakening Paradigm. *Front. Psychol.* **2013**, *4*, 55505. [CrossRef]
33. Feinberg, I. Changes in Sleep Cycle Patterns with Age. *J. Psychiatr. Res.* **1974**, *10*, 283–306. [CrossRef]
34. Pace-Schott, E.F.; Spencer, R.M.C. Age-Related Changes in the Cognitive Function of Sleep. *Prog. Brain Res.* **2011**, *191*, 75–89. [CrossRef]
35. Vitiello, M.V. Sleep in Normal Aging. *Sleep. Med. Clin.* **2006**, *1*, 171–176. [CrossRef]
36. Casagrande, M.; Forte, G.; Favieri, F.; Corbo, I. Sleep Quality and Aging: A Systematic Review on Healthy Older People, Mild Cognitive Impairment and Alzheimer's Disease. *Int. J. Env. Res. Public. Health* **2022**, *19*, 8457. [CrossRef]
37. Heft, M.W.; Robinson, M.E. Age Differences in Suprathreshold Sensory Function. *Age* **2014**, *36*, 1–8. [CrossRef] [PubMed]
38. Heft, M.W.; Robinson, M.E. Somatosensory Function in Old Age. *J. Oral. Rehabil.* **2017**, *44*, 327–332. [CrossRef] [PubMed]
39. Doty, R.L.; Kamath, V. The Influences of Age on Olfaction: A Review. *Front. Psychol.* **2014**, *5*, 72845. [CrossRef] [PubMed]
40. Murphy, C.; Gilmore, M.M. Quality-Specific Effects of Aging on the Human Taste System. *Percept. Psychophys.* **1989**, *45*, 121–128. [CrossRef] [PubMed]
41. Owsley, C. Aging and Vision. *Vis. Res.* **2011**, *51*, 1610–1622. [CrossRef] [PubMed]
42. Roth, T.N.; Hanebuth, D.; Probst, R. Prevalence of Age-Related Hearing Loss in Europe: A Review. *Eur. Arch. Oto-Rhino-Laryngol.* **2011**, *268*, 1101–1107. [CrossRef]
43. Schredl, M.; Reinhard, I. Gender Differences in Dream Recall: A Meta-Analysis. *J. Sleep Res.* **2008**, *17*, 125–131. [CrossRef]
44. Murray, M.M.; Eardley, A.F.; Edginton, T.; Oyekan, R.; Smyth, E.; Matusz, P.J. Sensory Dominance and Multisensory Integration as Screening Tools in Aging. *Sci. Rep.* **2018**, *8*, 8901. [CrossRef]
45. Hecht, D.; Reiner, M. Sensory Dominance in Combinations of Audio, Visual and Haptic Stimuli. *Exp. Brain Res.* **2009**, *193*, 307–314. [CrossRef] [PubMed]
46. Meaidi, A.; Jennum, P.; Ptito, M.; Kupers, R. The Sensory Construction of Dreams and Nightmare Frequency in Congenitally Blind and Late Blind Individuals. *Sleep. Med.* **2014**, *15*, 586–595. [CrossRef] [PubMed]
47. Okada, H.; Wakasaya, K. Dreams of Hearing-Impaired, Compared with Hearing, Individuals Are More Sensory and Emotional. *Dreaming* **2016**, *26*, 202–207. [CrossRef]
48. Schredl, M.; Kälberer, A.; Zacharowski, K.; Zimmermann, M. Pain Dreams and Dream Emotions in Patients with Chronic Back Pain and Healthy Controls. *Open Pain. J.* **2017**, *10*, 65–72. [CrossRef]
49. Doty, R.L. Olfactory Dysfunction in COVID-19: Pathology and Long-Term Implications for Brain Health. *Trends Mol. Med.* **2022**, *28*, 781–794. [CrossRef] [PubMed]
50. Llana, T.; Mendez, M.; Zorzo, C.; Fidalgo, C.; Juan, M.C.; Mendez-Lopez, M. Anosmia in COVID-19 Could Be Associated with Long-Term Deficits in the Consolidation of Procedural and Verbal Declarative Memories. *Front. Neurosci.* **2022**, *16*, 1082811. [CrossRef]

51. Hu, S.; Zhang, S.; You, Y.; Tang, J.; Chen, C.; Wang, C.; Wen, W.; Wang, M.; Chen, J.; Luo, L.; et al. Olfactory Dysfunction after COVID-19: Metanalysis Reveals Persistence in One-Third of Patients 6 Months after Initial Infection. *J. Infect.* **2023**, *86*, 516–519. [CrossRef] [PubMed]
52. Saletin, J.M.; Goldstein, A.N.; Walker, M.P. The Role of Sleep in Directed Forgetting and Remembering of Human Memories. *Cereb. Cortex* **2011**, *21*, 2534–2541. [CrossRef]
53. Feld, G.B.; Born, J. Sculpting Memory during Sleep: Concurrent Consolidation and Forgetting. *Curr. Opin. Neurobiol.* **2017**, *44*, 20–27. [CrossRef]
54. Sara, S.J.; Poe, G.R. Sleep Is for Forgetting. *J. Neurosci.* **2017**, *37*, 464–473. [CrossRef]
55. Langille, J.J. Remembering to Forget: A Dual Role for Sleep Oscillations in Memory Consolidation and Forgetting. *Front. Cell Neurosci.* **2019**, *13*, 439284. [CrossRef]

Article

Quantitative–Qualitative Assessment of Dream Reports in Schizophrenia and Their Correlations with Illness Severity

Gianluca Ficca [1], Oreste De Rosa [1,2], Davide Giangrande [1], Tommaso Mazzei [1,3], Salvatore Marzolo [3,4], Benedetta Albinni [1], Alessia Coppola [1], Alessio Lustro [1] and Francesca Conte [1,*]

[1] Department of Psychology, University of Campania "L. Vanvitelli", 81100 Caserta, Italy; gianluca.ficca@unicampania.it (G.F.); oreste.derosa@unicampania.it (O.D.R.); davide.giangrande@studenti.unicampania.it (D.G.); tommaso.mazzei@studenti.unicampania.it (T.M.); benedettaalbinni@gmail.com (B.A.); alessiacoppola001@gmail.com (A.C.); alessio.lustro@studenti.unicampania.it (A.L.)

[2] Dermatology Unit, Department of Mental and Physical Health and Preventive Medicine, University of Campania "L. Vanvitelli", 80138 Naples, Italy

[3] Residential Community for Therapy and Rehabilitation "Al di là dei sogni (Beyond Dreams)", 81037 Sessa Aurunca, Italy; salvatore.marzolo@aslcaserta.it

[4] Mental Health Unit 15, Local Health Authority 1, 81016 Piedimonte Matese, Italy

* Correspondence: francesca.conte@unicampania.it; Tel.: +39-0823-274790

Abstract: Positive symptoms of schizophrenia have been proposed to be an intrusion of dreaming in wakefulness; conversely, psychotic patients' abnormal cognitive and behavioral features could overflow into sleep, so that their dreams would differ from those of healthy people. Here we assess this hypothesis by comparing dream features of 46 patients affected by schizophrenic spectrum disorders to those of 28 healthy controls. In patients, we also investigated correlations of dream variables with symptom severity and verbal fluency. Overall, patients reported fewer and shorter dreams, with a general impoverishment of content (including characters, settings, interactions) and higher spatiotemporal bizarreness. The number of emotions, mainly negative ones, was lower in patients' reports and correlated inversely with symptom severity. Verbal fluency correlated positively with dream report length and negatively with perceptive bizarreness. In conclusion, our data show a significant impoverishment of dream reports in psychotic patients versus controls. Future research should investigate to what extent this profile of results depends on impaired verbal fluency or on impaired mechanisms of dream generation in this population. Moreover, in line with theories on the role of dreaming in emotion regulation, our data suggest that this function could be impaired in psychoses and related to symptom severity.

Keywords: dreams; schizophrenia; emotion regulation

Citation: Ficca, G.; De Rosa, O.; Giangrande, D.; Mazzei, T.; Marzolo, S.; Albinni, B.; Coppola, A.; Lustro, A.; Conte, F. Quantitative–Qualitative Assessment of Dream Reports in Schizophrenia and Their Correlations with Illness Severity. *Brain Sci.* **2024**, *14*, 568. https://doi.org/10.3390/brainsci14060568

Academic Editors: Luigi De Gennaro, Sergio A. Mota-Rolim, Brigitte Holzinger and Michael Nadorff

Received: 1 May 2024
Revised: 27 May 2024
Accepted: 29 May 2024
Published: 3 June 2024

1. Introduction

> "I had an historical dream! I was at Berlinguer's funeral. There was this Pertini character trying to escape by bike, I chased him."
>
> Patient n.17

Schizophrenia is a psychotic disorder characterized by disturbances in cognition, emotional responsiveness and behavior. According to the first criterion for the diagnosis of schizophrenia in the DSM-5 [1], at least two of the following symptoms must be present for most of the time for at least one month: (1) delusions; (2) hallucinations; (3) disorganized speech; (4) grossly disorganized or catatonic behavior; and (5) negative symptoms such as reduced volition or emotional expression. Several phenomenological similarities have often been underlined between dream features and these symptoms (e.g., [2]). As reviewed in Limosani et al. [3], besides sensory perceptions in absence of external stimulations which are shared by dreams and psychosis (i.e., hallucinations), cognition is characterized, in both states, by disorganized thought and unrealistic ideational contents, accompanied by bizarre experiences for which the subject shows an impairment of reality testing and, subjectively, a very intense emotional involvement.

These similarities clearly bear implications for both psychopathology and research on dream processes. Indeed, it has been suggested that dreams may represent a natural model for psychosis (e.g., [4]) and that schizophrenia could be a sort of "trapped state" between waking and dreaming, with the encroachment of experiences usually occurring in dreams into wakefulness [5]. For dream researchers, however, the dream–psychosis relationship is also extremely interesting the other way round, i.e., addressing the influence of wakefulness on sleep mentation and the possibility of making predictions on psychotic patients' dreams given the peculiar characteristics of their disorder. As a matter of fact, there are solid research lines trying to understand whether and to what extent waking experience is reflected in dreaming, in line with the widely held "continuity hypothesis" according to which a continuity would exist between waking-life experiences and dreams [6].

Furthermore, the relationships between wakefulness and dreams would be reflected not only in dream content, but also in its associated emotions [7]. Notably, a primary function of sleep for emotion regulation has been repeatedly proposed (e.g., [8,9]), also in light of REM sleep's peculiar neurotransmitter balance, which is believed to provide optimal conditions for offline processing of affects (see [10] for a review). Our own group has recently shown that poor sleep quality might impair sleep-related processes of affect regulation [11]. Thus, we deem it extremely interesting to look at dream features in the disturbances of the schizophrenic spectrum, where emotional dysregulation is a key characteristic.

According to what was said so far, waking-life psychotic symptoms could be directly linked to specific dream characteristics. When coming to schizophrenia spectrum disorders, this would imply that psychiatric symptoms could be predictors of differences in the dreams of psychotic patients relative to those of controls. However, probably due to the methodological difficulties of collecting dream reports from this kind of patient, the data available on this topic are still very sparse and foggy.

Dream reports in schizophrenia were found to be shorter in a number of rather old studies (e.g., [12–14]), but their results were obtained with different methodologies and did not clarify whether they were accounted for by an actual reduction in dream generation or by patients' reduced ability to recall and report their dreams, e.g., due to the impairment of verbal fluency repeatedly shown in schizophrenia [15].

Concerning their qualitative features, schizophrenic patients' dream reports, compared to healthy controls', were occasionally displaying reduced emotional involvement and emotional expression [16,17], less affect and less change in dream scenery [18], more frequent presence of familiar people [19] or of strangers [20,21], fewer words referring to the semantic field of hearing and a less active role of the dreamer [17].

Cognitive bizarreness—defined as "impossibility or improbability in the domains of dream plot, cognition and affect" [22]—has been considered as a cognitive marker shared by psychotic waking and dreaming state, but to what extent the high bizarreness in schizophrenic patients' waking ideation is maintained during dreams is still an open issue. Early studies found less bizarreness in schizophrenic patients' dreams in comparison to the dreams of a normal control population [12,23,24], whereas the more recent literature seems to point to equal [21] or even higher bizarreness scores in schizophrenic patients [25]. Interestingly, a study by Scarone et al. [26] showed that a comparable degree of formal cognitive bizarreness was shared by the waking cognition of schizophrenic subjects and the dream reports of both normal controls and schizophrenics.

In sum, the sparse and contrasting literature does not allow to draw clear conclusions on the relationships between dream features and psychosis. Therefore, here we compare dream reports of schizophrenic patients to those of healthy controls with regard both to quantity (Dream Recall Frequency, from now on DRF) and quality (length, content), in order to provide further data to enlighten the issue of whether psychotic symptomatology is reflected in dream content. Within the frame of this general objective, we specifically intend to focus on a few issues that have been covered very little, if at all, by previous research: (a) the possible role of lexical access ability, indexed by verbal fluency performance, in affecting dream report length; (b) the amount and types of emotions reported in dreams in the clinical vs. the control group; (c) the relationship of dream features with illness severity, as measured through the Brief Psychiatric Rating Scale (from now on, BPRS).

2. Materials and Methods

2.1. Participants

We recruited a convenience sample of 46 patients with diagnosed psychotic symptoms (F 10, M 36, age range: 19–54 years) at a residential facility (Rehabilitation Community "Beyond dreams", Sessa Aurunca (Caserta), Italy, n = 26) and a day-treatment center ("Integrazioni", Casoria (Napoli), Italy, n = 20) and 28 volunteer healthy participants (F 17, M 11, age range: 20–59 years), who were psychology students of the University of Campania "L. Vanvitelli" (Caserta, Italy) and their relatives and friends.

The main inclusion criterion for the patients' group (PG) was having a diagnosis of schizophrenia or any other psychotic disorder according to DSM-V criteria [1], including schizoaffective disorder. All patients were being treated with combined individual and group integrative psychotherapy. Family therapy was also followed by 24% of the subjects. Pharmacological treatments, which had to have been stable for at least three weeks before the study, were distributed as follows: no medication, 10.9%; mono therapy with antipsychotics, 32.6%; polytherapy with antipsychotics and benzodiazepines (only one patient was treated with zolpidem instead of benzodiazepines), 56.5%. Other inclusion criteria were (a) absence of comorbidity with other psychiatric or neurological disorders; (b) no evidence of mental retardation; (c) for inpatients in the residential facility, having stayed there for not less than 1 month and not more than 18 months.

As for the healthy control group (CG), inclusion criteria were (a) age between 18 and 60; (b) absence of any history of somatic and/or psychiatric disturbances; (c) absence of any history of sleep disorders; (d) regular sleep habits, evaluated through the Pittsburgh Sleep Quality Index, Italian Version [27]. Also, only subjects who reported recalling at least one dream per week were recruited.

All demographic characteristics of the two samples, including clinical diagnosis, therapies for the patients' group, are listed in Table 1.

Table 1. Clinical features in the patients' group.

Variable		
Diagnosis (N, %)		
	Paranoid Schizophrenia	16 (34.80)
	Disorganized Schizophrenia	10 (21.70)
	Schizotypic Personality Disorder	8 (17.40)
	Unspecified Schizophrenia	6 (13.00)
	Schizoaffective Disorder	6 (13.00)
Pharmacotherapy (N, %)		
	No Drugs	5 (10.90)
	Monotherapies	15 (32.60)
	Politherapies	26 (56.50)
Brief Psychiatric Rating Scale Total Score (m ± sd)		82.19 ± 20.46
BPRS Severity groups (N, %)	Absent	0 (0)
	Mild	5 (10.90)
	Moderate	10 (21.70)
	Moderately severe	18 (39.10)
	Severe	11 (23.90)
	Very severe	2 (4.30)
	Extremely severe	0 (0)
Verbal Fluency		
	Phonemic (m ± sd)	5.84 ± 3.81
	Semantic (m ± sd)	11.7 ± 3.75

2.2. Procedure

The study was approved by the Ethical Committee of the Department of Psychology, University of Campania (Italy). After providing information about the study, consent forms from both patients and healthy controls were obtained.

Before dream report collection, the psychopathological severity of each PG participant was evaluated through the BPRS, Expanded Edition 4.0 [28] during a one-hour individual therapy session. Moreover, a verbal fluency test [29] was administered to the same group to evaluate lexical access ability.

Participants were requested, 5 days a week (over a period of 30 days for the patients' group and of 15 days for the healthy controls), to report, immediately at spontaneous awakening, the mental activity they had memory of through the following classical instruction (presented in written form): "Please tell me everything you can remember of what was going through your mind before you woke up." [30]. For patients in the residential facility, it was the facility staff who solicited them to report the dreams at awakening, whereas patients from the day-treatment center and normal controls were instructed to write down or audio-record their dreams first thing after awakening. The patients were also requested to fill in a diary of their daily activities to control that they kept their daily routines stable, without any peculiar experience that might influence their dreams.

2.3. Instruments

For psychotic symptom severity assessment, we administered the BPRS [28], in its Italian version [31]. The BPRS has shown good internal consistency, with a Cronbach's alpha of 0.87 [32]. The severity of each one of the 24 symptoms is rated on a scale from 1 to 7, ranging from 1 (Absent) to 7 (Extremely severe). Ratings are based both on the patient's answers to the interviewer's questions and on the observed behavior during the interview. According to the total BPRS score, psychotic subjects were assigned to one of seven severity groups (Absent 0–24, Very Mild 25–48, Mild 49–72, Moderate 73–96, Moderately severe 97–120, Severe 121–144, Extremely severe 145–168).

The Verbal Fluency Test used in our study [29] consists of two tasks: Semantic fluency and Phonemic fluency. Subjects are given 1 min to produce as many words as possible within three semantic categories (i.e., "Car brands", "Fruits", "Animals") or starting with three given letters (i.e., "P", "F", "L"), respectively. Total scores for both tasks correspond to the total number of words generated in each task. These scores are then corrected for age and education in order to obtain equivalent scores from 0 to 4, where 0 is considered "pathological" and 4 "above normal". The test has shown good internal consistency, with a Cronbach's alpha of 0.83, and test–retest reliability within acceptable limits (i.e., r = 0.74) [33].

Finally, the diary of daily activities administered to patients consisted of two questions: (1) What were the main activities of your day? (2) What kind of emotions did you experience (positive, neutral, negative)?

2.4. Dream Analysis

Dream reports were evaluated by two independent raters, with a third rater, blind to the design and aims of the research, called to resolve possible disagreements. Whenever the subjects claimed that they had made "more than one dream", if they referred them to different bouts of sleep, only the report of the dream preceding the awakening was taken into account. Otherwise, the different dreams were considered as a single report (as in [34]).

Dream Recall Frequency (DRF) is defined as the percentage of days in which a dream report was obtained over the whole number of days of the protocol.

Length of reports was measured in temporal units (TUs) according to Foulkes and Schmidt's method [35]. A TU is assigned whenever (a) a character performs an action that, in waking life, could not be performed synchronically with his/her previous action; (b) a character responds to another character or event; (c) there is a topical change in the dream report.

Following Occhionero and Cicogna [36], type of Self-representation was coded into six categories:

(1)   Presence of Self as a pure thinking agent;

(2)   Total or partial Self body image, more or less associated with proprioceptive, kinesthetic, agreeable or painful sensations;

(3)   Representation of Self as a passive observer of the dream events;

(4)   A precise hallucination of both mind and body, analogous to wakefulness;

(5)   Identification with other characters in the dream;

(6)   A double representation of Self with two distinct and relatively active roles.

A seventh category—(0) Absence of Self-representation both as a physical entity and as thinking subjectivity—was added here to describe reports in which it is not possible to identify any type of Self-representation.

A number of dream content dimensions (as in [37]) were analyzed through a set of dichotomous categorical variables (presence/absence of that feature in the dream report):

-   Continuity, scored as present when the report's narrative structure did not show sudden interruptions or changes of main settings or characters (when the report was described as containing more than one dream, no continuity was assigned).

-   Impossibility/Implausibility Bizarreness, referring to events whose occurrence is implausible during wake;

-   Space/Time Bizarreness, referring to spatiotemporal distortions;

-   Perceptive Bizarreness, referring to images, characters or objects with distorted shapes, colors or dimensions;

-   Emotions, referring to spontaneously verbalized emotions which are clearly expressed by the subject and felt by the dreamer himself during the dream (other characters' emotions reported by the dreamer were not included);

-   Positive Emotions;

-   Negative Emotions;

-   Somatic Sensations, referring to spontaneously verbalized somatic sensations clearly expressed by the dreamer;

-   Non-Self Characters, referring to any additional character besides the Self;

-   Unknown Characters, referring to strangers or unfamiliar characters, appearing as single individuals or undefined groups;

-   Interactions, referring to direct (Self) and indirect (Others) interactions between characters;

-   Friendly Interactions, referring to friendly direct (Self) and indirect (Others) interactions between characters;

-   Aggressive Interactions, referring to aggressive direct (Self) and indirect (Others) interactions between characters;

-   Sexual Interactions, referring to sexual direct (Self) and indirect (Others) interactions between characters;

-   Setting, referring to a specific, clearly identifiable setting in which the oneiric scene takes place.

Three dream content dimensions were also assessed as continuous variables:

(a)   emotions (only those spontaneously verbalized by the subject in the dream report are included in scoring these variables): total number of emotions (including both the dreamer's and other characters' emotions), number of Self (dreamer's) emotions, number of non-Self (other characters') emotions;

(b)   characters: number of characters (these were scored only when they appeared as single individuals: both familiar and unknown characters were included, but the dreamer and undefined groups were excluded);

(c)   interactions (all actions identified by verbs clearly referring to interactions): total number of interactions, number of Self-interactions (those between the dreamer and other characters), number of non-Self interactions (those between two or more non-Self characters).

These continuous variables were all analyzed both as absolute numbers and as percentages over the number of TUs.

2.5. Statistics

A Kruskall–Wallis ANOVA was conducted to test between-group (CG vs. PG) differences in dream variables. Except for the analysis of between-group differences in (DRF) and white reports frequency, patients producing 0 dream reports over the whole data collection period were excluded from the analysis (N = 19). Spearman's correlation coefficient (rr) was used to detect possible correlations between age, global score at the BPRS, verbal fluency (phonemic and semantic), and dream variables in PG. Between-group differences in age and years of education were assessed with the Mann–Whitney U test, whereas differences in gender distribution with the Chi-Square (χ^2) test.

All analyses were performed with Jamovi 2.3.21 (The Jamovi Project, 2023), and the statistical significance level was set at $p \leq 0.05$. Descriptive statistics are reported as mean $\pm$ standard deviation.

3. Results

3.1. Characteristics of the Sample

The two groups did not differ in age (total sample: 35.7 $\pm$ 10.6; CG: 33.0 $\pm$ 6.8; PG: 36.0 $\pm$ 11; U = 99.5, p = 0.634) but differed in gender (total sample: F 27, M 47; CG: F 17, M 11; PG: F 10, M 36; χ^2 = 6.81, p = 0.009) and years of education (CG: 16.0 $\pm$ 2.7; PG: 11.4 $\pm$ 2.3, U = 28.0, p = 0.003).

More than half of the cases (56.5%) are schizophrenias of the paranoid and disorganized types, whereas the remaining 43.5% is divided between unspecified schizophrenia, schizoaffective disorder, and schizotypic personality disorders. Only 5 patients are drug-free and 26 of them are taking benzodiazepines and/or hypnotic drugs in addition to antipsychotic medications. Table 1 summarizes the patients' characteristics derived from the initial clinical assessment, namely diagnoses, pharmacotherapies, BPRS score and Verbal Fluency.

3.2. Inter-Rater Agreement

Inter-rater agreement turned out to be satisfactorily high both for the analysis of temporal units (r = 0.94) and for that of content variables (r = 0.92).

3.3. Diary of Daily Activities of Patients

Overall, patients carried out different activities during the data collection period, the most frequent of which were recreational activities (28%), training–work activities (23%), individual and/or group therapy (19%), meeting with family members (19%), and outdoor activities (11%). The most reported emotions during activities were neutral (50%), followed by negative (30%), and positive (20%).

3.4. Dream Recall Frequency

Nineteen patients did not produce any dream report across the entire data collection period (non-recallers: 41.3%). Instead, all the healthy controls produced at least one dream report. A total of 159 dream reports were obtained from the patients' group over the 30-day study period while controls' dream reports (collected over a 15-day period) were 111 overall. Therefore, DRF in the patients' group was significantly lower than in controls (18% $\pm$ 1.33 vs. 27% $\pm$ 4.25, χ^2_1 = 9.30, p = 0.002). In PG, non-recallers did not differ from recallers in age, years of education, and gender distribution, but significantly differed in BPRS global score, reporting higher severity of psychopathological symptoms (Table 2).

Table 2. Characteristics of dream recaller and non-recaller patients.

Variable	Recallers	Non Recallers	Statistics
Age	35.30 ± 10.83	37.10 ± 11.36	$U = 235, p = 0.631$
Sex	22 M 5 F	14 M 5 F	$\chi^2_1 = 0.4, p = 0.528$
Years of education	11.2 ± 2.06	11.7 ± 2.56	$U = 213, p = 0.288$
BPRS	77.10 ± 21.30	89.40 ± 17.26	$U = 162, p = 0.036$ *

Notes. Significant *p*-values are marked with an asterisk.

DRF did not significantly differ between inpatients (21.61 ± 1.71) and patients treated at the day-care facility (i.e., sleeping at home) (13.23 ± 2.26; $\chi^2_1 = 3.8, p = 0.060$).

A significantly higher proportion of contentless dreams, i.e., "white reports", was reported by patients compared with controls (PG: 4.00 ± 0.40 vs. CG: 0.00 ± 0.00, $\chi^2_1 = 8.12$, $p = 0.004$).

3.5. Between-Group Differences in Dream Report Features

Overall, PG reported significantly shorter dream reports than CG, and dream report length was higher in the inpatients group (2.70 ± 1.20) relative to the day-care facility patients (1.62 ± 0.77; $\chi^2_1 = 4.48, p = 0.034$).

Compared with CG, dream reports in PG were characterized by significantly fewer emotions (Table 3, Figure 1), fewer non-Self, unknown, and total characters, fewer interactions (all types except sexual ones), and a lower frequency of a precise setting. Furthermore, PG reported higher space/time bizarreness (see Table 3 for the complete results).

Table 3. Distribution of dream content features in the two groups.

Variable	CG	PG	χ^2_1	p
Temporal Units (n)	4.85 ± 2.35	2.43 ± 1.17	3.56	<0.0001 *
Continuity (%)	75.95 ± 8.50	84.28 ± 2.47	1.56	0.211
Bizarreness (%)	38.54 ± 2.56	41.45 ± 3.13	0.00	0.952
Implausibility Bizarreness (%)	33.75 ± 2.61	37.80 ± 3.34	0.02	0.898
Space/Time Bizarreness (%)	7.29 ± 1.55	14.13 ± 1.53	5.31	0.021 *
Perceptive Bizarreness (%)	11.90 ± 1.76	7.63 ± 1.26	0.53	0.468
Emotions Frequency in Dreams (%)	56.03 ± 2.84	35.52 ± 0.30	5.76	0.016 *
Emotions (n)	0.85 ± 0.42	0.42 ± 0.39	12.49	<0.001 *
Positive Emotions (%)	11.61 ± 1.86	13.85 ± 1.93	0.47	0.492
Negative Emotions (%)	58.68 ± 5.82	18.39 ± 2.18	17.56	<0.001 *
Somatic Sensations (%)	15.00 ± 2.14	15.77 ± 1.82	0.33	0.562
Non-Self Characters (%)	95.71 ± 0.94	82.21 ± 1.82	10.74	0.001 *
Unknown Characters (%)	73.00 ± 2.46	36.39 ± 3.24	15.00	<0.001 *
Characters (n)	1.98 ± 0.73	0.71 ± 0.65	24.81	<0.001 *
Interactions (%)	94.46 ± 1.20	52.79 ± 3.33	25.65	<0.001 *
Interactions (n)	1.89 ± 0.70	0.71 ± 0.49	29.46	<0.001 *
Interactions Self (n)	1.45 ± 0.55	0.61 ± 0.41	25.75	<0.001 *
Interactions Others (n)	0.45 ± 0.30	0.09 ± 0.28	23.59	<0.001 *
Self-Friendly Interactions (%)	66.71 ± 2.26	32.37 ± 2.72	17.22	<0.001 *
Self-Aggressive Interactions (%)	24.05 ± 2.20	6.76 ± 1.37	11.35	<0.001 *
Self-Sexual Interactions (%)	2.67 ± 0.78	3.51 ± 0.93	0.16	0.684
Others Friendly Interactions (%)	16.19 ± 1.87	2.55 ± 0.53	10.02	0.002 *
Others Aggressive Interactions (%)	15.98 ± 1.71	2.25 ± 0.98	11.81	<0.001 *
Others Sexual Interactions (%)	2.68 ± 1.04	0.46 ± 0.02	0.36	0.549
Settings (%)	88.06 ± 1.77	53.42 ± 3.48	15.21	<0.001 *

Notes. Frequencies are expressed as the percentage of dreams in which the feature was present over the total number of dreams obtained from that group. CG: Control Group, PG: Patient Group. Significant *p*-values are marked with an asterisk.

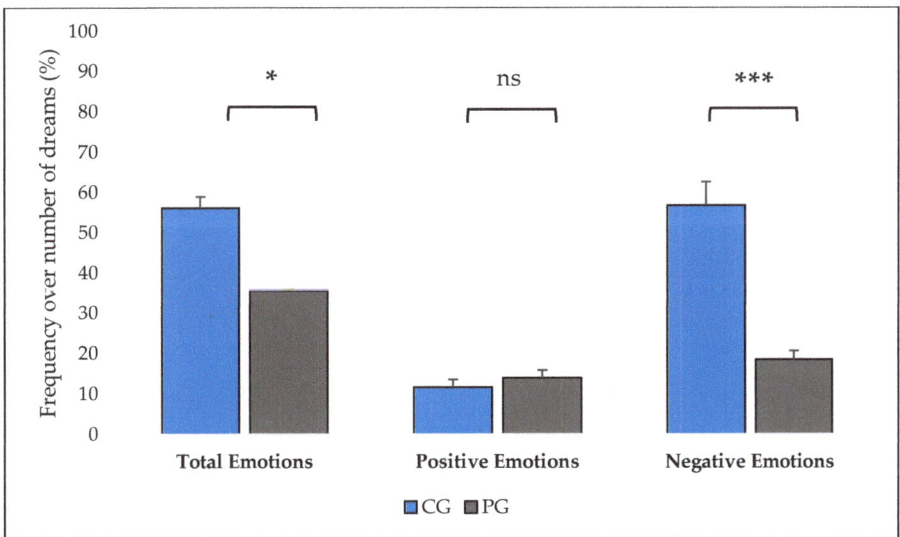

Figure 1. Differences between healthy controls (CG) and patients (PG) in total, positive and negative emotion frequency. *: $p < 0.05$; ***: $p < 0.001$; ns: not significant.

The two groups did not differ in the type of Self-representation in dreams (Table 4).

Table 4. Frequency distribution of different types of Self-representation in the two groups.

	Type 0	Type 1	Type 2	Type 3	Type 4	Type 5	Type 6
CG	2%	1%	1%	1%	90%	1%	1%
PG	7%	0%	1%	3%	87%	1%	1%
χ^2_1	1.96	0.30	1.87	0.01	0.96	0.00	0.84
p	0.161	0.578	0.172	0.895	0.326	0.959	0.359

Notes. CG: Control Group; PG: Patient Group.

3.6. Associations of Dream Variables with Symptom Severity and Verbal Fluency in the Patient Group

The BPRS global score was positively correlated with the frequency of non-Self characters in dreams ($p = 0.028$) and negatively with the number of emotions ($p = 0.030$).

As for verbal fluency, phonemic fluency positively correlated with dream report length (i.e., temporal units; $p = 0.040$) and frequency of somatic sensations ($p = 0.037$). Furthermore, both phonemic and semantic fluency were negatively correlated with the frequency of perceptive bizarreness ($p = 0.036$ and $p = 0.011$ respectively). The full heatmap of correlations is depicted in Figure 2.

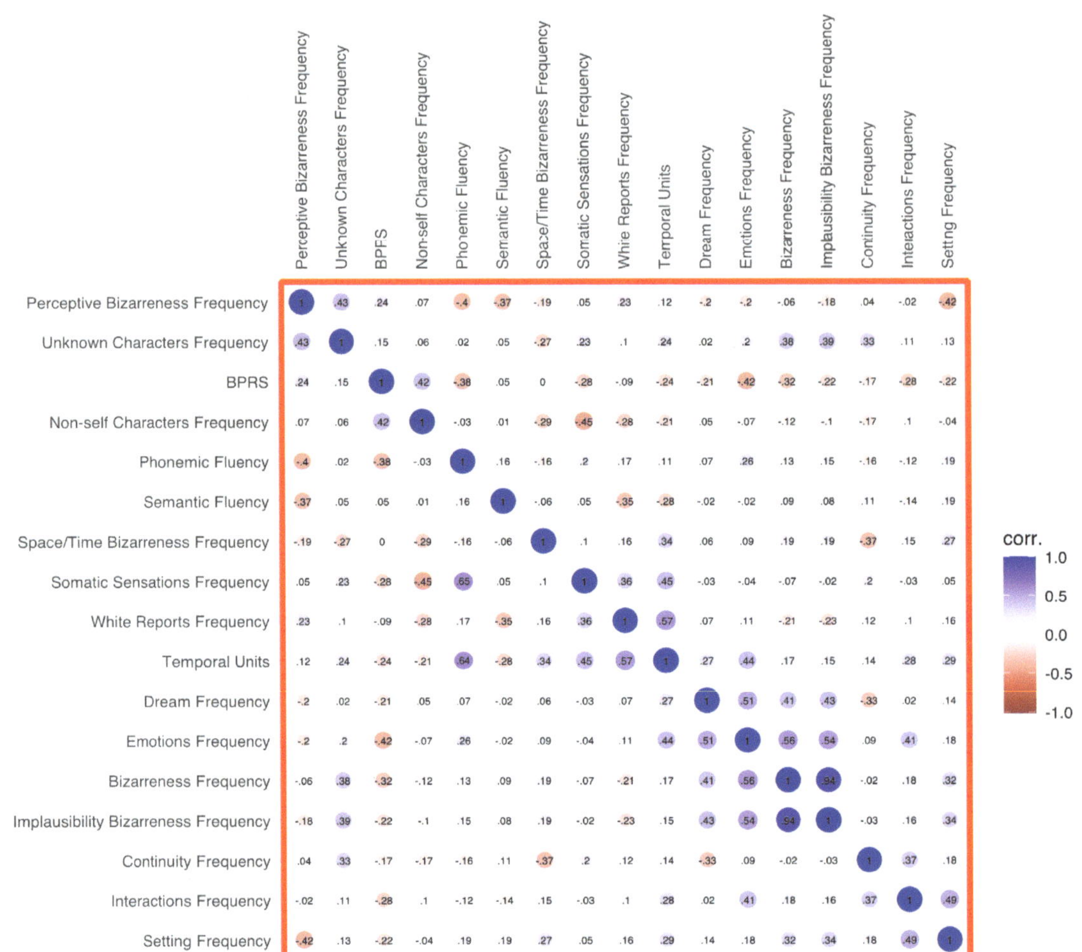

Figure 2. Heatmap of the correlations in the patient group.

4. Discussion

To the best of our knowledge, this is one of the very few studies analyzing the dream characteristics of a relatively wide sample of individuals with psychotic symptoms (compared to healthy subjects) in their habitual life context over an extended period of time (30 days) and assessing the relationships of their dream features with their illness severity and lexical ability.

As for quantitative aspects, our data are consistent with previous findings [12,20,23] showing a decreased frequency of dream recall in individuals suffering from psychotic disorders compared with healthy subjects. Additionally, dream reports appeared shorter in this group, which is also in agreement with a number of past results [12–14,26]. When interpreting these data, it is always very difficult to understand to what extent they depend on reduced oneiric production or, instead, on lower efficiency of dream retrieval processes in schizophrenic subjects. This is a well-known issue in all populations who show a similar quantitative reduction in dreams, such as the elderly [38]. Here, the overall low DRF in the patient population is largely accounted for by the rather high percentage of complete non-recallers, dramatically decreasing the DRF value, which would otherwise appear similar between patients and controls. On the other hand, the correlation between verbal fluency and dream report length, as well as the higher proportion of contentless reports (those mental activities the subjects are aware of but whose content they are unable to verbalize, also called "white reports") observed in the patient sample, support the hypothesis of an impairment in the process of verbally reporting dreams, whereas any conclusion on a possible reduction of oneiric production per se remains speculative. This idea is also coherent with our finding that non-recallers of the patient group showed significantly higher BPRS scores than recallers, suggesting that higher severity is accompanied by greater difficulties in dream recall processes.

Concerning dream content, a consistent profile emerges of global poverty in patients' dreams relative to those of healthy subjects. In the patients' reports, there are fewer characters, fewer settings, a lower number of total interactions, accounted for by the reduction of friendly and aggressive ones, both referring to first-person interactions or to other people's interactions witnessed as an observer. This profile suggests that dreams in psychotic patients are less rich and articulate than those of controls; however, again, considering that verbal fluency influenced dream report length, we cannot exclude that the scarcer contents of patients' dreams depend simply on their deficient lexical access abilities. As for bizarreness, it is not surprising that it was not found to be higher in PG, except for an increase of the space-time type. In line with this, as in Scarone et al. [26], we did not find any correlation between the BPRS severity score and dream bizarreness in the patient group. As suggested by the same authors [26], the gap between patients and controls in bizarre cognition during wakefulness appears to be filled during dreams.

One of the core results of this study refers to the rather impoverished emotional pattern of patients, with a decreased average number of emotions (paralleled by a lower number of reports in which emotions are expressed). Strikingly, this result, which is concordant with previous data [16], almost totally depends on the significant reduction of negative emotions. Apparently, the classical phenomenon observed in healthy subjects of negative dream emotions prevailing over positive ones [39,40] appears inverted in psychotic individuals. This phenomenon in healthy subjects has been interpreted as a possible index of the role of dreams in emotion regulation [7,11,41,42]: in fact, the negative emotions experienced more frequently or intensely in the period in which the dream occurs would be those in need of regulation during sleep, whereas positive emotions, requiring less modulation, would be underrepresented in the dream [7,11]; alternatively, the phenomenon could be due to a rebound of thoughts and related emotions suppressed during wakefulness [41,42]. Therefore, our result in the patient group comes as an impressive counterpart of their severe emotional dysregulation during wakefulness and warrants further investigations, e.g., in terms of its magnitude as a function of illness severity, given that, in our sample, we found that the number of dream emotions decreased as a function of illness severity.

In this regard, the negative association that emerged between BPRS scores and emotions is in line with data from Schredl and Engelhardt [43], who found correlations between the scores on the Symptom Checklist-90-R scales and dream features in patients with several mental disorders, including schizophrenia. These data suggest that it is the severity of specific symptoms (such as depressive mood or psychotic symptoms), rather than the diagnostic classification, that is primarily related to dream content.

Another interesting finding is the lack of differences between CG and PG in the representation of Self and one's own body. This feature has been occasionally described in healthy subjects [36,44,45], with data pointing to the presence of changes in the representation of the Self according to age and the sleep state from which the dream is recalled (being more similar to wakefulness in REM sleep dreams and more polymorphous in NREM sleep dreams). To our knowledge, this is the first time that Self-representation is assessed in a clinical population. Even in this aspect, there appear to be no particularly atypical characteristics transposed from wake symptomatology to the dreaming experience. The same can be observed for the frequency of somatic sensations, which also does not differentiate PG from CG.

A final speculation should be reserved to the issue of whether the present findings are in favor of the "continuity hypothesis" [6], whose original formulation might be usefully reported here: "(. . .) *dreams are continuous with waking life; the world of dreaming and the world of waking are one. The dream world is neither discontinuous nor inverse in its relationship to the conscious world. We remain the same person, the same personality with the same characteristics, and the same basic beliefs and convictions whether awake or asleep. The wishes and fears that determine our actions and thoughts in everyday life also determine what we will dream about)".* On a theoretical level, the hypothesis of continuity between waking and dream features seems particularly suited to the psychotic population in light of the similarity of this population's cognitive, emotional and behavioral characteristics during wakefulness with those of dreams (e.g., sensory perceptions in the absence of external stimulations, disorganized thought, unrealistic ideational contents, intense emotional involvement). In addition, these phenomena appear to be based on common neuroanatomical and neurochemical mechanisms of the waking and dreaming states. For instance, vivid sensory and motor imagery is related to a specific kind of neurotransmission imbalance, with lowered serotoninergic and noradrenergic tone relative to an increase of the cholinergic one [4]; disorganized thoughts are probably linked to the drastic reduction of functional brain connectivity, involving the majority of brain areas in dreams [46] and the thalamocortical circuits in schizophrenia [47], whereas bizarre experiences and the absence of reality testing are most likely dependent on a reduction of frontal cortex activity, especially in the dorsolateral areas [48]. Finally, the physiological basis for emotional involvement could be the remarkable activation of the amygdala and other limbic areas, clearly shown by functional neuroanatomy in both REM sleep [46,49] and schizophrenia [50].

However, it still seems hazardous to draw conclusions on the continuity hypothesis in psychotic individuals in a straightforward manner. In our study, the massive misperceptions, hallucinations, bizarre and disorganized thoughts and behaviors that are commonly associated with the psychotic experience do not seem to emerge in dreaming, at least not more pronouncedly than in healthy subjects. Actually, greater continuity appears to emerge with regard to the impoverishment of dream content, which parallels negative rather than positive psychotic symptomatology. Indeed, the manifestations of psychoses, especially in patients who have a long history of the disease and are chronically medicated, are often characterized by a prevalence of cognitive symptoms, such as remarkable deficits in working memory and attention [51]. Therefore, to what extent the impoverished dream reports found in our sample are a result of "continuity" with waking clinical features remains an open issue, which could prompt, in the future, the characterization of patients in a more refined way, e.g., through the use of instruments assessing day-by-day symptomatology in the positive, negative and cognitive dimensions. In addition, the correlation discussed above between verbal fluency measures and dream report length does not allow the ruling out of the possibility that greater continuity between waking and dreaming features would emerge if lexical access were not impaired.

Importantly, this same observation has to be kept in mind when interpreting our data (as well as any other data on dreaming in psychotic populations) in light of current theories on dreaming and psychosis. Specifically, our data do not appear to be compatible with the recent Defensive Activation Theory [52], which posits that REM sleep dreams and positive neurological symptoms play a "defensive" role against the rapid expansion of functional brain areas over understimulated ones (the visual cortex during sleep and functionally impaired networks in neurological syndromes) [53]. According to this hypothesis, we should have observed, in the dreams of the psychotic sample, a richness of elements even greater than that of healthy controls, reflecting the activation of a double protective process, i.e., that exerted by REM sleep dreaming for the benefit of the visual cortex (as in controls) and that exerted by more general overstimulation processes (reflected in positive symptomatology during wakefulness) in favor of wider brain networks impaired by psychiatric pathology. Instead, our data are coherent with another recent theoretical account, which interprets psychotic symptoms in light of the Active Inference Theory [54]:

specifically, positive symptoms are considered as an attempt of the central nervous system to overcome the mismatch between predictions and outcomes occurring when top-down attentional processes generate expectations but not predictions of their content [53]. As for negative symptoms, they have been interpreted as arising from uncertainty in social prediction brain circuits [55]: the general dearth of elements in our patients' dreams, as well as specifically the paucity of characters, interactions and emotions, could be seen as reflecting this kind of mechanism.

The findings in this study have to be interpreted cautiously in light of a few limitations. The first regards a crucial methodological choice, i.e., collecting dream reports at spontaneous awakening in the participants' habitual living environment. This approach is somewhere between the high experimental rigor achieved through report collection at provoked awakenings in the lab—which is, however, an extremely difficult procedure for clinical samples as demanding and delicate to deal with as psychotic patients—and retrospective interviews, much easier to collect but very unreliable in terms of waking interference and content bias [21]. Therefore, here experimental control is partially sacrificed in favor of an ecological design, determining both meaningful advantages and limitations. On the limitations side, it was not possible to control for the actual adherence of the participants sleeping at home (control subjects and the patients at the day care center) to the instruction of reporting their dream immediately at awakening. Also, we could not collect prompted reports, which would have solicited dream recall with more completeness than solely spontaneous reports, allowing us to better clarify the role of memory retrieval impairments in the scarcity and poorness of dream reports. Moreover, we could not obtain objective sleep data, making it impossible to relate dream report variables to the sleep states during which the dream was produced. On the other hand, our study design permits ruling out the insertion of a non-familiar environment in dreams (see the interesting discussion on that in [56]), since patients' and controls' habitual sleep habits and environments were respected. Furthermore, although our sample size is limited and participants were selected based on geographical proximity and accessibility, our research protocol allowed us to reach a sample of patients that is quite wide compared with most of the literature on the topic, as well as to obtain a fairly extended collection period (several weeks), which would have been scarcely feasible with a laboratory design (this usefully enlarged the total number of dream transcripts despite the low expected DRF).

Another limitation regards the disparity we found between groups in gender proportion and years of education. Regarding the latter, the difference was expected since educational attainment is generally lower in schizophrenia [57], probably due to the fact that the age of onset for this kind of pathology peaks around 20 years. As for the gender differences, studies on this topic consistently show that these differences mainly involve dream thematic content (e.g., sexuality, aggressive behaviors, etc.), whereas length, realism and level of interaction are similar [58,59]. Therefore, we are confident that the gender disparity between groups did not significantly affect our data, since our main result essentially regarded a general impoverishment of patients' dreams, affecting most of the dream elements.

Furthermore, it must be taken into account that most participants in the patient group were being treated with psychotropic medications (mostly antipsychotics, but also benzodiazepines and hypnotics), which have been shown to have some effect on dream features (especially DRF, although data are sparse and conflicting) [60].

Finally, for the patient group, although we collected and analyzed BPRS scores, we lack precise information on the presence and severity of positive and negative symptoms, which could have been useful to discuss our findings in light of recent theories on dreaming and psychosis. However, as mentioned above, because of the influence of impaired verbal fluency on dream report production, any interpretation in accordance with these theories would still have been speculative.

5. Conclusions

In conclusion, our data show significant quantitative and qualitative differences between dream reports of psychotic patients and those of healthy controls, collected over extended periods of time (several weeks) in the participants' habitual living environments. Specifically, patients show lower dream frequency as well as shorter dream reports. Moreover, healthy participants' dreams appear much richer and more articulate in terms of several content aspects, including number of characters, settings, interactions and emotions. The lower amount of negative emotions reported by patients and the correlation found between number of emotions and illness severity scores are in line with theories on the role of dreaming in emotion regulation, suggesting that this function could be impaired in psychoses and related to symptom severity. Finally, our data appear compatible with current theoretical accounts on dreaming and psychosis (such as the continuity hypothesis [6] or the Active Inference Theory [54]). Further research conducted within the framework of these hypotheses is warranted to investigate to what extent the general impoverishment of psychotic patients' dreams depends on their impaired verbal fluency or on impaired mechanisms of dream generation.

Author Contributions: Conceptualization, G.F. and F.C.; methodology, O.D.R., A.L. and G.F.; formal analysis, O.D.R., D.G., T.M., A.C., B.A. and A.L.; investigation, D.G., T.M., B.A. and S.M.; writing—original draft preparation, A.C., G.F. and F.C.; writing—review and editing, G.F., F.C., O.D.R. and A.L.; supervision, G.F. and F.C. All authors have read and agreed to the published version of the manuscript.

Funding: This research received no external funding.

Institutional Review Board Statement: The study was conducted in accordance with the Declaration of Helsinki and approved by the Ethics Committee of the University of Campania "L. Vanvitelli" (protocol code 1/2017; minute n. 6, Department Council of 21 March 2017).

Informed Consent Statement: Informed consent was obtained from all subjects involved in the study.

Data Availability Statement: The original contributions presented in the study are included in the article; further inquiries can be directed to the corresponding author due to privacy reasons.

Conflicts of Interest: The authors declare no conflicts of interest.

References

1. American Psychiatric Association. *Diagnostic and Statistical Manual of Mental Disorders*, 5th ed.; American Psychiatric Publishing: Arlington, VA, USA, 2013. [CrossRef]
2. Waters, F.; Blom, J.D.; Dang-Vu, T.T.; Cheyne, A.J.; Alderson-Day, B.; Woodruff, P.; Collerton, D. What is the link between hallucinations, dreams, and hypnagogic–hypnopompic experiences? *Schizophr. Bull.* **2016**, *42*, 1098–1109. [CrossRef] [PubMed]
3. Limosani, I.; D'Agostino, A.; Manzone, M.L.; Scarone, S. The dreaming brain/mind, consciousness and psychosis. *Conscious Cogn.* **2011**, *20*, 987–992. [CrossRef] [PubMed]
4. Gottesmann, C. The dreaming sleep stage: A new neurobiological model for schizophrenia. *Neurosci. J.* **2006**, *21*, 1105–1115. [CrossRef] [PubMed]
5. Llewellyn, S. In two minds: Is schizophrenia a state trapped between waking and dreaming? *Med. Hypotheses* **2009**, *73*, 572–579. [CrossRef] [PubMed]
6. Hall, C.S.; Nordby, V.J. *The Individual and his Dreams*; New American Library: New York, NY, USA, 1972.
7. Conte, F.; Cellini, N.; De Rosa, O.; Caputo, A.; Malloggi, S.; Coppola, A.; Albinni, B.; Cerasuolo, M.; Giganti, F.; Marcone, R.; et al. Relationships between dream and previous wake emotions assessed through the Italian modified Differential Emotions Scale. *Brain Sci.* **2020**, *10*, 690. [CrossRef] [PubMed]
8. Walker, M.P.; van der Helm, E. Overnight Therapy? The Role of Sleep in Emotional Brain Processing. *Psychol. Bull.* **2009**, *135*, 731–748. [CrossRef] [PubMed]
9. Levin, R.; Nielsen, T. Nightmares, Bad Dreams, and Emotion Dysregulation. *Curr. Dir. Psychol. Sci.* **2009**, *18*, 84–88. [CrossRef]
10. Scarpelli, S.; Bortolacci, C.; D'Atri, A.; Gorgoni, M.; De Gennaro, L. The functional role of dreaming in emotional processes. *Front. Psychol.* **2019**, *10*, 459. [CrossRef] [PubMed]
11. Conte, F.; Cellini, N.; De Rosa, O.; Rescott, M.L.; Malloggi, S.; Giganti, F.; Ficca, G. The effects of sleep quality on dream and waking emotions. *Int. Jou. Envir. Res. Pub. Health* **2021**, *18*, 431. [CrossRef]
12. Okuma, T.; Sunami, Y.; Fukuma, E.; Takeo, S.; Motoike, M. Dream content study in chronic schizophrenics and normals by REMP-awakening technique. *Folia Psychiatr. Neurol. Jp.* **1970**, *24*, 151–162.

13. Cartwright, R.D. Sleep fantasy in normal and schizophrenic persons. *J. Abnorm. Psychol.* **1972**, *80*, 275–279. [CrossRef]
14. Debieve, J.; Bedoret, J.M.; Meaux, J.P.; Fontan, M. The onset of dreams in schizophrenics by awakening during polygraphic registering of sleep. *Lille Med.* **1977**, *22*, 132–139.
15. Henry, J.D.; Crawford, J.R. A meta-analytic review of verbal fluency performance following focal cortical lesions. *Neuropsychology* **2004**, *18*, 284–295. [CrossRef]
16. Hadjez, J.; Stein, D.; Gabbay, U.; Bruckner, J.; Meged, S.; Barak, Y.; Elizur, A.; Weizman, A.; Rotenberg, V.S. Dream content of schizophrenic, nonschizophrenic mentally ill, and community control adolescents. *J. Adolesc.* **2003**, *38*, 331–342.
17. Zanasi, M.; Calisti, F.; Di Lorenzo, G.; Valerio, G.; Siracusano, A. Oneiric activity in schizophrenia: Textual analysis of dream reports. *Conscious. Cogn.* **2011**, *20*, 337–348. [CrossRef]
18. Richardson, G.A.; Moore, G.A. On the manifest dream in schizophrenia. *J. Am. Psychoanal Assoc.* **1963**, *11*, 281–290. [CrossRef] [PubMed]
19. Khazaie, H.; Tahmasian, M ; Younesi, G.; Schwebel, D.C.; Rezaei, M.; Rezaie, L.; Mohamadi, M.; Ghanbari, A. Evaluation of Dream Content among Patients with Schizophrenia, their Siblings, Patients with Psychiatric Diagnoses other than Schizophrenia, and Healthy Control. *Iran. J. Psychiatry* **2012**, *7*, 26–30. [CrossRef] [PubMed]
20. Kramer, M. Manifest dream content in normal and psychopathologic states. *Arch. Gen. Psychiatry* **1970**, *22*, 149–156. [CrossRef]
21. Lusignan, F.A.; Zadra, A.; Dubuc, M.J.; Daoust, A.M.; Mottard, J.P.; Godbout, R. Dream content in chronically-treated persons with schizophrenia. *Schiz. Res.* **2009**, *12*, 164–173. [CrossRef]
22. Hobson, J.A.; Hoffman, S.A.; Helfand, R.; Kostner, D. Dream bizarreness and the activation-synthesis hypothesis. *Hum. Neurobiol.* **1987**, *6*, 157–164.
23. Dement, W. Dream recall and eye movements during sleep in schizophrenics and normals. *J. Nerv. Ment. Dis.* **1955**, *122*, 263–269. [CrossRef] [PubMed]
24. Carrington, P. Dreams and schizophrenia. *Arch. Gen. Psychiatry* **1972**, *26*, 343–350. [CrossRef] [PubMed]
25. Noreika, V. Dreaming and waking experiences in schizophrenia: How should the (dis)continuity hypotheses be approached empirically? *Conscious. Cogn.* **2011**, *20*, 349–354. [CrossRef] [PubMed]
26. Scarone, S.; Manzone, M.L.; Gambini, O.; Kantzas, I.; Limosani, I.; D'Agostino, A.; Hobson, J.A. The dream as a model for psychosis: An experimental approach using bizarreness as a cognitive marker. *Schizophr. Bull.* **2008**, *34*, 515–522. [CrossRef] [PubMed]
27. Curcio, G.; Tempesta, D.; Scarlata, S.; Marzano, C.; Moroni, F.; Rossini, P.M.; Ferrara, M.; De Gennaro, L. Validity of the Italian Version of the Pittsburgh Sleep Quality Index. *J. Neurol. Sci.* **2013**, *34*, 511–519. [CrossRef] [PubMed]
28. Ventura, J.; Lukoff, D.; Nuechterlein, K.H.; Liberman, R.P.; Green, M.; Shaner, A. Appendix 1: Brief Psychiatric Rating Scale BPRS Expanded Version 4.0 scales. *Int. J. Methods Psychiatr. Res.* **1993**, *3*, 227–243. [CrossRef]
29. Novelli, G.; Papagno, C.; Capitani, E.; Laiacona, M. Tre test clinici di ricerca e produzione lessicale. Taratura su soggetti normali [Three clinical tests to research and rate the lexical performance of normal subjects]. *Arch. Di Psicol. Neurol. E Psichiatr.* **1986**, *47*, 477–506.
30. Cavallero, C.; Foulkes, D.; Hollifield, M.; Terry, R. Memory Sources of REM and NREM Dreams. *Sleep* **1990**, *13*, 449–455. [CrossRef] [PubMed]
31. Morosini, P.; Casacchia, M. Brief Psychiatric Rating Scale (BPRS): Versione 4.0 ampliata: Manuale di istruzioni. In *Scale, Definizioni Operative dei Livelli di Gravità dei Sintomi e Domande Dell'intervista*; 1995; pp. 1–6. Available online: https://www.riabilitazionepsicosociale.it/wordpress/wp-content/uploads/2016/04/BPRS-4.0-ampliata-Brief-Psychiatric-Rating-Scale.pdf (accessed on 20 October 2023).
32. Hofmann, A.B.; Schmid, H.M.; Jabat, M.; Brackmann, N.; Noboa, V.; Bobes, J.; Garcia-Portilla, M.P.; Seifritz, E.; Vetter, S.; Egger, S.T. Utility and validity of the Brief Psychiatric Rating Scale (BPRS) as a transdiagnostic scale. *Psychiatry Res.* **2022**, *314*, 114659. [CrossRef]
33. Tombaugh, T.N.; Kozak, J.; Rees, L. Normative data stratified by age and education for two measures of verbal fluency: FAS and Animal Naming. *Arch. Clin. Neuropsychol.* **1999**, *14*, 167–177. [CrossRef]
34. Schredl, M. Stability and Variability of dream content. *Percept. Mot. Ski.* **1998**, *86*, 733–734. [CrossRef] [PubMed]
35. Foulkes, D.; Schmidt, M. Temporal sequence and unit composition in dream reports from different stages of sleep. *Sleep* **1983**, *6*, 265–280. [CrossRef] [PubMed]
36. Occhionero, M.; Cicogna, P.C. Autoscopic phenomena and one's own body representation in dreams. *Conscious. Cogn.* **2011**, *20*, 1009–1015. [CrossRef] [PubMed]
37. Cicogna, P.C.; Natale, V.; Occhionero, M.; Bosinelli, M. A comparison of mental activity during sleep onset and morning awakening. *Sleep* **1998**, *21*, 462–470. [CrossRef] [PubMed]
38. De Padova, V.; Aiello, P.; Cocozza, G.; Ficca, G. Dream Recall in Elderly Subjects after Spontaneous Awakening in laboratory. In Proceedings of the World Association of Sleep Medicine, 1st Congress, Berlin, Germany, 15–18 October 2005; pp. 35–40.
39. Nielsen, T.A.; Deslauriers, D.; Baylor, G.W. Emotions in dream and waking event reports. *Dreaming* **1991**, *1*, 287. [CrossRef]
40. Merritt, J.M.; Stickgold, R.; Pace-Schott, E.; Williams, J.; Hobson, J.A. Emotion Profiles in the Dreams of Men and Women. *Conscious. Cogn.* **1994**, *3*, 46–60. [CrossRef]
41. Taylor, F.; Bryant, R.A. The tendency to suppress, inhibiting thoughts, and dream rebound. *Behav. Res. Ther.* **2007**, *45*, 163–168. [CrossRef]

42. Kröner-Borowik, T.; Gosch, S.; Hansen, K.; Borowik, B.; Schredl, M.; Steil, R. The effects of suppressing intrusive thoughts on dream content, dream distress and psychological parameters. *J. Sleep Res.* **2013**, *22*, 600–604. [CrossRef]
43. Schredl, M.; Engelhardt, H. Dreaming and psychopathology: Dream recall and dream content of psychiatric inpatients. *Sleep Hypn.* **2001**, *3*, 44–54.
44. Bosinelli, M.; Cavallero, C.; Cicogna, P. Self Rappresentation in Dream Experiences During Sleep Onset and REM Sleep. *Sleep* **1982**, *5*, 290–299. [CrossRef]
45. Mcnamara, P.; Mclaren, D.; Durso, K. Rappresentation of the Self in REM and NREm dreams. *Dreaming* **2007**, *17*, 113–126. [CrossRef] [PubMed]
46. Maquet, P.; Péters, J.; Aerts, J.; Delfiore, G.; Degueldre, C.; Luxen, A.; Franck, G. Functional neuroanatomy of human Rapid-Eye-Movement sleep and dreaming. *Nature* **1996**, *383*, 163–166. [CrossRef] [PubMed]
47. Chen, P.; Ye, E.; Jin, X.; Zhu, Y.; Wang, L. Association between thalamocortical functional connectivity abnormalities and cognitive deficits in schizophrenia. *Sci. Rep.* **2019**, *9*, 2952. [CrossRef] [PubMed]
48. Muzur, A.; Pace-Schott, E.F.; Hobson, J.A. The prefrontal cortex in sleep. *Tren. Cogn. Sci.* **2002**, *6*, 475–481. [CrossRef] [PubMed]
49. Dessailles, M.; Dang-Vu, T.; Sterpenich, V.; Schwartz, S. Cognitive and emotional processes during dreaming: A neuroimaging view. *Cons. Cogn.* **2011**, *20*, 998–1008. [CrossRef] [PubMed]
50. Mier, D.; Lis, S.; Zygrodnik, K.; Sauer, C.; Ulferts, J.; Gallhofer, B.; Kirsch, P. Evidence for altered amygdala activation in schizophrenia in an adaptive emotion recognition task. *Psychiatry Res.* **2014**, *221*, 195–203. [CrossRef]
51. Heilbronner, U.; Samara, M.; Leucht, S.; Falkai, P.; Schulze, T.G. The Longitudinal Course of Schizophrenia Across the Lifespan: Clinical, Cognitive, and Neurobiological Aspects. *Harv. Rev. Psychiatry* **2016**, *24*, 118–128. [CrossRef]
52. Eagleman, D.M.; Vaughn, D.A. The Defensive Activation Theory: REM sleep as a mechanism to prevent takeover of the visual cortex. *Front. Neurosci.* **2021**, *15*, 632853. [CrossRef] [PubMed]
53. Antonioni, A.; Raho, E.M.; Sensi, M.; Di Lorenzo, F.; Fadiga, L.; Koch, G. A new perspective on positive symptoms: Expression of damage or self-defence mechanism of the brain? *Neurol. Sci.* **2024**, *45*, 2347–2351. [CrossRef]
54. Friston, K.; FitzGerald, T.; Rigoli, F.; Schwartenbeck, P.; Pezzulo, G. Active inference: A process theory. *Neural. Comput.* **2017**, *29*, 1–49. [CrossRef]
55. Jeganathan, J.; Breakspear, M. An active inference perspective on the negative symptoms of schizophrenia. *Lancet Psychiatry* **2021**, *8*, 732–738. [CrossRef] [PubMed]
56. Schultz, H.; Salzarulo, P. Methodological approaches to problems in experimental dream research: An introduction. *J. Sleep. Res.* **1993**, *2*, 1–3. [CrossRef]
57. Crossley, N.A.; Alliende, L.M.; Czepielewski, L.S.; Aceituno, D.; Castañeda, C.P.; Diaz, C.; Iruretagoyena, B.; Mena, C.; Mena, C.; Ramirez-Mahaluf, J.P.; et al. The enduring gap in educational attainment in schizophrenia according to the past 50 years of published research: A systematic review and meta-analysis. *Lancet Psychiatry* **2022**, *9*, 565–573. [CrossRef]
58. Schredl, M.; Sahin, V.; Schäfer, G. Gender differences in dreams: Do they reflect gender differences in waking life? *Pers. Individ. Diff.* **1998**, *25*, 433–442. [CrossRef]
59. Schredl, M.; Reinhard, I. Gender differences in nightmare frequency: A meta-analysis. *Sleep Med. Rev.* **2011**, *15*, 115–121. [CrossRef]
60. Nicolas, A.; Ruby, P.M. Dreams, Sleep and Psychotropic Drugs. *Front. Neurol.* **2020**, *11*, 507495. [CrossRef]

Article

Personality Functioning Improvement during Psychotherapy Is Associated with an Enhanced Capacity for Affect Regulation in Dreams: A Preliminary Study

Simon Kempe *, Werner Köpp and Lutz Wittmann

Department of Psychology, International Psychoanalytic University, Stromstraße 1, 10555 Berlin, Germany
* Correspondence: simon.kempe@ipu-berlin.de

Abstract: Background: Clinical case illustrations of patients with an impairment of personality functioning (IPF) have repeatedly reported that progress during psychotherapy is reflected by alterations in dream content. However, quantitative studies based on samples of psychotherapy patients are scarce. As a core component of both personality functioning and contemporary psychodynamic dream theory, the construct of affect regulation is of specific significance in this context. Aims: To test if improvement in personality functioning in the course of psychotherapy is associated with an increasing ability to regulate affects in dreams. Method: In a longitudinal design, affect regulation was compared in N = 94 unsolicited dream reports from the first vs. last third of long term psychotherapy of ten patients with initial IPF. Dream reports were transcribed from recordings of the sessions. Expert ratings of the level of personality functioning were obtained using the Scales of Psychological Capacities. The capacity for affect regulation was assessed using the Zurich Dream Process Coding System. Group differences were assessed using linear mixed models, controlling for dream length as well as the nested structure of this data set. Results: Patients demonstrated an increased capacity for affect regulation in dreams that was primarily evident in three core features: the complexity of dream elements (cf., e.g., parameter attributes, $p = 0.024$); the extent of affective involvement in the dream ego (cf., e.g., parameter subject feeling, $p = 0.014$); and the flexibility to regulate the dynamics of safety/involvement processes ($p \leq 0.001$). This pattern was especially prominent in a subgroup (n = 7) of patients with more pronounced improvements in personality functioning. Conclusion: These findings support the hypotheses that decreasing IPF during psychotherapy is associated with increases in the capacity for affect regulation in dreams. Thus, researchers and therapists can utilize dream reports to illuminate the important aspects of treatment progress in clinical practice.

Keywords: personality functioning; personality change; dreaming; affect regulation; Zurich Dream Process Coding System

check for **updates**

Citation: Kempe, S.; Köpp, W.; Wittmann, L. Personality Functioning Improvement during Psychotherapy Is Associated with an Enhanced Capacity for Affect Regulation in Dreams: A Preliminary Study. *Brain Sci.* **2024**, *14*, 489. https://doi.org/10.3390/brainsci14050489

Academic Editor: Heide Klumpp

Received: 12 April 2024
Revised: 4 May 2024
Accepted: 7 May 2024
Published: 11 May 2024

1. Introduction

1.1. Dreams during the Course of Psychotherapy

Although the functions of dreaming are not conclusively understood, a review of empirical research by Gazzillo et al. [1] shows that many researchers assume an adaptive function of REM sleep and dreaming. For instance, REM dreaming is considered to have a "quasi-therapeutic adaptive function" [2] (p. 235) or to "actively moderate mood overnight" [3] (p. 1). Moreover, working with dreams is a standard technique that is frequently applied in the daily practice of psychotherapy [4]. Empirical research supports four conceptual uses of dreams in psychotherapy. In clinical practice, working with dreams contributes to "(I) facilitate the therapeutic process, (II) facilitate patient insight and self-awareness, (III) provide clinically relevant and valuable information to therapists and (IV) provide a measure of therapeutic change" [5] (p. 255). The first three of these applications focus on the therapeutic use of dreams in psychotherapy, whereas the fourth stresses dreams as a (secondary) psychotherapy outcome measure. Several clinical case studies indicate that manifest dream content changes in parallel with therapeutic progress. Warner [6], for example, found that the dreams of patients who improved in psychotherapy changed "from being self-punitive or self-denying to allowing them some satisfactions and gratification" (p. 314), while dreams of patients with limited improvement in psychotherapy demonstrated little or no change. However, systematic sample research studies addressing dream content as an outcome parameter are scarce.

Glucksman and Kramer [7] report three studies in which changes in manifest dream reports between the beginning and the end of psychotherapy were found, suggesting clinical progress in such a way that the positive affect increased, as opposed to decreasing negative affect. Fischer and Kächele [8] examined 240 dreams of 8 patients from the first and last third of treatment, with respect to the theoretical orientation of the therapist (Freudian–/Jungian-Therapy). The results indicate an initial significant association between the treatment method and dream content (such as manifesting sexual contents (Freudian) or contents familiar from mythology (Jungian)) that diminishes as treatment progresses, supporting the hypothesis that patients became more independent of the therapist. In a qualitative study, Roesler [9] investigated 202 dreams of 15 patients with diverse types of diagnoses. In total, 5 dream patterns were identified, reflecting a continuum of self-efficacy and involvement (from an absent or threatened dream ego to social involvement). These dream patterns were found to be transformed towards more satisfying interactions of the dream ego when progress in symptom level and personality structure was reported by the therapist. Kuelz et al. [10] found that obsessive compulsive (OC) disorder themes decreased in 40 dreams of 9 OC inpatients during the first 5 days after the commencement of exposure therapy. Contrary to expectations, no increase in dream length or emotional intensity was evident. Taken together, evidence from the diverse types of studies that focus on different parameters suggest a continuity between improvement in waking mental health and dream experience during the course of psychotherapy.

Melstrom and Cartwright [11] compared laboratory-collected dreams before and after psychotherapy in ten patients and four control subjects. In contrast to the observations, e.g., those of Glucksman and Kramer [7], the negative affect (level of dream anxiety) was increased in patients with psychotherapy assessed as successful, whereas it barely changed in patients with unsuccessful psychotherapy outcomes and control subjects. The authors hypothesize that a successful psychotherapy outcome is reflected in a tolerance and capacity to deal with higher levels of dream anxiety, whereas lower levels indicate defensive coping or avoidance of dream anxiety.

In summary, previous research ascertained that dream content changes in the course of psychotherapy. However, several methodological shortcomings need to be reflected upon. Firstly, a lack of prospectively defined study endpoints results in a falsification problem: both increasing the positive [7] and negative [11] affect in dreams were interpreted as reflecting successful therapy. Moreover, several studies did not specify inclusion and exclusion criteria for sampling, thus the dream content analysis did not focus on specific mental disorders but on the general effects of psychotherapy.

1.2. Dreaming and Impairment of Personality Functioning

The level of personality functioning is a core criterion in a dimensional classification of personality pathology as introduced in the ICD-11 [12]. Moreover, the impairment of personality functioning (IPF) is also a key variable for capturing psychopathology across various mental disorders and symptoms [13] and can thus, be used to distinguish patients with personality disorders, mood, and anxiety disorders, and healthy controls [14]. As for the association of dream characteristics with IPF, research has typically focused on patients diagnosed with borderline personality disorder (BPD). Overall, there are very few sample research studies on the dreams of BPD patients, but a recurring finding was an increased frequency of nightmares and more negatively toned dreams [15]. As for the psychodynamic literature, a large range of clinical case studies, reviewed by Hau [16], focuses on the so-called borderline-dream. Overall, no consistent picture of typical borderline dream content emerged. Rather, two types of dreams can be differentiated. One group of case reports describes archaic forms of representation as, e.g., characterized by unintegrated rage, whereas other reports find that flat, realistic dreams are characteristic of BPD.

In addition to the quality of dream content in BPD patients, a dimension that has received attention from both research on IPF as well as dream research is the capacity of affect regulation. Affect regulation is typically defined as "the attempt to alter or control one's mood or emotional state so as to maximize pleasant experiences and minimize unpleasant ones [. . .]" including strategies such as "[. . .] cognitive techniques such as reframing and distraction, behavioral methods such as progressive relaxation and meditation, and unconscious processes such as denial and dissociation" [17], and is repeatedly found to be impaired in patients diagnosed with BPD (e.g., [18]).

With respect to dreams, affect regulation can be assessed by the Zurich Dream Process Coding System ([ZDPCS; [19,20]) quantifying processes in the dream narrative that balance the level between two opposing tendencies, the need for experience of safety, and the need for emotional involvement of the dream ego. Euler et al. [21] asked participants with different levels of personality functioning for dream reports within the frame of semi-standardized clinically interviews and investigated affect regulation in 62 collected dream reports applying the ZDPCS. Markers of the level of personality functioning were found to be related to the richness and complexity of the dream narrative. Kempe et al. [22] examined the affect regulation in 77 dreams reported by 20 patients with and without IPF in the first third of psychotherapy. Results indicated that patients with IPF have a limited capacity for affect regulation in dreams, reflected in three dimensions: (I) less complexity of the dream elements; (II) less involvement of the dream ego on the interactional level; and (III) less flexibility in the safety/involvement regulation of the dream dynamic. Taken together, these findings indicate that patients with IPF use more preventive affect avoidance strategies due to an increased need for security in dreaming. Noteworthy to Kempe et al.'s [22] results that impaired affect regulation in dreams is a central marker in the differentiation of patients with and without IPF at the baseline of psychotherapy, clinical case studies (e.g., the work presented in [23]) indicate that these parameters change in parallel with the therapeutic progress.

This study is premised on the hypothesis that an improvement in personality functioning during psychotherapy is reflected by an increasing capacity for affect regulation in dreams. Based on the differences in dream affect regulation between patients with and without IPF [22], it is expected that, as a result of successful psychotherapy, the affect regulation parameters of dreams from patients with initial IPF, will approach the levels shown by patients without initial IPF. The alterations in dream content in parallel to the therapeutic progress is expected to manifest in three dimensions as above (for details, see the Section 3). This study attempts to overcome the methodological limitations of previous approaches through an external assessment of therapeutic progress in personality functioning, a naturalistic clinical sample, clear inclusion and exclusion criteria for sampling, a standardized evaluation method for dream content, a falsifiable operationalization of the hypotheses, and the analysis of dream report series, rather than single dream reports in psychotherapy (which is required for an adequate basis of data to establish relationships with personality trait factors) [19,24].

2. Method

2.1. Procedure

Patients provided written informed consent and the research project was approved by the institutional review board of the IPU-Berlin (no. 2019-10). Patients were treated with (modified) psychoanalytic psychotherapy in an outpatient setting in a cooperating private practice by one experienced psychotherapist with standard audio or video recordings. The average therapy length was 238.3 sessions (SD = 90.4, range = 119–431). In one case, a cut off was made after 431 sessions because therapy had not been completed at the time of data analysis. In total, 96 sessions from the first and last third of psychotherapy, containing a dream report, were identified, based on the therapeutic documentation. Of these, eleven sessions (11.5%) were accidentally not recorded and, therefore, dreams were obtained from the therapeutic transcript made during the session.

2.2. Sample

Ten patients with IPF at the beginning of treatment were included into this study. Eight of these are identical with the sample in Kempe et al. [22], while two of these patients did not report dreams at the end of psychotherapy and were, therefore, replaced. The average age of the sample is 32.5 years (SD = 7.8, range = 24–47) at the time of contact (female sex = 90.0%). Inclusion criteria are an ICD-10 F60 diagnosis or a social behavior disorder (F91.2, n = 1) and a low level of personality functioning (see Section 3). On average, patients received 2.5 diagnoses (range = 1–5) according to the ICD-10 (see Table 1). In addition to psychotherapy, four patients received psychopharmacological treatment (1. patient: Fluoxetine, Amitriptyline, Methylphenidate, Candesartan, 2. patient: Escitalopram, 3. patient: St. John's Wort, 4. patient: Quetiapine).

Table 1. ICD-10 diagnoses of participants (N = 10).

Variable	N
Mood [affective] disorders (F3)	3
Neurotic, stress-related, and somatoform disorders (F4)	5
Behavioral syndromes associated with physiological disturbances and physical factors (F5)	6
Disorders of adult personality and behavior (F6)	9
Emotionally unstable personality disorder (F60.3-)	
Aggressive (F60.30)	1
Borderline (F60.31)	4
Other specific personality disorders (F60.8)	4
Behavioral and emotional disorders with onset usually occurring in childhood and adolescence (F9)	1
≥2 diagnoses	9

Note: ICD-10 number refers to patients with at least one diagnosis from the respective cluster.

3. Measures

3.1. Scales of Psychological Capacities (SPC; [25])

The SPC are an expert-rated assessment tool to quantify personality functioning on the basis of 17 scales (35 subscales in total), such as self-coherence or impulse regulation. Each scale captures the severity of the respective impairment and coping possibilities to handle stressors alone (level 1), with help (level 2), or unable to, despite help (level 3), with one intermediate step each (7-point scale). Therapeutic changes in personality functioning can validly be assessed by the SPC [26]. In its original form, SPC ratings are based on semi-standardized interviews. In this study, SPC were rated based on the recording of the first and last two therapy sessions each. The construct validity and inter-rater reliability (IRR) of this procedure were satisfactory [22].

3.2. Zurich Dream Process Coding System (ZDPCS; [19])

The ZDPCS assesses affect regulation in dreams by analyzing strategies and capacities for regulating the course of dreams. An extensive introduction into the ZDPCS, including a fully coded dream example, as well as a comparison of the ZDPCS to other analytic approaches to dream content can be found in Kempe et al.'s work [27]. Previously, the ZDPCS has been applied for the characterization of dream affect regulation in veterans with a diagnosis of post-traumatic stress or adjustment disorder [28], as well as manifestations of psychotherapeutic progress in dreams [23]. The ZDPCS is based on a dream generation theory that extends psychoanalytic dream theory with contributions from cognitive science and artificial intelligence research. Following French [29] in applying a problem-solving paradigm, the function of a dream is understood as an attempt to solve or adapt to a complex. Complexes are defined as representations of interpersonal experiences associated with strong anxiety or disappointment which has not been disaffectualized during the process of memory consolidation [30]. Complexes originating from long-term memory are supposed to be brought to a solution by transforming the stored affective complex information back into simulated relational reality within the dream. Thereby, the dream ego is caught between the need to have a good enough sense of security (safety principle) so that the current tolerable degree of involvement of the dream ego in interpersonal processes is not exceeded, as well as the need to recommit to interpersonal relational reality (involvement principle).

3.3. Coding System

The aim of dream coding using the ZDPCS is to trace the affect regulatory mechanisms over the course of a dream. The dynamic shifts of the dream process between the poles of involvement and safety throughout the dream plot can be depicted. Dream reports are firstly edited (translation into presence, the deletion of comments) and then segmented. In each segment (comparable to a screenplay for the dreamwork), dream elements are coded in 5 fields with coding options in 11 main categories and 161 subcodes [28]. For this study, three ZDPCS dimensions were examined, which significantly differed between patients with and without IPF at psychotherapy baseline [22] and can thus, be assumed to be central markers for the level of personality functioning in dreams. The full scoring system, as well as its theoretical background, can be found in Moser and Hortig [19]. The IRR of the ZDPCS was examined twice [22,28] and was found to be substantial to strong (κ-range = 0.70–0.85), according to Landis and Koch [31].

3.4. Dream Parameters

(I) The complexity of the dream elements: Dream length (the word count of edited reports and the number of segments) is assumed to reflect the specific dimensions of personality, the ability to simulate complex inner-psychic processes, and the capacity for introspection. Furthermore, it is mandatory to adjust for dream length when comparing other dream content parameters for differences between two conditions [32]. A complex dream narrative is reflected primarily at the visual–pictorial level in the position field (PF), registering all elements, plus their attributes. A comprehensive PF is an indicator of the multifariousness of the cognitive and affective representations of the dream complex. Specific subgroups, such as the numbers of attributes in the PF, are of particular interest by highlighting how multifaceted dream elements are displayed in the dream narrative.

(II) The involvement of the dream ego on an interactional level: The frequency and the quality of interactions are registered in the interaction field (IAF) and are assumed to represent the current tolerable degree of involvement of the dream ego in interpersonal processes. A comprehensive IAF is an indicator of strong involvement of the dream ego. In addition, a categorization is made between six levels which reflect an increasing affectualization. The level of affectualization, i.e., the highest point of affectualization in a dream report, can be determined, as well as the frequencies of the different forms of interactions [20].

(III) Flexibility in the safety/the involvement regulation of the dream dynamic: The regulation between the two central tendencies in dream reports, according to the ZDPCS, is assessed. Transformations within the dream dynamic are tracked from segment to segment and differentiated with respect to whether the affectualization of the dream is increased (involvement principle) or reduced (safety principle). More frequent changes between the two poles are assumed to reflect a higher capacity in affect regulation.

A detailed description of the parameters utilized in the present study can be found in the Supplemental Materials.

4. Data Analysis

Statistical analyses were performed using the jamovi statistical software (v2.0). An outlier analysis of dream length (word count) resulted in excluding two dreams that were more than three standard deviations above the mean score. Due to the naturalistic study design, the number of dreams differ between patients, and the dreams of the individual patients are not independent of each other (nested data). As pointed out by Schredl [33], mixed-model approaches are mandatory in order to consider this specific nested data structure of multiple dream reports. Therefore, linear mixed models are computed to test for differences in continuous parameters after adjusting for dream length. Random effects are defined by the patient variable. Fixed effects are time points (dreams at the beginning/end of psychotherapy), dream length (word count), and the interaction term of these two. The dream length is centered in the mixed model analysis to reduce multicollinearity when computing the interaction term. For the ordinal parameters, ordinal logistic regression is required. This, however, does not allow for modeling cluster-level variables and random effects. The Shapiro–Wilks test for normality was run on each parameter and was found to be significant several times. Thus, results should be interpreted with caution, although some deviations from this normality assumption seem to be uncritical [34]. All hypotheses were tested unilaterally at $p < 0.05$ according to their clear unidirectional formulation. An inspection of patterns of change, with respect to personality functioning, revealed two clearly distinguishable subsamples. Post hoc analyses were performed for the larger subsample (n = 7), with more pronounced personality functioning improvement.

5. Results

5.1. Improvement in Personality Functioning and Descriptives

As reflected by the decreased SPC total scores, patients' personality functioning was significantly improved at the end (M = 0.7, SD = 0.3) compared to the beginning of psychotherapy (M = 2.1, SD = 0.3) (Wilcoxon signed-rank test; z = −2.80, p (one-tailed) = 0.003). Patients reported 45 dreams during the first third and 51 dreams during the last third of the psychotherapy sessions. On average, 4.5 dreams were reported by patients in the beginning (SD = 1.4, range = 2–6) and 5.1 dreams in the end (SD = 1.1, range = 4–7).

5.2. Differences between Dreams at Beginning and End of Psychotherapy

Hypothesis 1 (Dreams are expected to be more complex): Dreams were significantly longer at the end of psychotherapy, reflected by the number of words per edited report (cf. Table 2). However, this difference was not reflected by a significantly elevated quantity of segments per dream ($p = 0.074$). Within the PF, significantly more attributes were found in dreams at the end of psychotherapy. As expected, the additional parameters of complexity were increased at the end of psychotherapy, but they missed statistical significance. In particular, dreams showed more overall codes in PF at the end of psychotherapy ($p = 0.057$), and more human object processors ($p = 0.084$).

Table 2. ZDPCS dream characteristics (N = 10).

	First Part of Psychotherapy		Last Part of Psychotherapy		β (SE)	p [a]
	M	SD (Range)	M	SD (Range)		
Dream length (word count)	154.7	113.8 (13–470)	193.7	115.6 (41–488)	38.0 (21.3)	0.039
Quantity of segments per dream	7.0	4.7 (1–22)	8.4	4.8 (3–23)	1.4 (0.9)	0.074
Position field	20.7	14.1 (2–59)	28.0	16.5 (4–91)	2.9 (1.8)	0.057
Human object processors	5.3	3.9 (0–19)	7.6	6.4 (0–29)	1.2 (0.9)	0.084
Inanimate cognitive elements	3.4	3.6 (0–17)	3.6	3.1 (0–14)	−0.5 (0.5)	0.159
Attributes	2.8	2.7 (0–12)	4.2	2.5 (0–14)	0.9 (0.4)	0.024
Static positioning of relations	0.4	0.7 (0–3)	0.5	0.8 (0–3)	0.1 (0.2)	0.383

Table 2. *Cont.*

	First Part of Psychotherapy		Last Part of Psychotherapy		β (SE)	p [a]
	M	SD (Range)	M	SD (Range)		
Interaction field	6.4	6.1 (0–30)	7.0	5.1 (0–23)	−0.6 (1.0)	0.271
Displacement relations	0.9	1.2 (0–5)	0.8	1.0 (0–5)	−0.2 (0.2)	0.193
Responsive interactions	0.7	1.1 (0–5)	0.8	1.2 (0–5)	0.0 (0.2)	0.425
Subject feeling	0.1	0.3 (0–1)	0.3	0.6 (0–2)	0.2 (0.1)	0.014
Alternation between safety/involvement processes	5.0	3.5 (0–13)	8.4	5.7 (1–26)	2.1 (0.5)	<0.001

Note: [a] = Linear mixed model (*p*-value), ZDPCS = Zurich Dream Process Coding System.

Hypothesis 2 (Dreams are expected to show more involvement of the dream ego): The overall frequency of codes in the IAF did not significantly differ between dreams at the beginning and end of psychotherapy. Within the IAF, significantly more subject feeling was found at the end of psychotherapy. Figure 1 presents the distribution of dreams with respect to the highest observed level of affectualization (six levels) at the beginning and end of psychotherapy, according to Moser and von Zeppelin [20]. At both time points, more dreams in the high categories than in the lower categories were found. Dreams in the higher categories were more frequent at the end of psychotherapy in contrast to more dreams in lower categories at the beginning of psychotherapy, although this difference did not reach statistical significance (p = 0.188).

Hypothesis 3 (Dream dynamic is expected to be regulated more flexibly): Dreams at the end of psychotherapy demonstrated significantly more changes between the safety and involvement processes when compared to dreams at the beginning of psychotherapy (cf. Table 2).

5.3. Post Hoc Analyses

The inspection of the therapy-related improvements in personality functioning (the difference of the SPC total score from the beginning to the end of psychotherapy) revealed two subsamples: seven patients with strong effects (mean SPC change = 1.6, SD = 0.3, range = 1.2–2.0) as opposed to three patients (n = 3 reported 12 dreams each at the beginning and end of psychotherapies.) with moderate change (mean SPC change = 0.8, SD = 0.1, range = 0.7–0.9). Thus, the first group improved by 5.3 SD units with respect to the baseline values, as compared to 2.7 SD units of the second group. As previous studies (e.g., [6,11]) reported change in dreams only for successful psychotherapies, analyses were repeated for this subsample. In addition to the affect regulation parameters which changed in the overall sample, further indices, reflecting the complexity of dream elements showed significant differences, as predicted by hypothesis 1 (the quantity of segments per dream, overall codes in the position field; cf. Supplementary Table S1). Furthermore, as predicted by hypothesis 2, significantly more involvement of the dream ego (responsive interactions) was observed, as well as significant higher levels of affectualization in dreams at the end of psychotherapy (β = 0.877, p = 0.023, OR = 2.4, Supplementary Figure S1). The decrease in displacement relations and inanimate cognitive elements did not reach statistical significance (both $p < 0.10$).

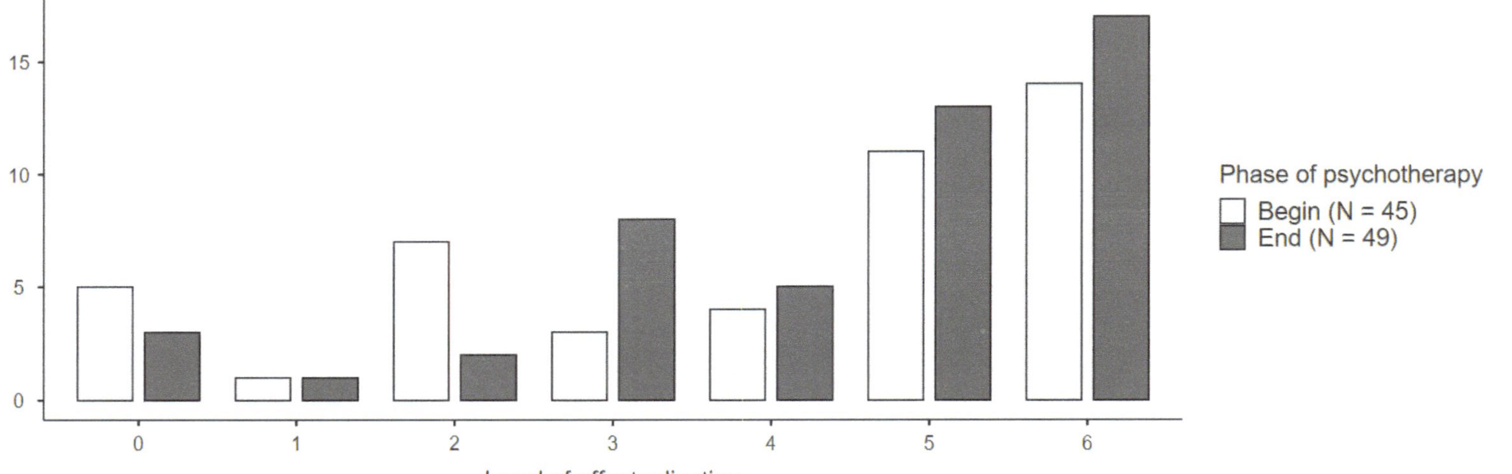

Figure 1. Distribution of the highest observed affectualization in dream reports (N = 10). *Note:* Level (L) 0: no interaction; L1: kinesthetic interactions; L2: displacement relations; L3: verbal relations; L4: constrained interactions; L5: resonant interactions; L6: responsive interactions.

6. Discussion

This analyses extends previous findings [22] of differences in affect regulation in dreams present at the baseline of psychotherapy between patients with and without IPF. Based on these findings, three specific hypotheses were formulated regarding the expected changes in dream affect regulation in response to therapy-related improvements in personality functioning. In the following section, we evaluate our hypotheses in light of the presented findings and methodological considerations.

6.1. Altering Dreams and Improving Personality Functioning

The results presented are largely in line with the first hypothesis that dreams are more complex at the end of psychotherapy. Dreams were significantly longer at the end of long term psychotherapy. This result contrasts with Kuelz et al. [10], who, however, examined dreams during the first five days of exposure treatment rather than long term psychotherapy, which allows for only limited comparability. The positive association between dream length and psychotherapy outcome may be explained by the enhanced ability to simulate complex inner-psychic processes. Alternatively, it could be assumed that patients described longer dreams as a result of working with dreams in psychotherapy, which could be explained by a stronger capacity for introspection or—alternatively improved dream recall. Importantly, in all other tested continuous parameters of dream content, dream length (word count) was used as a control variable, since it has been repeatedly shown that many scales correlate strongly with dream length (e.g., the work presented in [32]). Even though this effect of dream length on dream content is well known, this study is the first to methodologically adequately control for this parameter in research on therapeutic changes in dreams.

The presented results demonstrate that the complexity of dream content was increased at the end of psychotherapy. This was especially evident in significantly more attributes (ATTR) and longer dreams (word count). In addition, further parameters indicating the complexity of dream content pointed in the direction as expected: more overall codes in the position field (PF; $p = 0.057$), i.e., more equipment in the dream content and increased number of human object processors (OP; $p = 0.084$). Additionally, post hoc analyses in patients with strong personality functioning improvement showed a significant difference at the level of overall codes in the PF. The effect of significantly more ATTR and OP may be illustrated, for example, by the difference between dream content simply containing a house or a yellow house with a small balcony, round windows, and several people inside. Thus, cognitive elements in dreams at the end of psychotherapy offer more diverse affective points of contact because they are perceived as more multifaceted. Moreover, OP with a simulated inner life of their own reveal more potential to trigger affect compared to inanimate cognitive elements (CEU) [19]. An elevated number of OP at the end of psychotherapy thus, reflects an increased capacity for affect regulation. The only PF parameter which (non-significant, $p = 0.159$) decreased was the number of CEU. However, a decrease in CEU in favor of an increase in OP is in line with an increasing potential towards affect-intensive dreaming. These findings are in line with Moser and Hortig's [19] notion that an increased capacity for affect regulation enables the dream complex to be more comprehensively represented, or, to be shaped with broad affect without overstraining the security principle. Thus, more overall codes in the PF are a central marker for increased potential for more involvement in the dreams at the end of psychotherapy.

These further results are partially in line with the second hypothesis that dreams at the end of psychotherapy display more involvement of the dream ego. While the overall frequency of codes in the IAF did not significantly differ between both time points in psychotherapy, "Subject feeling" was found to be significantly increased at the end of psychotherapy. In addition, post hoc analyses showed a significant difference with respect to "Responsive interactions" and the highest observed degree of affective involvement of the dream ego within each dream report, in line with the results of Roesler [9]. Subject feeling is coded for the experience of self-efficacy. The finding that this was increased at the end of therapy is remarkable in the context of control-mastery theory, which stresses such adaptability as a function of dreams:

> "According to this model, dreams represent our unconscious attempts to find solutions to emotionally relevant problems. In dreams, people think about their main concerns, particularly those concerns that they have been unable to solve by conscious thought alone, and they try to develop and test plans and policies for dealing with them". [1] (p. 185)

Responsive interactions mark the distinct involvement and reciprocal regulation of interactions, which is said to be a primary marker of high affect tolerance [19]. The extension of affective intensity (according to the highest observed degree of affective involvement of the dream ego) at the end of psychotherapy is consistent with previous studies reporting higher levels of positive (e.g., the work presented in [7]) or negative affect (e.g., the work presented in [11]) in dreams as an outcome of psychotherapy. However, the ZDPCS operationalizes affective intensity by the extent of involvement and complexity of interpersonal experiences, rather than focusing on the extent of positive/negative aspects of affect in dream content analysis [27].

In accordance with the third hypothesis, dream work in patients at the end of psychotherapy was significantly more capable of oscillating between safety and involvement processes. This finding is in line with the clinical case studies of Döll-Hentschker [23], who observed a flexibilization of affect regulation in dreams in successful psychotherapies. Especially regarding patients with IPF (diagnosed with BPD), clinical case studies reported dream narratives during early stages of treatment to be simple and direct, without sudden shifts in contrast to an enrichment of the dream narratives in later stages [35], which is underlined by the present results.

In summary, the results show that, parallel to the improvement in personality functioning, the dream content approaches the parameters of patients without IPF at baseline of psychotherapy [22]. This development was visible in the overall sample, but more pronounced in the subsample (n = 7) with strong personality functioning improvement. In these successful psychotherapies (as indicated by SPC total scores changing from pre-treatment (M = 2.2, SD = 0.3) to post-treatment (M = 0.6, SD = 0.1)), the improvement in personality functioning corresponds to the patients group without IPF at baseline of psychotherapy (M = 0.6, SD = 0.1) [22] and is comparable to SPC-post-treatment-scores after psychoanalytic psychotherapy from another study (M = 0.60, SD = 0.25; [26]). The difference in therapy-related improvements in personality functioning between the two subsamples may be framed according to Kernberg's [36] concept of personality organization. According to this model, personality pathology is assessed on a continuum, ranging from a low to high level of borderline personality organization (BPO). The sample studied here with initial IPF, i.e., severe personality pathology (N = 10), can be located at a low level of BPO. Therefore, it is plausible that therapy-related improvements in personality functioning varied, enabling patients to achieve normal/neurotic personality organization (n = 7) or high BPO level (n = 3). Accordingly, dream parameters of the latter group do not display characteristics as known from samples without IPF [22].

Taken together, the results indicate an inter-relationship between personality functioning and affect regulation in dreams. Central markers of affect regulation in dreaming related to higher levels of personality functioning are: (I) how rich and multifaceted the dream equipment is shaped, (II) to what extent affective involvement can be permitted, (III) how flexibly the dream work can oscillate in the dynamics of safety/involvement regulation.

6.2. Implications for Clinical Practice and Methodological Aspects

The alteration in dream-inherent affect regulation as a function of therapy-related improvement in personality functioning has implications for clinical practice and research. As summarized above, changes in dream content have been suggested several times as an outcome measure for therapeutic improvement with regard to particular factors. This study stresses that the capacity for affect regulation in dreams could serve as an outcome measure for the level of personality functioning as a criterion in a dimensional classification of personality pathology as introduced in the ICD-11 [12]. The ZDPCS, as a secondary endpoint in psychotherapy research, offers a variety of starting points to generate further hypotheses, e.g., on personality functioning in diverse diagnostic categories.

As indicated above, working with dreams in psychotherapy can serve to promote the therapeutic progress and patient's ability for introspection [5]. Specifically at the beginning of psychotherapy, it has been noted several times that BPD patients are "usually reluctant or unable to give associations to a dream and to explore its meaning" [35] (p. 489). Thus, it is assumed that a lack of ability to symbolize leads to more concretistic forms of representation, which make an interpretation obsolete, because the meaning is openly revealed. Instead, the immediate communicative message and the expression of affect should be incorporated by the therapist. An approach such as the ZDPCS may help therapists to decide if a dream contains such signs of structural impairment or not, and which approach to dream interpretation, therefore, appears to be favorable. This is essential as our results show a rather large standard deviation and a range of affect regulation in dreams at both time points. In other words, dream reports of patients with IPF (and of those without) are by no means strictly homogenous.

6.3. Strengths and Limitations

To the best of our knowledge, this longitudinal study is the first to evaluate changes in unsolicited dream reports during long term psychotherapy in a systematic sample research approach based on standardized audio or video recordings, which has been performed rarely and in clinical case studies (e.g., the work presented in [23]). Previous studies were based on dream reports in psychotherapy documented by the attending clinician right after the session (e.g., the work presented in [8]) or in a sleep laboratory setting [11]. Furthermore, these results are particularly strengthened by a naturalistic clinical sample in which we focused on personality functioning. The data were analyzed retrospectively and thus, were not influenced by the longitudinal study design, which should be acknowledged as laboratory-based dream content is susceptible to experimenter bias or state factors such as participants' laboratory experience [37]. In addition, therapeutic progress was assessed by external assessment. Furthermore, the content validity of this study is particularly strengthened by analyzing dream report series, rather than single dream reports in psychotherapy. Assessing dream report series rather than single dream reports, (which have been frequently conducted in previous research) is of importance as dream content intrapersonally varies. Thus, single dream reports provide an inadequate basis of data to establish relationships with personality trait factors [19,24]. Importantly, the resulting nested structure of this dataset and dream length were controlled for the first time in a study of alteration of dreams in psychotherapy.

Notwithstanding these strengths, this study is not free of limitations that need to be considered for any interpretation of our results. Firstly, on the data level of patients (N = 10), the sample size is relatively small, lowering statistical power in the hierarchical analyses for detecting small effects in N = 94 dream reports. By entering patients as a clustering variable, it was not possible to additionally include psychotropic medication or different diagnostic categories as a further level in the linear mixed-models. A larger sample would also have allowed testing for differences between the two subsamples. Secondly, data analysis was not performed blind with regard to the time points. However, satisfactory reliability checks based on randomly chosen dream reports against a blind second rater indicated that the occurrence of a severe bias due to this shortcoming appears to be improbable. Lastly, but importantly, it remains unclear if and how patients select dreams to then report them in psychotherapy.

7. Conclusions

This study explored the changes in dream-inherent affect regulation in ten patients during long term psychotherapy. For this purpose, personality functioning and N = 94 dreams from the beginning and end of treatment were compared. Changing dream content was primarily evident in three core features: the complexity of dream elements, the extent of affective involvement of the dream ego, and the flexibility to regulate dynamics of safety/involvement processes. In summary, this study provides support to the idea that improved personality functioning at the end of psychotherapy is associated with an enhanced capacity for affect regulation in dreams. Results imply that dream-inherent affect regulation can serve as an (secondary) outcome measure for the level of personality functioning. Therapists can utilize dream reports to decide between different approaches to working with dreams in clinical practice depending on levels of personality functioning.

Supplementary Materials: The following supporting information can be downloaded at: https://www.mdpi.com/article/10.3390/brainsci14050489/s1, Table S1: ZDPCS dream characteristics (n = 7). Figure S1: Distribution of the highest observed affectualization in dream reports (n = 7).

Author Contributions: Conceptualization, S.K., W.K. and L.W.; methodology, S.K., W.K. and L.W.; software, S.K. and L.W.; validation, S.K., W.K. and L.W.; formal analysis, S.K.; investigation, S.K., W.K. and L.W.; resources, S.K., W.K. and L.W.; data curation, S.K., W.K. and L.W.; writing—original draft preparation, S.K., W.K. and L.W.; writing—review and editing, S.K., W.K. and L.W.; visualization, S.K., W.K. and L.W.; supervision, W.K. and L.W.; project administration, S.K., W.K. and L.W.; funding acquisition, S.K., W.K. and L.W. All authors have read and agreed to the published version of the manuscript.

Funding: This study was supported by the Heigl-Stiftung, the Köhler-Stiftung, and the Hamburger Stiftung zur Förderung von Wissenschaft und Kultur.

Institutional Review Board Statement: The study was conducted in accordance with the Declaration of Helsinki and approved by the Institutional Review Board of the International Psychoanalytic University (2019-10; 11 July 2019).

Informed Consent Statement: Informed consent was obtained from all subjects involved in the study.

Data Availability Statement: The data file can be shared on request. The participants of this study did not give written consent for their data to be shared publicly, so due to the sensitive nature of the research, supporting data are not available.

Conflicts of Interest: The authors report that there are no competing interests to declare.

References

1. Gazzillo, F.; Silberschatz, G.; Fimiani, R.; De Luca, E.; Bush, M. Dreaming and adaptation: The perspective of control-mastery theory. *Psychoanal. Psychol.* **2020**, *37*, 185–198. [CrossRef]
2. Hartmann, E. Nightmare after Trauma as Paradigm for All Dreams: A New Approach to the Nature and Functions of Dreaming. *Psychiatry* **1998**, *61*, 223–238. [CrossRef]
3. Cartwright, R.; Luten, A.; Young, M.; Mercer, P.; Bears, M. Role of REM sleep and dream affect in overnight mood regulation: A study of normal volunteers. *Psychiatry Res.* **1998**, *81*, 1–8. [CrossRef] [PubMed]
4. Hill, C.E.; Liu, J.; Spangler, P.; Sim, W.; Schottenbauer, M. Working with dreams in psychotherapy: What do psychoanalytic therapists report that they do? *Psychoanal. Psychol.* **2008**, *25*, 565–573. [CrossRef]
5. Eudell-Simmons, E.M.; Hilsenroth, M.J. A review of empirical research supporting four conceptual uses of dreams in psychotherapy. *Clin. Psychol. Psychother.* **2005**, *12*, 255–269. [CrossRef]
6. Warner, S.L. Can psychoanalytic treatment change dreams? *J. Am. Acad. Psychoanal.* **1983**, *11*, 299–316. [CrossRef] [PubMed]
7. Glucksman, M.L.; Kramer, M. The Manifest Dream Report and Clinical Change. In *Dream Research: Contributions to Clinical Practice*; Kramer, M., Glucksman, M.L., Eds.; Routledge: New York, NY, USA, 2015; pp. 106–122.
8. Fischer, C.; Kächele, H. Comparative analysis of patients' dreams in Freudian and Jungian treatment. *Int. J. Psychother.* **2009**, *13*, 34–40.
9. Roesler, C. Dream content corresponds with dreamer's psychological problems and personality structure and with improvement in psychotherapy: A typology of dream patterns in dream series of patients in analytical psychotherapy. *Dreaming* **2018**, *28*, 303–321. [CrossRef]
10. Kuelz, A.K.; Stotz, U.; Riemann, D.; Schredl, M.; Voderholzer, U. Dream recall and dream content in obsessive-compulsive patients: Is there a change during exposure treatment? *J. Nerv. Ment. Dis.* **2010**, *198*, 593–596. [CrossRef] [PubMed]
11. Melstrom, M.A.; Cartwright, R.D. Effects of Successful vs. Unsuccessful Psychotherapy Outcome on Some Dream Dimensions. *Psychiatry* **1983**, *46*, 51–65. [CrossRef] [PubMed]

12. WHO. *International Statistical Classification of Diseases and Related Health Problems*, 11th ed.; World Health Organization: Geneva, Switzerland, 2020.
13. Vierl, L.; Juen, F.; Benecke, C.; Hörz-Sagstetter, S. Exploring the associations between psychodynamic constructs and psychopathology: A network approach. *Pers. Ment. Health* **2023**, *17*, 40–54. [CrossRef] [PubMed]
14. Doubková, N.; Heissler, R.; Preiss, M.; Sanders, E. Differences in personality functioning impairment in mood, anxiety, and personality disorders: A cluster analysis. *BMC Psychiatry* **2022**, *22*, 315. [CrossRef] [PubMed]
15. Schredl, M. Dreams and nightmares in personality disorders. *Curr. Psychiatry Rep.* **2016**, *18*, 15. [CrossRef] [PubMed]
16. Hau, S. Traum und Sexualität bei Borderline-Patienten. In *Borderline-Störungen und Sexualität*; Dulz, B., Benecke, C., Richter-Appelt, M., Eds.; Schattauer: Stuttgart, Germany, 2009; pp. 251–266.
17. APA. *APA Dictionary of Psychology*, online ed.; American Psychological Association: Washington, DC, USA, 2022.
18. Conklin, C.Z.; Bradley, R.; Westen, D. Affect regulation in borderline personality disorder. *J. Nerv. Ment. Dis.* **2006**, *194*, 69–77. [CrossRef] [PubMed]
19. Moser, U.; Hortig, V. *Mikrowelt Traum. Affektregulierung und Reflexion*; Brandes & Apsel: Frankfurt am Main, Germany, 2019.
20. Moser, U.; von Zeppelin, I. *Der Geträumte Traum: Wie Träume Entstehen und Sich Verändern*; Kohlhammer: Stuttgart, Germany, 1996.
21. Euler, J.; Henkel, M.; Bock, A.; Benecke, C. Strukturniveau, Abwehr und Merkmale von Träumen. *Forum Psychoanal.* **2016**, *32*, 267–284. [CrossRef]
22. Kempe, S.; Köpp, W.; Blomert, E.; Wittmann, L. Low levels of personality functioning are associated with affect dysregulation in dreams. *Int. J. Dream Res.* **2024**, *17*, 60–67. [CrossRef]
23. Döll-Hentschker, S. *Die Veränderung von Träumen in Psychoanalytischen Behandlungen: Affekttheorie, Affektregulierung und Traumkodierung*; Brandes & Apsel: Frankfurt am Main, Germany, 2008.
24. Deserno, H.; Kächele, H. Traumserien. Ihre Verwendung in Psychotherapie und Therapieforschung. In *Der Traum*; Janta, B., Unruh, B., Walz-Pawlita, S., Eds.; Psychosozial: Gießen, Germany, 2013; pp. 233–244.
25. Huber, D.; Brandl, T.; Klug, G. The scales of psychological capacities: Measuring beyond symptoms. *Psychother. Res.* **2006**, *14*, 89–106. [CrossRef] [PubMed]
26. Huber, D.; Henrich, G.; Clarkin, J.; Klug, G. Psychoanalytic versus psychodynamic therapy for depression: A three-year follow-up study. *Psychiatry* **2013**, *76*, 132–149. [CrossRef]
27. Kempe, S.; Krome, A.-J.; Köpp, W.; Wittmann, L. Different keys unlock different doors: Analyzing content and affect regulation in dream reports. *Somnologie* **2023**, *27*, 192–197. [CrossRef]
28. Wittmann, L.; Kempe, S.; Anstadt, T.; Schredl, M.; Protić, S.; Höllmer, H.; Gorzka, R.-J. Characterizing veterans' dreams applying the Zurich dream process coding system. *Dreaming* **2022**, *32*, 331–344. [CrossRef]
29. French, T.M. The Integration of Behavior. In *The Integrative Process in Dreams*; University of Chicago Press: Chicago, IL, USA, 1954; Volume 2.
30. Stern, D.N. *The Interpersonal World of the Infant*; Basic Books: New York, NY, USA, 1985.
31. Landis, J.R.; Koch, G.G. The Measurement of Observer Agreement for Categorical Data. *Biometrics* **1977**, *33*, 159–174. [CrossRef] [PubMed]
32. Urbina, S.P. Methodological issues in the Quantitative Analysis of Dream Content. *J. Personal. Assess.* **1981**, *45*, 71–78. [CrossRef] [PubMed]
33. Schredl, M. Challenges in quantitative dream research: A commentary on Curtiss Hoffman's "Research articles in Dreaming: A review of the first 20 years". *Int. J. Dream Res.* **2013**, *6*, 136–138.
34. Schielzeth, H.; Dingemanse, N.J.; Nakagawa, S.; Westneat, D.F.; Allegue, H.; Teplitsky, C.; Réale, D.; Dochtermann, N.A.; Garamszegi, L.Z.; Araya-Ajoy, Y.G. Robustness of linear mixed-effects models to violations of distributional assumptions. *Methods Ecol. Evol.* **2020**, *11*, 1141–1152. [CrossRef]
35. Blechner, M.J. Changes in the Dreams of Borderline Patients. *Contemp. Psychoanal.* **1983**, *19*, 485–498. [CrossRef]
36. Kernberg, O.F. *Severe Personality Disorders: Psychotherapeutic Strategies*; Yale University Press: New Haven, CT, USA, 1984.
37. Picard-Deland, C.; Nielsen, T.; Carr, M. Dreaming of the sleep lab. *PLoS ONE* **2021**, *16*, e0257738. [CrossRef] [PubMed]

Article

The Dream Experience and Its Relationship with Morning Mood in Adolescents Hospitalized after a Suicide Attempt

Emma Desjardins [1], Lina Gaber [1], Emily Larkin [1], Antoine Benoit [1], Addo Boafo [2] and Joseph De Koninck [1,*]

[1] School of Psychology, University of Ottawa, 136 Jean-Jacques Lussier, Ottawa, ON K1N 6N5, Canada; edesj086@uottawa.ca (E.D.); lgabe060@uottawa.ca (L.G.); elark042@uottawa.ca (E.L.); abeno099@uottawa.ca (A.B.)

[2] Children's Hospital of Eastern Ontario, 401 Smyth Road, Ottawa, ON K1H 8L1, Canada; aboafo@cheo.on.ca

* Correspondence: jdekonin@uottawa.ca

Abstract: Suicidality in adolescents has been associated with emotional distress, stressful life events, relationship issues, and nightmares to name a few. This study explored the actual dream content and the mood at pre-sleep, during a reported dream, and in the morning in 33 adolescents admitted to the hospital on account of a suicide attempt. In all aspects, hospitalized adolescents were compared to 33 matched adolescents who had followed the same protocol. In accordance with the Continuity and the Threat Simulation theories of dream formation, it was hypothesized that the waking-life experiences of suicidal adolescents would transpire in both dream mood and content as well as in the frequency of nightmares. Dreams were analyzed by independent judges using traditional dream content scales, including for the presence of negative and destructive themes and types of interpersonal relationships. As predicted, more suicidal adolescents experienced frequent nightmares, which was significant. A higher negative mood at pre-sleep, within dreams, and at post-sleep was also observed. Furthermore, their dreams contained a higher prevalence of destructive themes and failures, as well as self-directed and death-resulting aggressions. Regression analyses indicated that morning mood was most accurately predicted by positive and negative dream mood in the normative adolescents, whereas only negative dream mood appeared to predict subsequent waking affect in suicidal participants. Our results underline the valuable potential of implementing nightmare-reducing therapies in the presence of suicidal adolescents who suffer from frequent nightmares.

Keywords: nightmares; dreams; mood; suicide; adolescents

Citation: Desjardins, E.; Gaber, L.; Larkin, E.; Benoit, A.; Boafo, A.; De Koninck, J. The Dream Experience and Its Relationship with Morning Mood in Adolescents Hospitalized after a Suicide Attempt. *Brain Sci.* **2024**, *14*, 804. https://doi.org/10.3390/brainsci14080804

Academic Editors: Sergio A. Mota-Rolim, Brigitte Holzinger, Michael Nadorff, Luigi De Gennaro and Serena Scarpelli

Received: 16 April 2024
Revised: 31 July 2024
Accepted: 31 July 2024
Published: 10 August 2024

1. Introduction

Suicide is a major concern and consistently ranks among the top causes of death for adolescents [1]. For example, Canada reports it as the second-leading cause of death for this age group [2]. In fact, about 90% of Canadian adolescent psychiatric inpatients display high-risk suicide ideation and behaviour [3]. It is thus crucial to develop identification, prevention, and intervention strategies to reduce this risk, and conduct further research to gain a better understanding of the surrounding factors, more specifically, those related to sleep [4].

Previous research has found significant associations between sleep disturbance and the risk of suicidality [5–8]. In order to more precisely study this phenomenon in real-life situations, we have initiated a series of studies with adolescents admitted to our local hospital psychiatric clinic following a suicide attempt who were willing to share dreams. Relevant to the current report, we have now studied a cohort with full polysomnography and event-related potentials (ERPs) [9,10].

Compared to a reference group, they had REM sleep latency abnormalities, shallower sleep, high REM density, and late night slow-wave sleep reduction [9]. ERP measures revealed poorer inhibition in response to emotional stimuli that correlated with shorter REM sleep latency, higher REM sleep, and more frequent nocturnal awakenings [10].

The next step is to explore the dreaming experience of this clinical group again when in crisis. Looking at the dreaming literature, a first observation is that several studies have shown that disturbing dreams, including nightmares and bad dreams, are also associated with a higher risk of suicidality [11–15]. More recent studies have confirmed that experiencing frequent nightmares during the COVID-19 pandemic increased the probability of suicidal ideation [16]. In the current research, we aim to focus beyond nightmares on the actual dream content of our clinical group. Indeed, in a previous study, we observed that frequent nightmare sufferers experience an overall significant difference in the presence of fear and anxiety in their day-to-day dreams that is combined with significantly more negative mood and higher stress levels before and after sleep [17].

Looking at the relevant literature, researchers have extensively studied the connection between waking life and dream content. The Continuity Theory (CT) suggests that daytime events, cognitions, and emotions are reflected in dreams in a distorted and selective way [18–20]. Further, mundane activities such as reading and writing are seldom represented in dreams, while emotional experiences like stress are preferentially incorporated [20–23]. Indeed, studies have consistently observed that dreams tend to be more negative than waking experiences, containing more negative emotions, interactions, and outcomes [24]. One of the most well-known studies regarding the negative tone of dream content was the American normative study from Hall and Van de Castle (1966) [25], which has since been reverified and reconfirmed to be present in other samples (more recently for example by Dale et al. 2016) [26].

One of the earliest clinical studies of dreams in suicidal patients is that of Gutheil (1948) [27], in which it was determined that dreams not only indicate a patient's intentions, but typically draw notion to a deeper mental mechanism involved. Gutheil also remarked that as suicide risk increases, dreams representing suicide or death are perceived more positively as a satisfying resolution to a painful situation or conflict [27].

Subsequent studies have found that the dreams of suicidal patients contain a high representation of aggressive, violent, and destructive themes and materials relating to death and dying [28–30]. Results from Firth et al. (1986) [31] found that although this was true, these destructive themes were most common in patients who were already more aggressive, impulsive, and severely depressed, demonstrating that the link between these dream themes and suicide attempts was indirect, as depression among patients could be attributed to the appearance of the death theme in dreams.

Further research has attempted to determine why negative dream content may be preferentially incorporated in suicidal patient's dreams. For instance, it has been observed that patients who reported lower levels of psychological well-being tended to have dreams that contained more aggressive than friendly interactions, more negative than positive emotions, and more failures and misfortunes than successes and good fortunes [32].

Ağargün et al. (1998) [12] explored the relationship between recurring and fearful dreams and suicidal tendencies in patients with major depression. Results showed that patients with repetitive nightmares, women specifically, had a higher mean score on the Schedule for Affective Disorders suicide subscale (SADS) than others. A follow-up study also showed a significant correlation between the incidence of nightmares and suicide risk [33]. Recent research has also linked frequent nightmares and depressive symptoms in adolescents. This association appeared to be fully mediated by the distress experienced by individuals due to their recurring nightmares [34].

Given that nightmares and waking emotional distress are linked with suicidality, it is not surprising that individuals with suicidal tendencies present both a predominance of negative affect during waking life and greater difficulty with emotion regulation [10,35–39].

Consistent with the CT, it is thus of interest to explore how other waking elements translate into the dreaming experience. Studies have demonstrated that family members are frequently represented characters in children's dreams. However, according to Foulkes (1982) [40], a general decline has been observed in dreaming about family members in adolescents between 13 and 15 years old [41,42]. However, using a cluster analysis, one study identified family relationships have a typical episode in the dreams of adolescents and young adults [43]. Additionally, Nöltner and Schredl (2023) [44] examined the frequency of core family members in dreams of students and their interactions with the dreamer. This study revealed that about twenty-one percent of the student's dreams included the mother, the father, or the parent unit as characters [44]. Furthermore, research has also found rejecting-neglecting parenting and low levels of parental care to be significantly associated with youth suicide attempts [45,46]. Using the Parental Acceptance-Rejection Questionnaires (PARQs), one study determined that adolescents who had attempted suicide tended to perceive their parents as more aggressive, neglecting, rejecting, and cold [47]. There is thus a gap in the literature as to how parental rejection and neglect might be translated into the dream experience of adolescents.

To further discuss the relationship between waking and dreaming, the Threat Simulation Theory (TST) postulates that dreaming has evolved as an adaptive survival mechanism to simulate threats drawn from waking-life experiences [48]. In a modern context, individuals who have a more active threat simulation system (TSS) (i.e., traumatized children) have been shown to report a higher number of and more severe dream threats [49].

Additionally, the frequency and severity of oneiric threats have been associated with reporting waking-life threatening events, daily stress levels, pre-sleep negative emotions, and recently the presence of COVID-19 pandemic references [50–52]. Evidence therefore suggests that the menacing qualities of dreams are influenced by emotional waking-life experiences. The link, however, with the representation of threats in dreams and its value in subsequent adaptation has so far not been established if not ignored, relegating the TST to a dream formation theory.

Moving on to emotions, according to Özdemir et al. (2017) [53], suicide attempters demonstrate greater difficulty in emotion regulation compared to non-attempters as demonstrated by higher scores across all subscales of the Difficulties in Emotion Regulation Scale. This leads to the Emotion Regulation Theory (ERT) of dreaming, which postulates that the purpose of dreams is to regulate emotions by providing a secure and controlled setting for processing emotional experiences from wakefulness, thereby reducing the intensity of negative emotions, like anxiety and fear [54–58]. None of the previous studies, however, have examined the relationship between dream content itself and morning psychological state or mood in order to test this notion.

Of further interest is the phenomenon known as the positivity bias, which has been observed in many studies, wherein individuals evaluating their own emotions in dreams tend to perceive them as more positive than independent judges who analyze their dream narratives [54,59–63]. Some have suggested that the discrepancy between the literal content of the dream narrative and the dreamer's subjective emotional experience contributes to the process of mood regulation [54]. Since most dreams that are recalled in the morning arise from REM sleep, it has been proposed that the muscle atonia associated with this phase of sleep may assist in desensitizing dreamers from the negative dream content [64].

While the literature so far provides very valuable information on dream content and suicidality, each study has focused on specific components such as dream themes and content, the fearful quality of dreams, and the frequency of nightmares. There is need to conduct a more complete analysis of the dream experience of suicidal adolescents.

The present study aims to contribute to this need by focusing on the dreams of a cohort of adolescents who attempted suicide. These participants were invited to answer a sleep questionnaire covering sleep habits, dream recall, and frequency of nightmares, and report their mood on one night before going to bed (pre-sleep) and upon waking up (post-sleep), and retrospectively write their dreams in the morning from the same night along with their dream mood. This allowed us to examine their dreams and determine the presence and frequency of specific themes in their content compared with a normative sample of adolescents that used the same method of dream collection, except that it was applied in the hospital setting. As established in previous studies, it was expected that the clinical participants would report experiencing a higher frequency of nightmares than observed in normative studies. Based on the Continuity hypothesis, it was further predicted that the participants who had previously attempted suicide would experience dreams that incorporate their waking-life concerns and suicidal ideations, with a greater occurrence of destructive themes, reduced positive social interactions, and a perception of parental neglect and rejection. Additionally, in line with the Threat Simulation Theory, it was expected that these participants would experience more oneiric threats. Regarding mood, both the Continuity hypothesis and TST would predict a positive correlation between pre-sleep mood and dream mood. For instance, the pre-sleep negative experience (i.e., on the previous day) would be expected to result in a more negative mood in the dream. Finally, in the context of the MRT, our objective was to examine the relationship between morning mood, dream mood, and pre-sleep mood. We also sought to determine whether the latter two variables were predictors of morning mood.

2. Methods

2.1. Participants

The participants were recruited upon admission at the Children's Hospital of Eastern Ontario (CHEO) inpatient psychiatry unit on account of a suicide attempt within a month of the current admission. All patients admitted to the inpatient psychiatry unit from 1 January 2022 to 30 June 2023 who met the eligibility criteria above were approached. There were 50 of them. Only participants who were willing and able to provide informed consent to take part in the study were included. Eleven declined to participate. Thirty-nine consented to take part in the study and completed it without dropping out. We do not have information as to how the group of 11 compares with the group of 39 as we did not have consent to review their information. Six participants were excluded from the analyses because they did not recall a dream. The suicide attempt group was composed of 33 anglophone adolescents (31 biological females and 2 biological males) between the ages of 13 and 17. It is noted that participants were grouped according to sex assigned at birth. Amongst the biological females, ten participants identified as transgender males, and one as a non-binary individual, and, in the biological males, one identified as a transgender female.

It should be noted that the high proportion of female participants is congruent with the typical observation that females are admitted to inpatient settings more frequently than male counterparts [65]. Indeed, females are more likely than males to experience emotional or sexual abuse, exhibit higher levels of anxiety and depressive symptoms, and have a greater risk for suicidal ideation and attempts [65]. This disparity can also be attributed to the fact that while male adolescents exhibit a higher suicide rate, ranging from twice to four times that of their female counterparts, females display a greater frequency of suicidal attempts, ranging from three to nine times that of males [66,67].

Comparator group participants were drawn from a normative study of Canadian dreams conducted at the University of Ottawa Sleep Laboratory; the data for this study were collected from 2004 to 2017. This normative study contains dream reports and surveys from Canadian males and females from the Ottawa–Gatineau area, both anglophones and francophones, from the ages of 12 to 85, divided into five age groups of approximatively 200 per group. It was used in several studies of the ontogenesis of dreams across age groups [68,69]. Each participant completed a daily recording of the day's events for 10 days or until two dreams were reported. Given the high stability of the dream norms over time, even within cultures, several studies of special populations have used normative data as a benchmark when no control groups were feasible (i.e., Sabourin et al., 2018) [70]. The protocol for dream collections and mood recordings was comparable to that of the normative study, with the exception of the hospital setting, which required consideration. In total, 33 adolescents (31 biological females and 2 biological males; 28 anglophones and 5 francophones) between the ages of 13 and 17 were selected from this normative study to match the mean age and biological sex of the suicide attempt participants. This selection was made according to an inclusion criterion of at least one reported dream from one night, with a dream length of more than 40 words, which is within the norm in dream research to ensure sufficient content. Moreover, the word count between the groups was comparable. No participants from the normative study identified as transgender. No consideration was made of nightmare frequency and actual dream narrative.

2.2. Procedure

The 33 adolescents in the suicide attempt group accepted to write their dreams for one night during their stay at the hospital. Patients were invited to participate in the study by a nurse on the ward, and after a verbal agreement, parents and youth signed the consent forms. Then, a psychiatrist interviewed the participants and carried out a diagnostic intake interview and the morning dream report.

Participants completed an adapted version of the Dream Diary questionnaire from the normative study which is a general questionnaire that explores the sources of dreams, sleep habits, ability to recall dreams, and the influence of dreams on waking life. They rated their mood prior to sleep, upon awakening in the morning, and their dream mood retrospectively. The dream narratives were analyzed by two independent judges for the presence of destructive themes by using Firth's scale (1986), content themes by using specific categories from the Hall and Van de Castle (HVDC) system of quantitative content analysis (1966), parental rejection by using the PARQ short form child version [71], and threats by using an adapted version of Revonsuo and Valli's Dream threat scale (DTS) (2000) [72]. Their mood, dream content, and general information were compared to those of the 33 normative adolescents matched for age and gender of the comparator group. This project received full approval from the Ethics Committee of the Children's Hospital of Eastern Ontario (CHEO) in January 2022.

2.3. General Information Questionnaire

The purpose of this questionnaire was to collect information regarding the participants' demographics (i.e., age, maternal language, sex, marital status, education, employment, and medications), sleep habits (usual sleep schedule, subjective insomnia, and daytime naps), and dream recall (frequency of dreams and nightmares). The frequency of nightmares was determined on a scale of one to six; each number was assigned to a possible frequency range (1—less than one a month; 2—approximately once a month; 3—approximately once per two weeks; 4—approximately one a week; 5—many times a week; 6—almost every night). It should be noted that data regarding sleep habits are not included in this report, as previous cohorts have already been studied and a separate report is currently in preparation. This new report will take into account the time of day or night of the suicide attempt in order to explore the potential influence of circadian rhythms and sleep duration.

2.4. Dream Journal

Upon awakening, the participants documented their recalled dreams, providing as much detail as possible to describe the dream's locations, events, characters, interactions, activities, feelings, and emotions. They also reported their mood both in the evening (prior to sleep), in the morning (after waking up), and, retrospectively, their dream mood. A list of different emotions was rated, and the mood was reported on a Likert scale from 0 to 3, 0 being not at all and 3 a lot. One positive emotion item (Happy) and one negative emotion item (Sad) were used for the temporal analysis of mood and positivity bias.

2.5. Destructive Themes

For this study, we selected relevant themes from Firth's scale (1986) [31], including death [30], destructive or violent hostility [30], separation or loss [73], helplessness [25], masochism [30], realistic versus unrealistic content [73], and hostility [74]. Each theme was evaluated for its presence or absence in the dream narrative and assigned a score of either 1 or 0.

2.6. Hall and Van de Castle's System of Content Analysis

The HDVC instrument is a frequently used and well-validated dream content rating system [75]. Both judges had previously received extensive training in the HDVC coding system and intercoder reliability was established (correlation of 0.80, percentage of perfect match was 92%, t-test proved no significant difference between ratings ($t(2203) = -0.079$, $p > 0.05$)). Participants' dreams were coded for the following categories of the HVDC system of quantitative content analysis: characters, aggression, friendliness, sexuality, success and failure, good fortune and misfortune, and emotions.

2.7. Dream Threat Scale

The methodology employed in this study was based on the approach utilized by Bradshaw et al. (2016) and replicated for the purposes of this research [51]. Threats were categorized following the different levels of threats' severity established by the DTS: 1.: Life-threatening event, 2.: Socially, psychologically, or financially severe threat, 3.: Physically severe threat, and 4.: Minor threat. The threatening intensity of the dreams was assessed using a 0 to 3 scale, where a score of 3 corresponded to the presence of at least one type 1 (life-threatening) threat as defined by the DTS. A score of 2 indicated the presence of at least one type 2 or 3 (physical or social/psychological/financial) threat, while a score of 1 indicated the presence of at least one type 4 (minor) threat. A score of 0 represented the lowest level of threatening intensity, indicating an absence of threats in the dream content.

2.8. Parental Acceptance-Rejection Questionnaire Short Form

The PARQ child version short form was used to determine the dreamer's perception of parental acceptance-rejection based on the presence of one or both parents (mother and/or father) in the dream narrative. In waking-life research, the PARQ is a self-report instrument, but in this instance of dream content analysis, a judge has used this instrument based on the parental acceptance-rejection reported by the dreamer in the dream narrative. This methodology is derived from previous dream research that used the Eysenck personality inventory to determine if the waking personality of the dreamer has a compensatory or continuous relationship with his oneiric personality [76]. The lowest possible score of the PARQ short form scale is 24 (revealing maximum perceived acceptance), the highest is 96 (revealing maximum perceived rejection), and the midpoint is 60.

2.9. Statistical Analysis

The difference in frequency of nightmares between the groups was investigated using a Chi-squared test. The hypotheses of a higher negative and a lower positive reported mood during the pre-sleep, dream, and post-sleep measures were verified using a repeated measure ANOVA. A multiple regression analysis and bivariate correlation analyses were also performed to examine the relationship between moods at the different time points. The expected greater occurrence of waking-life concerns, destructive themes, and oneiric threats and the prediction of reduced positive social interactions in the dreaming experience of the suicide attempt participants were examined using an independent sample t-test. Control for multiple comparisons was applied by dividing the error rate by the number of comparisons and adopting a family-wise error rate.

3. Results

3.1. Nightmare Frequency

Figure 1A illustrates the frequency distribution of participants in each of the six nightmare frequency ranges. The distribution of suicide attempt participants across the items is notably dispersed, with approximately half of them falling into the higher frequency of nightmare items. In contrast, the distribution of the comparator participants is heavily concentrated in the lower frequency of nightmare items. Although there is no frequency criterion for nightmare disorder in the Diagnostic and Statistical Manual of Mental Disorders, Fifth Edition (DSM-5), individuals who experience one or more episodes per week are categorized as presenting a moderate nightmare disorder [77]. Therefore, we established a threshold using this frequency to further differentiate the sample into two subgroups (i.e., frequent nightmare sufferers and non-frequent nightmare sufferers). In that context, a significant difference was observed between the suicide attempt group and the comparator group (Chi-squared = 9.587, p = 0.002) (see Figure 1B).

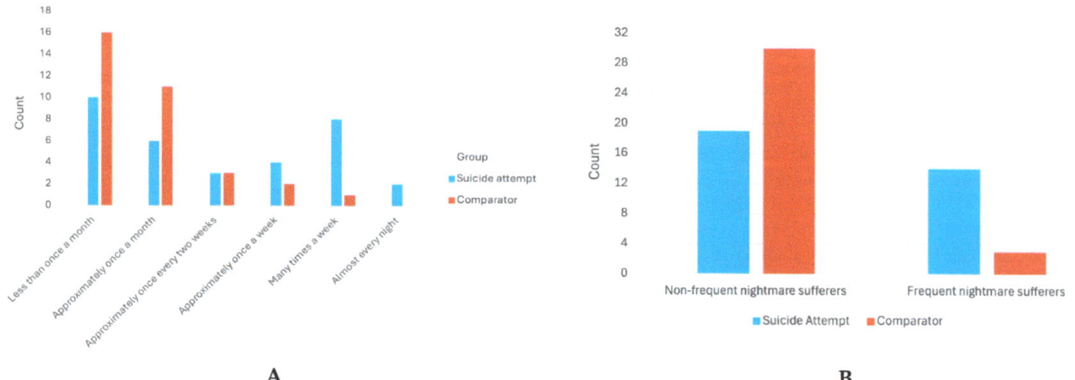

Figure 1. Nightmare frequency according to group. (**A**) illustrates the frequency distribution of participants in each six nightmare frequency ranges (1—less than one a month; 2—approximately once a month; 3—approximately once per two weeks; 4—approximately one a week; 5—many times a week; 6—almost every night) according to group. (**B**) illustrates the frequency distribution of participants in two frequency ranges according to group: those who experience frequent nightmares and those who do not.

3.2. Mood

The following results are derived from the negative and positive moods (Happy versus Sad) reported in the dream journal. Figure 2 illustrates their variation both in terms of negative (2A) and positive mood (2B) at pre-sleep, dreaming, and post-sleep in both groups.

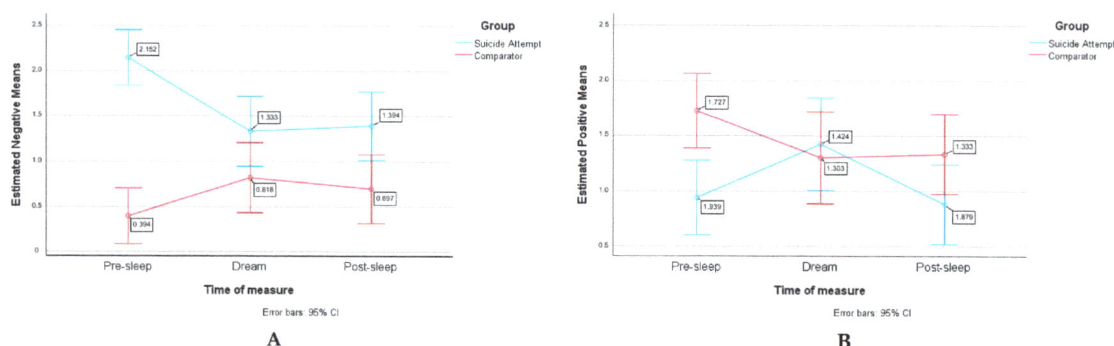

Figure 2. Negative (panel **A**) and positive (panel **B**) moods at pre-sleep, dreaming, and post-sleep in both groups.

The suicide attempt group showed a significantly higher combined negative mood (M_{SA} = 1.626) compared to the comparator group (M_C = 0.636) while simultaneously reporting significantly lower combined positive mood (M_{SA} = 1.081) compared to the comparator group (M_C = 1.455).

The results of a repeated measure ANOVA showed a significant main effect of time of measure on mood (F(2.63) = 3.874, p = 0.023, partial eta square = 0.057), confirming a difference in mood levels between pre-sleep, during dreaming, and post-sleep. There was also a significant interaction between mood and group ($F_{(1.64)} = 15.159$, $p < 0.001$, partial eta squared = 0.192), indicating a different pattern of mood level variation between the two groups. Although there was no significant interaction between time of measure and group ($F_{(2.128)} = 2.427$, $p = 0.092$, partial eta squared = 0.037), there was a significant interaction effect between mood, time of measure, and groups ($F_{(2.63)} = 12.030$, $p < 0.001$, partial eta squared = 0.158); the variation in mood level thus seems to depend on both the group and the time at which the mood was measured; the pattern of mood progression is thus significant and varies between groups.

To investigate the Continuity Theory hypothesis, bivariate correlation analyses were performed on the data of the comparator group (see Table 1) and determined that a positive mood in the evening was strongly and positively correlated with a positive mood during dreaming ($r = 0.605$, $p < 0.01$), and strongly and positively correlated with a positive mood in the morning ($r = 0.563$, $p < 0.01$). Additionally, a positive mood during dreaming was strongly and positively correlated with a positive mood in the morning ($r = 0.724$, $p < 0.01$). A positive mood in the morning was moderately and negatively correlated with a negative mood during dreaming ($r = -0.463$, $p < 0.01$).

Table 1. Correlation table for positive and negative moods of the comparator group at pre-sleep, dreaming, and post-sleep.

		Positive Mood			Negative Mood		
		Evening	Dreaming	Morning	Evening	Dreaming	Morning
Positive mood	Evening	1	0.605 **	0.563 **	−0.398 *	−0.326	−0.224
	Dreaming		1	0.724 **	−0.101	−0.425	−0.209
	Morning			1	−0.120	−0.463 **	−0.455 **
Negative mood	Evening				1	0.203	0.337
	Dreaming					1	0.495 **
	Morning						1

**. Correlation is significant at the 0.01 level (2-tailed). *. Correlation is significant at the 0.05 level (2-tailed).

Furthermore, fostering a negative mood during dreaming was moderately and positively correlated with a negative mood in the morning ($r = 0.495$, $p < 0.01$).

Bivariate correlation analysis was performed on the data of the suicide attempt group (see Table 2) and determined that a positive mood in the evening was moderately and positively correlated with a positive mood in the morning ($r = 0.371$, $p < 0.05$). Not surprisingly, a positive mood during dreaming was moderately and negatively correlated with a negative mood during dreaming ($r = -0.405$, $p < 0.05$). A positive mood in the morning was moderately and negatively correlated with a negative mood in the evening ($r = -0.362$, $p < 0.05$).

Table 2. Correlation table for positive and negative moods of the suicide attempt group at pre-sleep, dreaming, and post-sleep.

		Positive Mood			Negative Mood		
		Evening	Dreaming	Morning	Evening	Dreaming	Morning
Positive mood	Evening	1	0.279	0.371 *	−0.553 **	−0.177	−0.196
	Dreaming		1	0.283	−0.133	−0.405 *	−0.316
	Morning			1	−0.362 *	−0.082	−0.188
Negative mood	Evening				1	0.393 *	0.391 *
	Dreaming					1	0.406 *
	Morning						1

**. Correlation is significant at the 0.01 level (2-tailed). *. Correlation is significant at the 0.05 level (2-tailed).

Furthermore, fostering a negative mood in the evening was moderately and positively correlated with negative mood during dreaming ($r = 0.393$, $p < 0.05$) and moderately and positively correlated with negative mood in the morning ($r = 0.391$, $p < 0.05$). A negative mood during dreaming was moderately and positively correlated with a negative mood in the morning ($r = 0.406$, $p < 0.05$).

An objective of this study was to determine whether the pre-sleep and dream mood variables could be used as predictors of morning mood. To this end, a multiple regression analysis was performed. It revealed that among suicide attempt participants ($n_{SA} = 33$), reporting a positive mood in the evening significantly and positively predicted reporting a positive mood in the morning ($\beta = 0.316$, $p = 0.036$). However, reporting a positive mood during dreaming was not found to be a robust predictor of mood in the morning ($\beta = 0.195$, $p = 0.212$). Moreover, the results indicated that reporting a negative mood during dreaming was a moderate predictor of a negative mood in the morning ($\beta = 0.298$, $p = 0.045$) in this group.

Among the comparator group ($n_C = 33$), reporting a negative mood during dreaming moderately predicted reporting a negative mood in the morning ($\beta = 0.445$, $p = 0.008$), while a negative mood in the evening failed to predict a negative mood in the morning ($\beta = 0.247$, $p = 0.058$). Moreover, reporting a positive mood during dreaming had a strong significant positive predicting value of reporting a positive mood in the morning ($\beta = 0.605$, $p < 0.001$), yet there was no significant relationship between reporting a positive mood in the evening and in the morning ($\beta = 0.197$, $p = 0.189$) in the comparator group. Results of the multiple regression are illustrated in Figures 3 and 4.

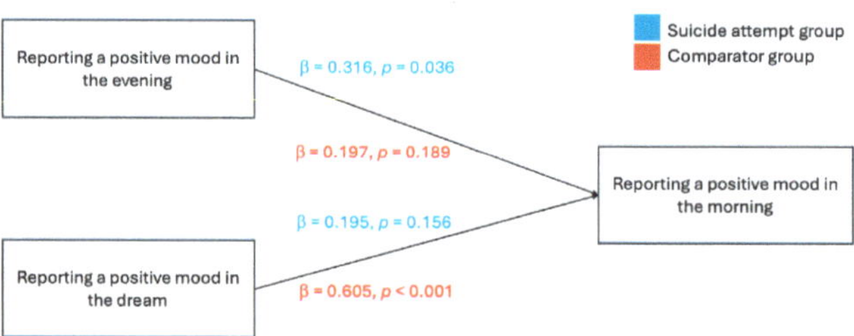

Figure 3. Illustration of a positive mood in the morning predicted by the report of a positive mood in the evening or in the dream.

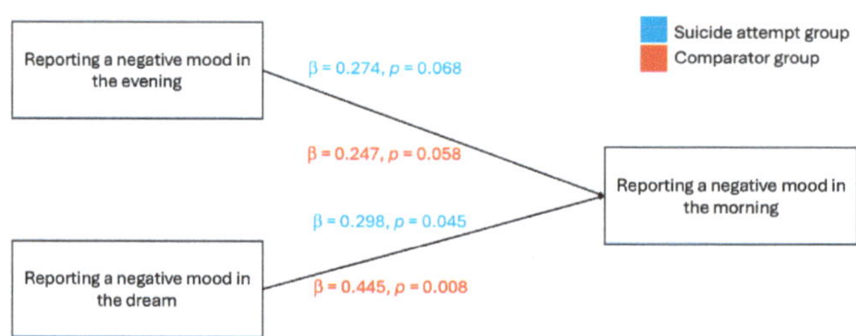

Figure 4. Report of a negative mood in the morning predicted by the report of a negative mood in the evening or in the dream. Notes. Standardized coefficients are presented. The estimates above each line correspond to the estimate for the suicide attempt group, while the estimates presented below each line correspond to the comparator group.

3.3. Positivity Bias

The result of an independent sample *t*-test comparing the suicide attempt group's average self-evaluated dream mood ($n_{SA} = 33$) to an external judge's evaluation of the dream mood revealed that there was a significant difference in terms of positive mood between the participant's self-evaluation and the external judge's evaluation of the dream mood ($t(60.520) = 3.427$, $p < 0.001$, Cohen's $d = 0.844$). To be more precise, the dream mood was evaluated on average more positively by participants from the suicide attempt group themselves than by an external judge ($M_{SA} = 1.424$, $M_{EJ} = 0.5152$). This is consistent with the previous literature on this phenomenon [54].

3.4. Dream Content

Destructive themes. Results from independent sample *t*-tests comparing the comparator group's dreams ($n_{SA} = 33$) to our suicide attempt group's dreams ($n_C = 33$) revealed that there was no significant difference in terms of the presence of destructive themes between the dream content of the comparator group and the suicide attempt group ($t(64) = 1.487$, $p = 0.071$). After applying control for multiple comparisons, the most statistically significant individual theme differences were those of death ($t(51.450) = 2.530$, $p = 0.007$, Cohen's $d = 0.623$) and masochism ($t(32) = 3.028$, $p = 0.002$, Cohen's $d = 0.745$) (see Table 3). To be more precise, the dream content of participants from the suicide attempt group featured on average more death ($M_{SA} = 0.273$, $M_C = 0.061$) and more masochism ($M_{SA} = 0.197$, $M_C = 0.00$) compared to the participants from the comparator group.

Table 3. Comparison of destructive themes' frequency in dream content between groups.

	t	*df*	*Significance (One-Sided p)*	*Cohen's d*
Death	2.530	51.450	0.007 **	0.623
Violent Hostility	0.593	64	0.278	-
Separation	−0.175	64	0.431	-
Helplessness	0.542	64	0.295	-
Masochism	3.028	32	0.002 **	0.745
Unrealistic content	0.000	64	0.500	-
Hostility	0.137	64	0.446	-
Sum	1.487	64	0.071	-

**. Correlation is significant at the 0.01 level (one-tailed), after control for multiple comparisons.

Hall and Van de Castle categories. When comparing the dreams of the comparator group ($n_C = 33$) to those of the suicide attempt group ($n_{SA} = 33$) using independent sample *t*-tests, we found a significant disparity in the number of self-directed aggressions ($t(32) = −1.936$, $p = 0.031$, Cohen's $d = 0.477$), friendly interactions ($t(48.468) = −2.387$, $p = 0.010$, Cohen's' $d = −0.588$), failures ($t(49.039) = 2.549$, $p = 0.007$, Cohen's $d = 0.628$), and level of happiness ($t(45.493) = 2.030$, $p = 0.024$, Cohen's $d = 0.500$) within the dream content (see Table 4). To be more precise, the dream content of participants from the suicide attempt group featured on average more self-directed aggressions ($M_{SA} = 0.242$, $M_C = 0.000$), fewer friendly interactions ($M_{SA} = 0.379$, $M_C = 1.000$), more failures ($M_{SA} = 0.288$, $M_C = 0.0601$), and a higher level of happiness ($M_{SA} = 0.333$, $M_C = 0.091$) compared to the participants from the comparator group. In the friendliness category, the sharing of pleasant social activities subcategory emerged as the most significant ($t(32) = −2.667$, $p = 0.006$, Cohen's' $d = −0.656$). Despite there being no statistically significant difference in the overall quantity of aggressions in the suicide attempt group's dream content ($t(64) = −1.168$, $p = 0.123$) when compared to the comparator group, there was a significantly higher incidence of death-resulting aggression (type 8 aggression) ($t(32) = −2.620$, $p = 0.007$, Cohen's' $d = 0.645$) within this group ($M_{SA} = 0.197$, $M_C = 0.000$).

Table 4. Comparison of Hall and Van de Castle categories' frequency in dream content between groups.

	t	*df*	*Significance (One-Sided p)*	*Cohen's d*
Characters	−1.334	64	0.093	-
Total Aggression	1.168	64	0.123	-
Death-resulting aggression	2.620	32	0.007 **	0.645
Self-directed aggression	1.936	32	0.031 *	0.477
Friendliness	−2.387	48.468	0.010 *	−0.588
Pleasant activity sharing	−2.667	32	0.006 **	−0.656
Friendliness towards dreamer	−1.782	58.428	0.040	−0.439
Sexuality	−0.632	64	0.265	-
Success	0.436	64	0.424	-
Failure	2.549	49.039	0.007 **	0.628
Misfortune	0.490	64	0.313	-
Good Fortune	−0.149	64	0.441	-
Negative emotions	1.445	64	0.077	-
Happiness	2.030	45.493	0.024 *	0.500

**. Correlation is significant at the 0.01 level (one-tailed). *. Correlation is significant at the 0.05 level (one-tailed).

Threats. As illustrated in Table 5, no significant difference in the total quantity of threats between the two groups (n_{SA} = 33, n_C = 33) was revealed by the statistical analyses, but a higher quantity of psychological/financial/social type threats (t(57.432)= 2.081, p = 0.021, Cohen's' d = 0.512) was observed in the suicide attempt group's dream content (M_{SA} = 0.636, M_C = 0.303). The dreams of this group also seemed to demonstrate a higher threatening intensity in terms of psychological and physical integrity (t(64) = 2.029, p = 0.023, Cohen's' d = 0.500) when compared to those of the comparator group (M_{SA} = 1.727, M_C = 1.212).

Table 5. Comparison of threat frequency in dream content between groups.

	t	*df*	*Significance (One-Sided p)*	*Cohen's d*
Total threats	0.942	64	0.175	-
Type 1—Life-threatening	−0.119	64	0.453	-
Type 2—Psycho./Financial/Social	2.081	57.432	0.021 *	0.512
Type 3—Physical	0.000	64	0.500	-
Type 4—Minor	−0.407	64	0.343	-
Threatening intensity	2.029	64	0.023 *	0.500

*. Correlation is significant at the 0.05 level (one-tailed). **. Correlation is significant at the 0.01 level (one-tailed).

Parental Acceptance-Rejection Questionnaire short form. Among the 33 suicidal dreams, 14 contained at least one parent character (42% of the dreams), whereas in the comparator group, the presence of parental figures was observed in only 9 out of the 33 dreams (27% of the dreams). Only three dream narratives from the suicide attempt group and one dream narrative from the comparator group included sufficient detailed interactions with a parental character to allow for analysis of the relationship using the PARQ short form. The average score of parental acceptance-rejection in the suicide attempt group was 72.667. As for the only comparative score for parental acceptance-rejection, the result was 45.

3.5. Dream Sample

In order to demonstrate how elements from waking life can be incorporated into dreams, we present examples of dreams shared by participants in the suicide attempt group:

"(. . .) another person I do not really remember who it was, but he was talking about like how his life was like falling apart and I saw him like self-harming a lot and there was a lot of blood and I remember that the blood was like dripping onto snow, below him so it is really easy to see because it was like bright red on the white snow (. . .)."

"(. . .) As a result of these emotions I abruptly said I was going to leave. I started packing my suitcase. My mom questioned why I was leaving and I said nobody wants me around that's why. We had a very verbal and loud argument about it. I remember thinking I hoped they cared enough to tell me to stay but they didn't. In this situation I felt like a burden to everyone."

4. Discussion

In the context of current theories, and clinical observations and interventions, the main goal of this study was to investigate the mood, frequency of nightmares, and presence and frequency of specific themes in the dreams of suicidal adolescents, as compared to a normative sample of adolescents. We comment here on the implications of the findings.

4.1. Nightmare Frequency

First, our findings demonstrate that the suicide attempt group had a greater number of participants who could be considered as frequent nightmare sufferers than those in the comparator group, with nearly fifty percent of the suicide attempt group reporting one or more nightmares a week. This finding is consistent with the existing literature, which identifies that frequent nightmares are a risk factor for suicidality [12,33]. It is important to note that the occurrence of frequent nightmares does not necessarily indicate the presence of a nightmare disorder, as additional DSM-5 criteria must be met [77]. Nevertheless, addressing nightmares may be an important component of suicide prevention and intervention efforts for adolescents.

The literature demonstrates that non-pharmacological interventions, such as Imagery Rehearsal Therapy and lucid dreaming, are effective in reducing nightmares in both children and adolescents [78,79]. For instance, Imagery Rehearsal Therapy has been linked to enhanced sleep quality and reductions in the severity of PTSD symptoms in sexual assault survivors [80]. Nielsen and Levin (2007) proposed the affective network dysfunction model, suggesting that nightmares are caused by a disruption in the adaptive function of fear memory extinction during dreaming [81]. Reducing nightmares could improve stress and emotional distress associated with fear memory and sleep quality, thus acting as a promising preventive strategy for suicide among adolescents who have major depression and experience frequent nightmares.

4.2. Mood

Next, as shown by our results, our suicide attempt group demonstrated a significantly higher combined negative mood, and a significantly lower combined positive mood, at night, in dreams, and in the morning, in comparison to the comparator group. This observation is consistent with prior research, and supports that negative affect is strongly associated with an increased risk for suicidality [37,39].

As illustrated in both panels of Figure 2, our results revealed a significant relationship between mood, time, and group, indicating that the mood varied when controlled for the time it was measured and the group it pertained to.

In terms of negative mood, the progression throughout time periods was different between the two groups, showing a reverse pattern. The comparator group exhibited a typical negative mood progression, with a low negative mood level at the pre-sleep measure and an increase in negative emotions during dreaming. Indeed, this pattern is consistent with previous research, wherein participants have typically reported a greater prevalence of negative emotions, or a "negative tone", in their dreams in comparison to their waking life [24–26,82]. This negative tone has been associated with the predominant activation of the amygdala, a brain structure involved in the processing of fear, during rapid eye movement (REM) sleep [24,83]. In our study, however, the suicide attempt group had baseline levels of evening negative mood so high that the dream and post-sleep moods were low in comparison.

It is of value to reflect on what the implication of these findings is in relation to the ERT. In our study, the levels of negative emotions in dreaming and in waking reported by suicide attempt participants differ from the pattern seen in the comparator participants and previous research. This may suggest an alteration in the processing of emotion regulation during dreaming, and reflect the observed difficulties in regulating emotions during wakefulness among individuals with suicidal tendencies [53]. This observation is also consistent with our previous findings regarding disrupted rapid eye movement (REM) sleep and waking emotional processing in a different cohort of the same patient group, as mentioned in our introduction [9,10]. Despite this difference in pattern, the suicide attempt group still reported a lower negative mood at the post-sleep measure than at the pre-sleep measure, indicating some improvement between the two measures of time. This is consistent with previous research that has demonstrated enhanced emotional adjustment, mood, and well-being after intact sleep, particularly REM sleep, even when disrupted [84].

Further, it is of interest to examine the progression of positive mood with the suicide attempt group, who reported lower positive mood levels than the comparator group before and after sleep, but not in the dream. More specifically, in the comparator group, positive mood was higher at pre-sleep, decreased during dreaming, and then remained stable at the post-sleep measure. Suicidal participants started with low positive mood levels at pre-sleep, which increased during dreaming to rise above the level of the comparator group and decreased at the post-sleep measure. A similar observation emerged in the study of Punamäki (1999), where they found that a negative evening mood in children residing in traumatic environments was associated with the report of positive dreams [85]. The author attributed this phenomenon to a model of compensatory dream function, whereas the presence of these positive emotions could be the result of a psychological mechanism that attempts to protect the dreamer's well-being by providing a relief from a painful reality [85–87]. This would align with the disruption-avoidance-adaptation model of dreaming, where disruption facilitates mastery over a stressful stimulus, while avoidance protects the dreamer against this disruption to ultimately allow adaptation [86].

Furthermore, it is interesting to note that the progression of positive mood in the suicide attempt group is similar to the progression of negative mood in the comparator group with an increase in mood during the dream. Also, the progression of negative mood in the suicide attempt group is similar to the progression of positive mood in the comparator group with a decrease in mood during the dream. Thus, it would seem that the progression in mood is reversed according to the valence between the groups. This is also consistent with the disruption-avoidance-adaptation model of dreaming. From this perspective, the observed increase in positivity among adolescents who have attempted suicide could be indicative of avoidance, whereas the increase in negativity among the comparators could be indicative of mastery.

The findings from the bivariate correlation on the comparator group's mood partially aligned with the CT of dreaming. Specifically, a positive mood in the evening was associated with a positive mood during the dream and in the morning. Indeed, previous studies have demonstrated that emotional experiences and mood states in wakefulness can influence the nature of dreams, as positive emotional experiences tend to lead to more positive and pleasant dreams, whereas negative emotional experiences tend to lead to more negative and unpleasant dreams [88–90].

In addition, a positive mood during the dream was strongly linked to a positive mood in the morning and a negative mood during dreaming was moderately associated to a negative mood in the morning. Conversely, a negative mood during the dream was moderately and negatively linked to a positive mood in the morning. These findings are consistent with the discussion by Mallett et al. (2021) proposing that dream experiences can influence subsequent waking affect [91].

The suicide attempt group's bivariate correlation results were also partially consistent with the CT of dreaming. Specifically, a negative mood in the evening was linked to a negative mood during dreaming and in the morning. Additionally, a negative mood during dreaming was moderately linked with a negative mood in the morning. These findings collectively indicate that the reported negative moods were consistently related ted across the different times of measure.

The multiple linear regression results suggest that both positive and negative dream mood is a predictor of morning mood in the comparator group. This is consistent with the hypothesis that dreaming plays a role in regulating emotions upon waking, as proposed by the ERT [54].

Contrary to the results of the comparator group, the positive dream mood of the suicide attempt group was not found to be an indicator of the morning mood. Instead, only the positive evening mood was found to be an indicator of the morning mood. Nevertheless, a negative dream mood was found to be a moderate predictor of a negative mood in the morning in this group. Once more, the emotion regulation process hypothesized by the ERT may be altered in this group as only the negative dream mood seems to predict the subsequent waking affect.

Based on these preliminary results, it can be hypothesized that there is a tentative compensation effect in dreaming, as evidenced by the increased positive mood. However, this effect does not appear to be sufficient to ameliorate mood in the morning. This supports the recommendation that improving dream experience may improve mood in adolescents at risk for suicidality.

4.3. Positivity Bias

Moreover, the suicide attempt group exhibited a discrepancy between the participants' and the judge's evaluations regarding the positive mood in the dream. Thus, despite the potential alteration of emotion regulation in dreaming discussed before, this observation demonstrates that the suicide attempt group exhibited a positive bias. Such a bias has been hypothesized to aid in the desensitization process that may occur during dreams [54] and may contribute to the lower post-sleep negative mood observed in the suicide attempt group.

4.4. Dream Content

Destructive themes. Our research findings corroborate the conclusions drawn by Firth et al. (1986), in that the occurrence rate of dreams concerning death and dying and featuring the theme of masochism was considerably greater among individuals with suicidal tendencies than in the comparator group [31]. Negative self-concept has previously been associated with suicidal ideation and attempts in youth [92–94]. This relationship between negative self-concept and suicidal tendencies may explain why both themes, masochism and death, appeared more frequently in the dreams of suicidal adolescents. Research has shown that helplessness and hopelessness, which are closely related strong predictors of suicidality [95,96]. In a study conducted by Glucksman and Kramer (2017), these factors have also been observed as significant themes in the dream content of clinically depressed potentially suicidal patients [28]. However, in our study, when examined individually, the other destructive themes did not exhibit any notable differences between the two groups. Indeed, the analyzed theme of helplessness did not display a significant difference in frequency between the groups. This could potentially be attributed to the challenge that the judges faced in inferring from the dream narrative whether the dreamer experienced a feeling of helplessness.

4.5. Hall and Van de Castle Categories

Friendliness. Positive social relationships, such as social support and friendships with others, tend to be protective factors against suicidal behavior [97,98]. Our finding that suicidal adolescents reported significantly fewer friendly interactions in their dream is consistent with the Continuity Theory of dreaming, in that it mirrors their waking deficit in positive social interactions in their dreams compared to normative adolescents.

Aggression. Self-directed aggressions are any form of aggression from the HVDC scale accomplished by and directed toward the dreamer. This category is very similar in concept to the masochism theme analyzed earlier. Consistent with our prior presented results, the prevalence of self-directed aggression on the HVDC scale confirms the higher frequency of negative self-thoughts or actions in the dream content of the suicide attempt group.

Prior studies have demonstrated that aggressivity can be considered a predictor of suicide attempts or even a feature of suicide capability [99,100]. Our results showed no statistically significant difference in the overall quantity of aggressions between the two groups. However, there was a significantly higher incidence in dreams of death-resulting aggression within the suicide attempt group. This category is established as the highest intensity of aggressive interactions on the HVDC scale. This result could be associated with the prevalence of death-related themes in dream content, thus indicating that not only is death more frequent but also more aggressively portrayed.

Failure. Our analysis indicates that the suicide attempt group had a significantly higher quantity of failures compared to the comparator group. In the context of dream content, "failure" is defined by the HVDC scale as the inability of a character to achieve their desired goal due to personal limitations and inadequacies [25]. This definition can be associated with the negative self-concept, helplessness, and hopelessness discussed earlier. This result can suggest that suicidal dreamers perceive a lack of control, which prevents them from reaching their goals.

Happiness. The higher level of expressed written feelings of happiness in the dream content of the suicide attempt group is consistent with the higher reported positive mood associated with the dream experience in the suicide attempt group when compared to the comparator group, and the increase in positive mood observed from evening to dream in the suicide attempt group, as depicted in Figure 2B. The same elements as discussed in the mood section of this paper in relation to the possibility of a different emotion regulation process during dreaming in the clinical sample could be hypothesized from this result. Also consistent with this observation is the fact that the positivity bias was observed in the suicide attempt group and interpreted as a tentative manifestation of mood regulation. In addition, this combination of positive with negative content has been observed recently by Vallat et al. (2018) [101] and Barbeau et al. (2022b) [54] and interpreted as possible desensitization attempts. However, further replication and investigation of this relationship is necessary, and the utilization of text analysis tools (text mining software) could yield valuable insights on this matter.

4.6. Threats

Our findings showed that the number of threats in the dream content of suicidal adolescents was not higher than the one of normative adolescents. Nevertheless, the higher quantity of type 3 threats or psychological/financial/social type threats and a higher threatening intensity in terms of psychological and physical integrity observed in the suicide attempt group's dream content are consistent with prior research on waking clinical correlates for suicidal ideation; suicide has indeed been closely associated with psychological distress, depression, and higher levels of stress (i.e., traumatic, interpersonal) in adolescents [102–104]. Threatening and negative dreams have also been associated with rumination and intrusive thoughts, which are closely related to depression, hopelessness, and suicidality [105,106].

4.7. Parental Acceptance-Rejection Questionnaire Short Form

There was a slightly higher frequency of the presence of parental characters in the dreams of suicidal adolescents, with 42% compared to 27% in control participants. The details about the relationship with the parent also seemed higher in the suicide attempt group, as three dream narratives allowed for analysis with the PARQ short form as opposed to one in the comparator group. The average parental-acceptance score was higher in the suicide attempt group than in the comparator group, suggesting that suicidal adolescents conveyed a more rejecting parental image in their dream narrative. This result is consistent with prior research in which suicide attempters tended to perceive their parents to be more rejecting [47]. A potential avenue for further investigation could be to assess the perceived level of parental acceptance-rejection during waking and correlate the result to the inferred one from the dream narrative of suicidal adolescents. Completing the PARQ short form during the day could also trigger parental presence in the dreams of adolescents, as elements from salient or emotional recent waking-life experiences can be integrated into dreams [107,108].

4.8. Transgender and Non-Binary Participants

Finally, it is important to note that among the thirty-three suicidal participants in this study, twelve individuals identified as a gender different from that assigned at birth. Specifically, ten participants identified as transgender males, one as a transgender female, and one as non-binary. This percentage is representative of, and in accordance with, the prior established prevalence of suicidality in this population. With estimates ranging from 31% to 64%, transgender individuals exhibit a significantly higher risk for depression and anxiety than the general population [109]. Research shows that transgender adolescents are more likely to report suicide risk than cisgender adolescents [110]. Stigma, LGBT victimization, and gender non-conformity have been associated with self-harm and/or suicidal ideation [111]. There may be differences in dream content between transgender and cisgender individuals in terms of interactions and identity. Conducting a larger study with more participants could yield valuable insights into how gender identity manifests in dream content.

4.9. Limitations

Several Factors May Influence the Implications of the Results of this Study

Dream recall. Dream recall is known to be imperfect as dream content can be reinterpreted and rationalized upon waking [112]. Dream reports often fail to capture the full scope and intensity of the dream experience. Indeed, dreamers tend to provide minimal details and overlook the grander aspects of their dreams when recounting them. As a result, the experiences described in dream reports can be quite different from the actual dream experience [113]. Furthermore, the situation of hospitalization, the medication, and multiple other factors can affect dream recall [24]. Although the dream diary method is a superior approach to questionnaires, it is still far from ideal [24]. Additionally, the field of brain recording of dream activity is still in its infancy [24].

Convenience sample. The clinical sample for this study was recruited through an emergency hospital inpatient unit on a volunteer basis and represents a convenience sample. Consequently, the conclusions presented in this article should be considered preliminary and may not be representative of a more general population and, in particular, of non-hospitalized adolescents who have attempted suicide. Nevertheless, these findings are highly relevant as they were obtained using the current standard of dream collection and warrant further investigation with larger samples. Finally, as we mentioned earlier, the sleep habit data were not included in this report since they were studied in our previous cohort and are included with a larger sample of circadian and sleep factors as well as timing of suicide attempt.

5. Conclusions

Our study provides support for the CT of dreaming, as it unveils that the dreams of suicidal adolescents reflected different aspects of their waking life. These findings align with the emotional dysregulation and distress, negative affect, negative self-concept, parental rejection, lack of social support, and suicidal ideation previously associated with at-risk suicidal adolescents. Our results underline the value in addressing nightmares, and support utilizing intervention techniques and therapies to mitigate the negative experience of dream content in relation to suicidal tendencies. By allowing adolescents who experience frequent nightmares and suicidal ideations to undergo therapies aimed at reducing nightmares, emotional distress could be alleviated, sleep quality could be improved, and overall psychological well-being could be enhanced. The observed association of positivity bias and morning mood suggests that further studies should explore this relation and the empirical value of the ERT in the context of adolescent suicidality. It would also be of interest to investigate the documented emotion regulation value of REM sleep in this population [114].

Author Contributions: E.D. completed an honours thesis on this study and participated in the final review and write-up. L.G. completed an honours thesis on part of this study. E.L., E.D., and L.G. participated in the dream scoring. A.B. (Antoine Benoit) executed statistical analyses. A.B. (Addo Boafo) performed the intake of the participants, collected the dreams and questionnaires, and oversaw the hospitalization of participants. J.D.K. obtained funding, and supervised the entire study, write-up, and submission. All authors have read and agreed to the published version of the manuscript.

Funding: This study was partially funded by The Social Sciences and Humanities Research Council of Canada (SSHRC), grant number 435-2020-0851.

Institutional Review Board Statement: Research Ethic Board (REB) of the Children's hospital of Eastern Ontario (CHEO) REB Protocol No: 21/137X, Protocol Title: CHEOREB# 21/137X—Sleep and Dreams in Adolescence Approval Date: January 2022 and renewed to December 2024. REB@cheo.on.ca.

Informed Consent Statement: Informed consent was obtained from all subjects involved in the study.

Data Availability Statement: The data base of our manuscript will be available upon request, respecting the confidentiality and the consent of participants in accordance with the ethic approval of the project. Copies of questionnaires are also available.

Conflicts of Interest: The authors declare no conflict of interest.

References

1. World Health Organization Suicide. Available online: https://www.who.int/news-room/fact-sheets/detail/suicide (accessed on 24 May 2024).
2. Government of Canada, Statistics Canada. Health Fact Sheets. Available online: https://www150.statcan.gc.ca/n1/pub/82-625-x/2015001/article/14296-eng.htm# (accessed on 24 May 2024).
3. Greenham, S.L.; Bisnaire, L. An Outcome Evaluation of an Inpatient Crisis Stabilization and Assessment Program for Youth. *Resid. Treat. Child. Youth* **2008**, *25*, 123–143. [CrossRef]
4. Andrews, S.; Hanna, P. Investigating the Psychological Mechanisms Underlying the Relationship between Nightmares, Suicide and Self-Harm. *Sleep Med. Rev.* **2020**, *54*, 101352. [CrossRef]
5. Liu, X.; Yang, Y.; Liu, Z.-Z.; Jia, C.-X. Bidirectional Associations between Sleep Problems and Suicidal Thought/Attempt in Adolescents: A 3-Wave Data Path Analysis. *J. Affect. Disord.* **2024**, *350*, 983–990. [CrossRef] [PubMed]
6. Goldstein, T.R.; Bridge, J.A.; Brent, D.A. Sleep Disturbance Preceding Completed Suicide in Adolescents. *J. Consult. Clin. Psychol.* **2008**, *76*, 84–91. [CrossRef] [PubMed]
7. Pigeon, W.R.; Pinquart, M.; Conner, K. Meta-Analysis of Sleep Disturbance and Suicidal Thoughts and Behaviors. *J. Clin. Psychiatry* **2012**, *73*, e1160–e1167. [CrossRef] [PubMed]
8. Boafo, A.; Dion, K.; Greenham, S.; Barrowman, N.; Reddy, D.; De Koninck, J.; Robillard, R. Sleep Problems and Complexity of Mental Health Needs in Adolescent Psychiatric Inpatients. *J. Psychiatr. Res.* **2021**, *139*, 8–13. [CrossRef]
9. Boafo, A.; Armitage, R.; Greenham, S.; Tavakoli, P.; Dale, A.; Nixon, A.; Lafrenière, A.; Ray, L.B.; De Koninck, J.; Robillard, R. Sleep Architecture in Adolescents Hospitalized during a Suicidal Crisis. *Sleep Med.* **2019**, *56*, 41–46. [CrossRef] [PubMed]
10. Tavakoli, P.; Lanthier, M.; Porteous, M.; Boafo, A.; De Koninck, J.; Robillard, R. Sleep Architecture and Emotional Inhibition Processing in Adolescents Hospitalized during a Suicidal Crisis. *Front. Psychiatry* **2022**, *13*, 920789. [CrossRef] [PubMed]

11. Russell, K.; Rasmussen, S.; Hunter, S.C. Insomnia and Nightmares as Markers of Risk for Suicidal Ideation in Young People: Investigating the Role of Defeat and Entrapment. *J. Clin. Sleep Med.* **2018**, *14*, 775–784. [CrossRef]

12. Ağargün, M.Y.; Çilli, A.S.; Kara, H.; Tarhan, N.; Kincir, F.; Öz, H. Repetitive and Frightening Dreams and Suicidal Behavior in Patients with Major Depression. *Compr. Psychiatry* **1998**, *39*, 198–202. [CrossRef]

13. Bernert, R.A.; Joiner, T.E.; Cukrowicz, K.C.; Schmidt, N.B.; Krakow, B. Suicidality and Sleep Disturbances. *Sleep* **2005**, *28*, 1135–1141. [CrossRef] [PubMed]

14. Cukrowicz, K.C.; Otamendi, A.; Pinto, J.V.; Bernert, R.A.; Krakow, B.; Joiner, T.E. The Impact of Insomnia and Sleep Disturbances on Depression and Suicidality. *Dreaming* **2006**, *16*, 1–10. [CrossRef]

15. Liu, X.; Liu, Z.-Z.; Chen, R.-H.; Cheng, X.-Z.; Bo, Q.-G.; Wang, Z.-Y.; Yang, Y.; Fan, F.; Jia, C.-X. Nightmares Are Associated with Future Suicide Attempt and Non-Suicidal Self-Injury in Adolescents. *J. Clin. Psychiatry* **2019**, *80*, 12806. [CrossRef] [PubMed]

16. Bolstad, C.J.; Holzinger, B.; Scarpelli, S.; De Gennaro, L.; Yordanova, J.; Koumanova, S.; Mota-Rolim, S.; Benedict, C.; Bjorvatn, B.; Chan, N.Y.; et al. Nightmare Frequency Is a Risk Factor for Suicidal Ideation during the COVID-19 Pandemic. *J. Sleep Res.* **2024**, e14165. [CrossRef] [PubMed]

17. Antunes-Alves, S.; De Koninck, J. Pre- and Post-Sleep Stress Levels and Negative Emotions in a Sample Dream among Frequent and Non-Frequent Nightmare Sufferers. *Arch. Psychiatry Psychother.* **2012**, *14*, 11–16.

18. Schredl, M. Continuity Hypothesis of Dreaming. In *Dreams: Understanding Biology, Psychology, and Culture*; ABC-CLIO: Santa Barbara, CA, USA, 2019; pp. 88–94.

19. Hall, C.S. *The Meaning of Dreams*; Harper & Row: New York, NY, USA, 1953.

20. Domhoff, G.W. *The Emergence of Dreaming: Mind-Wandering, Embodied Simulation, and the Default Network*; Oxford University Press: Oxford, UK, 2017; ISBN 978-0-19-067343-7.

21. Koulack, D. Dreams and Adaptation to Contemporary Stress. In *The Functions of Dreaming*; State University of New York Press: New York, NY, USA, 1993; pp. 321–340.

22. Picchioni, D.; Goeltzenleucher, B.; Green, D.N.; Convento, M.J.; Crittenden, R.; Hallgren, M.; Hicks, R.A. Nightmares as a Coping Mechanism for Stress. *Dreaming* **2002**, *12*, 155–169. [CrossRef]

23. Malinowski, J.; Horton, C.L. Evidence for the Preferential Incorporation of Emotional Waking-Life Experiences into Dreams. *Dreaming* **2014**, *24*, 18–31. [CrossRef]

24. De Koninck, J. Sleep, Dreams, and Dreaming. In *Handbook of Sleep and Sleep Disorders*; Morin, C., Epsie, C., Eds.; Oxford University Press: Oxford, UK, 2012; pp. 150–171. [CrossRef]

25. Hall, C.S.; Van de Castle, R.L. *The Content Analysis of Dreams*; Hall, C.S., Van de Castle, R.L., Eds.; Century Psychology series; Appleton-Century-Crofts: New York, NY, USA, 1966.

26. Dale, A.; Lortie-Lussier, M.; Wong, C.; De Koninck, J. Dreams of Canadian Students: Norms, Gender Differences, and Comparison with American Norms. *J. Cross-Cult. Psychol.* **2016**, *47*, 941–955. [CrossRef]

27. Gutheil, E.A. Dream and Suicide. *Am. J. Psychother.* **1948**, *2*, 197–362. [CrossRef]

28. Glucksman, M.L.; Kramer, M. Manifest Dream Content as a Predictor of Suicidality. *Psychodyn. Psychiatry* **2017**, *45*, 175–185. [CrossRef]

29. Farberow, N.L. Personality Patterns of Suicidal Mental Hospital Patients. *Genet. Psychol. Monogr.* **1950**, *42*, 3–79. [PubMed]

30. Raphling, D.L. Dreams and Suicide Attempts. *J. Nerv. Ment. Dis.* **1970**, *151*, 404–410. [CrossRef] [PubMed]

31. Firth, S.T.; Blouin, J.; Natarajan, C.; Blouin, A. A Comparison of the Manifest Content in Dreams of Suicidal, Depressed and Violent Patients. *Can. J. Psychiatry* **1986**, *31*, 48–53. [CrossRef] [PubMed]

32. Pesant, N.; Zadra, A. Dream Content and Psychological Well-Being: A Longitudinal Study of the Continuity Hypothesis. *J. Clin. Psychol.* **2006**, *62*, 111–121. [CrossRef]

33. Tanskanen, A.; Tuomilehto, J.; Viinamäki, H.; Vartiainen, E.; Lehtonen, J.; Puska, P. Nightmares as Predictors of Suicide. *Sleep* **2001**, *24*, 845–848. [CrossRef]

34. Wang, Z.; Zhang, K.; He, L.; Sun, J.; Liu, J.; Hu, L. Associations between Frequent Nightmares, Nightmare Distress and Depressive Symptoms in Adolescent Psychiatric Patients. *Sleep Med.* **2023**, *106*, 17–24. [CrossRef]

35. Sheftall, A.H.; Bergdoll, E.E.; James, M.; Bauer, C.; Spector, E.; Vakil, F.; Armstrong, E.; Allen, J.; Bridge, J.A. Emotion Regulation in Elementary School-Aged Children with a Maternal History of Suicidal Behavior: A Pilot Study. *Child Psychiatry Hum. Dev.* **2020**, *51*, 792–800. [CrossRef] [PubMed]

36. Zlotnick, C.; Donaldson, D.; Spirito, A.; Pearlstein, T. Affect Regulation and Suicide Attempts in Adolescent Inpatients. *J. Am. Acad. Child Adolesc. Psychiatry* **1997**, *36*, 793–798. [CrossRef] [PubMed]

37. Stein, D.; Apter, A.; Ratzoni, G.; Har-Even, D.; Avidan, G. Association between Multiple Suicide Attempts and Negative Affects in Adolescents. *J. Am. Acad. Child Adolesc. Psychiatry* **1998**, *37*, 488–494. [CrossRef]

38. Levin, R.; Nielsen, T.A. Disturbed Dreaming, Posttraumatic Stress Disorder, and Affect Distress: A Review and Neurocognitive Model. *Psychol. Bull.* **2007**, *133*, 482–528. [CrossRef]

39. Rubio, A.; Oyanedel, J.C.; Bilbao, M.; Mendiburo-Seguel, A.; López, V.; Páez, D. Suicidal Ideation Mediates the Relationship Between Affect and Suicide Attempt in Adolescents. *Front. Psychol.* **2020**, *11*, 524848. [CrossRef] [PubMed]

40. Foulkes, D. *Children's Dreams: Longitudinal Studies*; Wiley: New York, NY, USA, 1982; ISBN 978-0-471-08181-4.

41. Honig, A.S.; Nealis, A.L. What Do Young Children Dream About? *Early Child Dev. Care* **2012**, *182*, 771–795. [CrossRef]

42. Colace, C. Children's Dreaming: A Study Based on Questionnaires Completed by Parents. *Sleep Hypn.* **2006**, *8*, 19–32.

43. Maggiolini, A.; Morelli, M.; Falotico, E.; Montali, L. Dream Contents of Early Adolescents, Adolescents, and Young Adults: A Cluster Analysis with T-LAB. *Dreaming* **2016**, *26*, 221–237. [CrossRef]

44. Nöltner, S.; Schredl, M. Interactions with Family Members in Students' Dreams. *Dreaming* **2023**, *33*, 19–31. [CrossRef]

45. Donath, C.; Graessel, E.; Baier, D.; Bleich, S.; Hillemacher, T. Is Parenting Style a Predictor of Suicide Attempts in a Representative Sample of Adolescents? *BMC Pediatr.* **2014**, *14*, 113. [CrossRef]

46. Boričević Maršanić, V.; Margetić, B.A.; Zečević, I.; Herceg, M. The Prevalence and Psychosocial Correlates of Suicide Attempts Among Inpatient Adolescent Offspring of Croatian PTSD Male War Veterans. *Child Psychiatry Hum. Dev.* **2014**, *45*, 577–587. [CrossRef]

47. Wong, J.C.M.; Chang, C.L.; Shen, L.; Nyein, N.; Loh, A.S.W.; Yap, N.H.; Kroneman, L.M.; Feng, L.; Tan, C.H. Temperament, Parenting, Mental Disorders, Life Stressors and Help-Seeking Behavior of Asian Adolescent Suicide Attempters: A Case Control Study. *Front. Psychiatry* **2022**, *13*, 999089. [CrossRef]

48. Revonsuo, A. The Reinterpretation of Dreams: An Evolutionary Hypothesis of the Function of Dreaming. *Behav. Brain Sci.* **2000**, *23*, 877–901. [CrossRef]

49. Valli, K.; Revonsuo, A.; Pälkäs, O.; Ismail, K.H.; Ali, K.J.; Punamäki, R.-L. The Threat Simulation Theory of the Evolutionary Function of Dreaming: Evidence from Dreams of Traumatized Children. *Conscious. Cogn.* **2005**, *14*, 188–218. [CrossRef]

50. Loukola, V.; Tuominen, J.; Kirsilä, S.; Kyyhkynen, A.; Lahdenperä, M.; Parkkali, L.; Ranta, E.; Malinen, E.; Vanhanen, S.; Välimaa, K.; et al. Viral Simulations in Dreams: The Effect of the COVID-19 Pandemic on Threatening Dream Content in a Finnish Sample of Diary Dreams. *Conscious. Cogn.* **2024**, *119*, 103651. [CrossRef]

51. Bradshaw, S.; Lafrenière, A.; Amini, R.; Lortie-Lussier, M.; De Koninck, J. Threats in Dreams, Emotions and the Severity of Threatening Experiences in Waking. *Int. J. Dream Res.* **2016**, *9*, 102–109.

52. Lafrenière, A.; Lortie-Lussier, M.; Dale, A.; Robidoux, R.; De Koninck, J. Autobiographical Memory Sources of Threats in Dreams. *Conscious. Cogn.* **2018**, *58*, 124–135. [CrossRef]

53. Özdemir, P.G.; Kefeli, M.C.; Özcan, H. Emotion Regulation Difficulties, Personality Traits and Coping Styles in First-Time Suicide Attempters. *J. Psychol. Psychother. Res.* **2017**, *4*, 27–34. [CrossRef]

54. Barbeau, K.; Turpin, C.; Lafrenière, A.; Campbell, E.; De Koninck, J. Dreamers' Evaluation of the Emotional Valence of Their Day-to-Day Dreams Is Indicative of Some Mood Regulation Function. *Front. Behav. Neurosci.* **2022**, *16*, 947396. [CrossRef]

55. Kramer, M. The Selective Mood Regulatory Function of Dreaming: An Update and Revision. In *The Functions of Dreaming*; SUNY series in dream studies; State University of New York Press: Albany, NY, USA, 1993; pp. 139–195, ISBN 978-0-7914-1297-8.

56. Moffitt, A.; Kramer, M.; Hoffmann, R. *The Functions of Dreaming*; State University of New York Press: New York, NY, USA, 1993; ISBN 978-1-4384-1339-6.

57. Kramer, M. *The Dream Experience: A Systematic Exploration*; Routledge: New York, NY, USA, 2006; ISBN 978-0-203-94370-0.

58. Scarpelli, S.; Bartolacci, C.; D'Atri, A.; Gorgoni, M.; De Gennaro, L. The Functional Role of Dreaming in Emotional Processes. *Front. Psychol.* **2019**, *10*, 459. [CrossRef] [PubMed]

59. Schredl, M.; Doll, E. Emotions in Diary Dreams. *Conscious. Cogn.* **1998**, *7*, 634–646. [CrossRef]

60. Sikka, P.; Valli, K.; Virta, T.; Revonsuo, A. I Know How You Felt Last Night, or Do I? Self- and External Ratings of Emotions in REM Sleep Dreams. *Conscious. Cogn.* **2014**, *25*, 51–66. [CrossRef]

61. Sikka, P.; Feilhauer, D.; Valli, K.; Revonsuo, A. How You Measure Is What You Get: Differences in Self- and External Ratings of Emotional Experiences in Home Dreams. *Am. J. Psychol.* **2017**, *130*, 367–384. [CrossRef]

62. Sikka, P.; Revonsuo, A.; Sandman, N.; Tuominen, J.; Valli, K. Dream Emotions: A Comparison of Home Dream Reports with Laboratory Early and Late REM Dream Reports. *J. Sleep Res.* **2018**, *27*, 206–214. [CrossRef]

63. Sikka, P.; Valli, K.; Revonsuo, A.; Tuominen, J. The Dynamics of Affect across the Wake-Sleep Cycle: From Waking Mind-Wandering to Night-Time Dreaming. *Conscious. Cogn.* **2021**, *94*, 103189. [CrossRef]

64. Perlis, M.L.; Nielsen, T.A. Mood Regulation, Dreaming and Nightmares: Evaluation of a Desensitization Function for REM Sleep. *Dreaming* **1993**, *3*, 243–257. [CrossRef]

65. Stewart, S.L.; Semovski, V.; Lapshina, N. Adolescent Inpatient Mental Health Admissions: An Exploration of Interpersonal Polyvictimization, Family Dysfunction, Self-Harm and Suicidal Behaviours. *Child Psychiatry Hum. Dev.* **2024**, *55*, 963–974. [CrossRef]

66. Wunderlich, U.; Bronisch, T.; Wittchen, H.-U.; Carter, R. Gender Differences in Adolescents and Young Adults with Suicidal Behaviour. *Acta Psychiatr. Scand.* **2001**, *104*, 332–339. [CrossRef]

67. Blumenthal, S.J. *Suicide over the Life Cycle: Risk Factors, Assessment, and Treatment of Suicidal Patients*; Suicide over the life cycle: Risk factors, assessment, and treatment of suicidal patients; American Psychiatric Association: Arlington, VA, USA, 1990; p. 799, ISBN 978-0-88048-307-0.

68. Dale, A.; Lortie-Lussier, M.; De Koninck, J. Ontogenetic Patterns in the Dreams of Women across the Lifespan. *Conscious. Cogn.* **2015**, *37*, 214–224. [CrossRef] [PubMed]

69. Dale, A.; Lafrenière, A.; De Koninck, J. Dream Content of Canadian Males from Adolescence to Old Age: An Exploration of Ontogenetic Patterns. *Conscious. Cogn.* **2017**, *49*, 145–156. [CrossRef]

70. Sabourin, C.; Robidoux, R.; Pérusse, A.D.; De Koninck, J. Dream Content in Pregnancy and Postpartum: Refined Exploration of Continuity between Waking and Dreaming. *Dreaming* **2018**, *28*, 122–139. [CrossRef]

71. Rohner, R. Parental Acceptance-Rejection Questionnaire (PARQ): Test Manual. In *Handbook for the Study of Parental Acceptance and Rejection*; Rohner Research Publications: Storrs, CT, USA, 2005; pp. 43–106.

72. Revonsuo, A.; Valli, K. *Dream Threat Scale*; American Psychological Association: Washington, DC, USA, 2000.

73. Langs, R.J. Manifest Dreams from Three Clinical Groups. *Arch. Gen. Psychiatry* **1966**, *14*, 634. [CrossRef] [PubMed]

74. Saul, L.; Sheppard, E.; Selby, D.; Lhamon, W.; Sachs, D.; Master, R. The Quantification of Hostility in Dreams with Reference to Essential Hypertension. *Science* **1954**, *119*, 382–383. [CrossRef]

75. Domhoff, G.W. *The Scientific Study of Dreams: Neural Networks, Cognitive Development, and Content Analysis*; American Psychological Association: Washington, DC, USA, 2003; ISBN 978-1-55798-935-2.

76. Samson, H.; De Koninck, J. Continuity or Compensation between Waking and Dreaming: An Exploration Using the Eysenck Personality Inventory. *Psychol. Rep.* **1986**, *58*, 871–874. [CrossRef]

77. American Psychiatric Association (APA). *Diagnostic and Statistical Manual of Mental Disorders: DSM-5*, 5th ed.; American Psychiatric Publishing: Washington, DC, USA, 2013; ISBN 978-0-89042-554-1.

78. Ellis, J.G.; De Koninck, J.; Bastien, C.H. Managing Insomnia Using Lucid Dreaming Training: A Pilot Study. *Behav. Sleep. Med.* **2021**, *19*, 273–283. [CrossRef] [PubMed]

79. St-Onge, M.; Mercier, P.; De Koninck, J. Imagery Rehearsal Therapy for Frequent Nightmares in Children. *Behav. Sleep. Med.* **2009**, *7*, 81–98. [CrossRef]

80. Krakow, B.; Hollifield, M.; Schrader, R.; Koss, M.; Tandberg, D.; Lauriello, J.; McBride, L.; Warner, T.D.; Cheng, D.; Edmond, T.; et al. A Controlled Study of Imagery Rehearsal for Chronic Nightmares in Sexual Assault Survivors with PTSD: A Preliminary Report. *J. Trauma. Stress* **2000**, *13*, 589–609. [CrossRef] [PubMed]

81. Nielsen, T.; Levin, R. Nightmares: A New Neurocognitive Model. *Sleep Med. Rev.* **2007**, *11*, 295–310. [CrossRef] [PubMed]

82. Nielsen, T.A.; Deslauriers, D.; Baylor, G.W. Emotions in Dream and Waking Event Reports. *Dreaming* **1991**, *1*, 287–300. [CrossRef]

83. Dang-Vu, T.T.; Schabus, M.; Desseilles, M.; Schwartz, S.; Maquet, P. Neuroimaging of REM Sleep and Dreaming. In *The New Science of Dreaming: Volume 1. Biological Aspects*; Praeger perspectives; Praeger Publishers/Greenwood Publishing Group: Westport, CT, USA, 2007; pp. 95–113, ISBN 978-0-275-99046-6.

84. Vandekerckhove, M.; Cluydts, R. The Emotional Brain and Sleep: An Intimate Relationship. *Sleep Med. Rev.* **2010**, *14*, 219–226. [CrossRef]

85. Punamäki, R.-L. The Relationship of Dream Content and Changes in Daytime Mood in Traumatized vs. Non-Traumatized Children. *Dreaming* **1999**, *9*, 213–233. [CrossRef]

86. Wright, J.; Koulack, D. Dreams and Contemporary Stress: A Disruption-Avoidance-Adaptation Model. *Sleep* **1987**, *10*, 172–179. [CrossRef] [PubMed]

87. Stewart, D.W.; Koulack, D. The Function of Dreams in Adaptation to Stress over Time. *Dreaming* **1993**, *3*, 259–268. [CrossRef]

88. Barbeau, K.; Lafrenière, A.; Ben Massaoud, H.; Campbell, E.; De Koninck, J. Dissociated Effects of Age and Recent Troubling Experiences on Nightmares, Threats and Negative Emotions in Dreams. *Front. Psychiatry* **2022**, *13*, 770380. [CrossRef]

89. Gilchrist, S.; Davidson, J.; Shakespeare-Finch, J. Dream Emotions, Waking Emotions, Personality Characteristics and Well-Being--A Positive Psychology Approach. *Dreaming* **2007**, *17*, 172–185. [CrossRef]

90. Schredl, M.; Reinhard, I. The Continuity between Waking Mood and Dream Emotions: Direct and Second-Order Effects. *Imagin. Cogn. Personal.* **2010**, *29*, 271–282. [CrossRef]

91. Mallett, R.; Picard-Deland, C.; Pigeon, W.; Wary, M.; Grewal, A.; Blagrove, M.; Carr, M. The Relationship Between Dreams and Subsequent Morning Mood Using Self-Reports and Text Analysis. *Affect. Sci.* **2022**, *3*, 400–405. [CrossRef] [PubMed]

92. Orbach, I.; Gilboa-Schechtman, E.; Sheffer, A.; Meged, S.; Har-Even, D.; Stein, D. Negative Bodily Self in Suicide Attempters. *Suicide Life. Threat. Behav.* **2006**, *36*, 136–153. [CrossRef] [PubMed]

93. Kaplan, H.B.; Pokorny, A.D. Self-Derogation and Suicide—I: Self-Derogation as an Antecedent of Suicidal Responses. *Soc. Sci. Med. 1967* **1976**, *10*, 113–118. [CrossRef]

94. Lewinsohn, P.M.; Rohde, P.; Seeley, J.R. Psychosocial Risk Factors for Future Adolescent Suicide Attempts. *J. Consult. Clin. Psychol.* **1994**, *62*, 297–305. [CrossRef] [PubMed]

95. Kuo, W.-H.; Gallo, J.J.; Eaton, W.W. Hopelessness, Depression, Substance Disorder, and Suicidality: A 13-Year Community-Based Study. *Soc. Psychiatry Psychiatr. Epidemiol.* **2004**, *39*, 497–501. [CrossRef] [PubMed]

96. Brown, G.K.; Beck, A.T.; Steer, R.A.; Grisham, J.R. Risk Factors for Suicide in Psychiatric Outpatients: A 20-Year Prospective Study. *J. Consult. Clin. Psychol.* **2000**, *68*, 371–377. [CrossRef] [PubMed]

97. Van Meter, A.R.; Paksarian, D.; Merikangas, K.R. Social Functioning and Suicide Risk in a Community Sample of Adolescents. *J. Clin. Child Adolesc. Psychol.* **2019**, *48*, 273–287. [CrossRef]

98. Hirsch, J.K.; Barton, A.L. Positive Social Support, Negative Social Exchanges, and Suicidal Behavior in College Students. *J. Am. Coll. Health* **2011**, *59*, 393–398. [CrossRef]

99. Brokke, S.S.; Landrø, N.I.; Haaland, V.Ø. Impulsivity and Aggression in Suicide Ideators and Suicide Attempters of High and Low Lethality. *BMC Psychiatry* **2022**, *22*, 753. [CrossRef] [PubMed]

100. Wang, L.; He, C.Z.; Yu, Y.M.; Qiu, X.H.; Yang, X.X.; Qiao, Z.X.; Sui, H.; Zhu, X.Z.; Yang, Y.J. Associations between Impulsivity, Aggression, and Suicide in Chinese College Students. *BMC Public Health* **2014**, *14*, 551. [CrossRef] [PubMed]

101. Vallat, R.; Chatard, B.; Blagrove, M.; Ruby, P. Correction: Characteristics of the Memory Sources of Dreams: A New Version of the Content-Matching Paradigm to Take Mundane and Remote Memories into Account. *PLoS ONE* **2018**, *13*, e0193440. [CrossRef] [PubMed]

102. Wu, J.; Wang, X.; Zheng, Y.; Shi, J.; Liu, J.; Yang, L.; Kang, C.; Yang, W.; Guan, T.; Liu, S.; et al. Sex Differences in the Prevalence and Clinical Correlates of Suicidal Ideation in Adolescent Patients with Depression in a Large Sample of Chinese. *Res. Sq.* **2022**. [CrossRef]

103. Avenevoli, S.; Swendsen, J.; He, J.-P.; Burstein, M.; Merikangas, K.R. Major Depression in the National Comorbidity Survey–Adolescent Supplement: Prevalence, Correlates, and Treatment. *J. Am. Acad. Child Adolesc. Psychiatry* **2015**, *54*, 37–44.e2. [CrossRef] [PubMed]

104. Lee, D.; Jung, S.; Park, S.; Hong, H.J. The Impact of Psychological Problems and Adverse Life Events on Suicidal Ideation among Adolescents Using Nationwide Data of a School-Based Mental Health Screening Test in Korea. *Eur. Child Adolesc. Psychiatry* **2018**, *27*, 1361–1372. [CrossRef] [PubMed]

105. Feng, X.; Wang, J. Presleep Ruminating on Intrusive Thoughts Increased the Possibility of Dreaming of Threatening Events. *Front. Psychol.* **2022**, *13*, 809131. [CrossRef] [PubMed]

106. Morrison, R.; O'Connor, R.C. A Systematic Review of the Relationship Between Rumination and Suicidality. *Suicide Life. Threat. Behav.* **2008**, *38*, 523–538. [CrossRef] [PubMed]

107. Arnulf, I.; Grosliere, L.; Le Corvec, T.; Golmard, J.-L.; Lascols, O.; Duguet, A. Will Students Pass a Competitive Exam That They Failed in Their Dreams? *Conscious. Cogn.* **2014**, *29*, 36–47. [CrossRef]

108. Schredl, M. Continuity between Waking and Dreaming: A Proposal for a Mathematical Model. *Sleep Hypn.* **2003**, *5*, 38–52.

109. Reisner, S.L.; Poteat, T.; Keatley, J.; Cabral, M.; Mothopeng, T.; Dunham, E.; Holland, C.E.; Max, R.; Baral, S.D. Global Health Burden and Needs of Transgender Populations: A Review. *Lancet* **2016**, *388*, 412–436. [CrossRef] [PubMed]

110. Johns, M.M.; Lowry, R.; Andrzejewski, J.; Barrios, L.C.; Demissie, Z.; McManus, T.; Rasberry, C.N.; Robin, L.; Underwood, J.M. Transgender Identity and Experiences of Violence Victimization, Substance Use, Suicide Risk, and Sexual Risk Behaviors Among High School Students—19 States and Large Urban School Districts, 2017. *MMWR Morb. Mortal. Wkly. Rep.* **2019**, *68*, 67–71. [CrossRef] [PubMed]

111. Liu, R.T.; Mustanski, B. Suicidal Ideation and Self-Harm in Lesbian, Gay, Bisexual, and Transgender Youth. *Am. J. Prev. Med.* **2012**, *42*, 221–228. [CrossRef] [PubMed]

112. Rosen, M.G. What I Make up When I Wake up: Anti-Experience Views and Narrative Fabrication of Dreams. *Front. Psychol.* **2013**, *4*, 514. [CrossRef] [PubMed]

113. Strauch, I.; Meier, B. *In Search of Dreams: Results of Experimental Dream Research*; SUNY Press: Albany, NY, USA, 1996; ISBN 978-0-7914-2759-0.

114. Murkar, A.L.A.; De Koninck, J. Consolidative Mechanisms of Emotional Processing in REM Sleep and PTSD. *Sleep Med. Rev.* **2018**, *41*, 173–184. [CrossRef]

Article

How COVID-19 Affected Sleep Talking Episodes, Sleep and Dreams?

Milena Camaioni [1], Serena Scarpelli [1], Valentina Alfonsi [1], Maurizio Gorgoni [1,2], Rossana Calzolari [1,3], Mina De Bartolo [1], Anastasia Mangiaruga [1], Alessandro Couyoumdjian [1] and Luigi De Gennaro [1,2,*]

[1] Department of Psychology, Sapienza University of Rome, Via dei Marsi 78, 00185 Rome, Italy; milena.camaioni@uniroma1.it (M.C.); serena.scarpelli@uniroma1.it (S.S.); valentina.alfonsi@uniroma1.it (V.A.); maurizio.gorgoni@uniroma1.it (M.G.); debartolo.1847715@studenti.uniroma1.it (M.D.B.); alessandro.couyoumdjian@uniroma1.it (A.C.)

[2] Body and Action Lab, IRCCS Fondazione Santa Lucia, Via Ardeatina 306, 00179 Rome, Italy

[3] Department of General Psychology, University of Padova, Via Venezia 8, 35131 Padova, Italy

* Correspondence: luigi.degennaro@uniroma1.it

Abstract: Background: The COVID-19 pandemic increased symptoms of stress and anxiety and induced changes in sleep quality, dream activity, and parasomnia episodes. It has been shown that stressful factors and/or bad sleep habits can affect parasomnia behaviors. However, investigations on how COVID-19 has affected sleep, dreams, and episode frequency in parasomnias are rare. The current study focuses on the impact of the pandemic on a specific parasomnia characterized by speech production (sleep talking, ST). Methods: We selected 27 participants with frequent ST episodes (STs) during the pandemic and compared them with 27 participants with frequent STs from a previous study conducted during a pre-pandemic period. All participants performed home monitoring through sleep logs and recorded their nocturnal STs for one week. Results: We observed a higher frequency of STs in the pandemic group. Moreover, STs were related to the emotional intensity of dreams, independent of the pandemic condition. The pandemic was associated with lower bizarreness of dreams in the pandemic group. There were no differences in sleep variables between the two groups. Conclusion: Overall, these results suggest a stressful effect of COVID-19 on the frequency of STs. Both the pandemic and the frequency of STs affect qualitative characteristics of dreams in this population.

Keywords: sleep talking; parasomnia; dreaming

Citation: Camaioni, M.; Scarpelli, S.; Alfonsi, V.; Gorgoni, M.; Calzolari, R.; De Bartolo, M.; Mangiaruga, A.; Couyoumdjian, A.; De Gennaro, L. How COVID-19 Affected Sleep Talking Episodes, Sleep and Dreams? *Brain Sci.* **2024**, *14*, 486. https://doi.org/10.3390/brainsci14050486

Academic Editor: Heide Klumpp

Received: 11 April 2024
Revised: 7 May 2024
Accepted: 9 May 2024
Published: 11 May 2024

1. Introduction

The strong countermeasures to control the COVID-19 pandemic have had a profound impact on sleep and mental health worldwide, as evidenced by numerous studies [1–4]. Specifically, empirical evidence in Italy showed poor sleep quality, increased time spent in bed, and delayed sleep schedules [3,5,6]. Moreover, some studies have reported high levels of stress, depression, and anxiety symptoms [7–10].

However, the effects of these restrictions are complex. In fact, research has shown a mixed picture, with some studies reporting both beneficial and destructive effects on sleep [5,6,11–14]. On the one hand, stress and environmental circumstances of staying at home, such as reduced exposure to natural light and physical activity and increased screen exposure [15], led to increased sleep difficulties. On the other hand, individuals were able to adapt their sleep patterns to their natural circadian tendencies, resulting in longer sleep and better daytime functioning [6].

Interestingly, longitudinal studies across the different COVID-19 waves have pointed to a general improvement in sleep disturbances, depression, and anxiety symptoms during the post-lockdown period [6,16,17]. Specifically, some studies have found a high percentage of poor sleepers (51%) in the general Italian population during the initial lockdown phase, followed by a partial reversal in the subsequent follow-up phases (44% poor sleepers) [6]. This dynamic pattern of sleep changes underscores the evolving nature of the pandemic's impact on sleep and mental health.

The COVID-19 pandemic not only affected sleep patterns but also led to significant changes in dream activity [18,19]. Studies comparing the pre-pandemic period with the lockdown phases of the first wave of the pandemic have revealed an increase in dream recall frequency in both healthy adults [18–22] and COVID-19 patients [23]. Additionally, dream contents were characterized by greater emotional intensity with more negative valence [20,24] compared to the pre-pandemic period. However, longitudinal studies have shown variations in dream activity and emotional features throughout the different waves of the pandemic [18], with some studies reporting a reduction in dream recall and emotional oneiric features during the second and third waves compared to the initial phase [25,26].

An additional factor of interest is the increase in nightmares observed during the pandemic [26,27]. A higher frequency of nightmares was reported during the first wave than in the pre-pandemic period, which decreased during the third wave [21,26–28]. This increase was associated with poor sleep and psychological symptoms (i.e., anxiety) [21,28]. In the parasomnia literature, it has been observed that stress or fragmented sleep can increase the frequency of nightmares or other altered nighttime sleep disorders [29].

Some studies have also shown the impact of COVID-19 on other parasomnias. In particular, the International COVID-19 Sleep Study (ICOSS) collaborative study showed a strikingly high prevalence of dream enactment behaviors (DEBs) in the general population during the pandemic [30]. Moreover, sleep phenomena such as sleep talking (ST), sleep maintenance problems, and symptoms of REM Behavior Disorder (RBD) were associated with high dream recall during the pandemic [30]. These findings are consistent with the arousal–retrieval model [31], supporting the notion that frequent arousals during sleep facilitate memory storage of dream content [19,32].

However, it remains unclear to what extent COVID-19 affects the frequency of parasomniac episodes and whether it generates a change in sleep and dreams. Studies have focused mainly on nightmares, with little investigation into other kinds of altered nighttime behaviors.

Our research aims to expand knowledge on this topic considering a specific parasomnia: sleep talking (ST), or somniloquy. ST is a common parasomnia in the general population, characterized by the production of linguistic vocalizations that can occur during all stages of sleep [33–35]. A recent study showed an association between verbal episodes and increased sleep fragmentation [35]. In addition, ST also appears to influence some dream characteristics, such as emotional intensity [35]. It has been also observed that sleep talking is often associated with other sleep disorders, psychiatric or medical conditions, and stress [36,37]. A recent study on young adults who self-identified as habitual sleep talkers showed a frequent co-occurrence of ST with other parasomnias, including nightmares (94%) and RBD (47%) [34]. According to this finding, we hypothesized that COVID may have also affected somniloquy in its isolated form.

Our study takes a unique approach by comparing two groups of sleep talkers recorded during the pre-pandemic and pandemic periods. This comparative analysis aims to identify differences in the frequency of the phenomenon. In addition, we investigated whether changes in oneiric characteristics could be explained by variables related to the pandemic or the parasomniac phenomenon, providing a comprehensive understanding of the topic.

2. Materials and Methods

2.1. Participants

We selected two groups of participants with frequent ST episodes (verbal activations—STs), paired by gender and age: the pandemic group (PN) (N $n = 27$; F = 21, mean age: 23.63, sd: ±2.83) and the pre-pandemic group (pre-PN) ($n = 27$; F = 21, mean age: 23.74, sd: ±2.85).

The PN was selected from a study conducted during the pandemic from January 2021 to October 2021 [35]. Meanwhile, the pre-PN was identified in a previous study performed during a pre-pandemic period, specifically from 2016 to 2018 [34].

Both groups were recruited using an identical procedure consisting of two steps (Figure 1): an online survey and home monitoring. An online survey was administered to the general population via digital platforms (e.g., Facebook, WhatsApp, Instagram). Subjects who reported frequent sleep talk were selected on the basis of this survey and were contacted to participate in home monitoring to verify the actual presence of nighttime speech. The two selection steps are discussed in detail in the following sections.

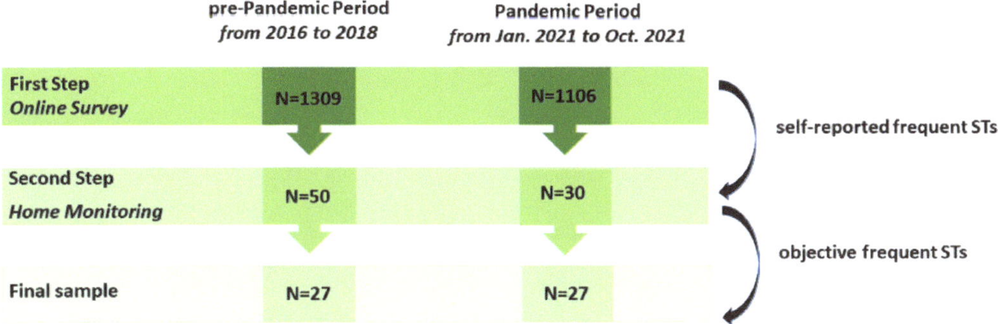

Figure 1. The recruitment process of the two groups.

2.1.1. Online Survey

The online survey consisted of an ad hoc questionnaire of case history to assess general health, the Pittsburgh Sleep Quality Index (PSQI) [38], and the Munich Parasomnia Questionnaire (MUPS) [39].

The PSQI consists of 19 self-report items that measure subjective sleep quality. The questionnaire produces a global score ranging from 0 to 21. A global score greater than 5 indicates poor sleep quality [38].

The MUPS is a 22-item self-report questionnaire assessing the presence and frequency of altered nocturnal behaviors on a Likert scale from 1 ("Never") to 7 ("Very frequently—every or nearly every night") [39]. We considered very frequent behaviors to be those that received a score greater than 5.

For the first step, the inclusion criteria were:

1. Participants who self-reported frequent STs, as indicated by the MUPS item related to STs ("How often do you experience the following behaviors? talking during sleep") ranging from 5 ("Sometimes—one or more times per month) to 7 ("Very frequently—every or nearly every night");
2. Age range: 18–35 years;
3. Absence of medical conditions and psychiatric disorders (this was investigated with a specific question in the questionnaire on case history: "Do you have acute or chronic health problems (medical or psychological problems? (If yes, please specify in 'other')");
4. Absence of other sleep disorders except ST;
5. No drug or alcohol abuse.

During the pre-pandemic period, 1309 subjects responded to the online survey. From this sample, $n = 50$ met the inclusion criteria of the first phase and were recruited for home monitoring. Similarly, of the 1106 questionnaires collected during the pandemic period, $n = 30$ subjects were eligible for home monitoring (Figure 1).

2.1.2. Home Monitoring

In both groups, the participants selected in the first step performed 7 days of home monitoring (see below in the procedure section).

Although the selected participants had subjectively reported a high ST frequency on the MUPS item, we wanted to ensure that they were objectively frequent sleep talkers. Therefore, only subjects who produced at least one ST episode during home monitoring were included in the final sample.

Thus, the sample selected for analysis consisted of $n = 27$ STs in the pre-PN and $n = 27$ STs in the PN.

Participants were informed of the procedure and signed a written informed consent. The protocol complied with the Declaration of Helsinki and was approved by the Institutional Review Board of the Department of Psychology of Sapienza University of Rome (protocol number of 04/02/2020: 0000226).

2.2. Procedure

The home monitoring procedure was the same for both groups. Participants used an open-source voice-activated recording app installed on their smartphones (Dream Talk Recorder or Voice Activated Recorder, based on iOS or Android operating system, respectively) during sleep to record ST episodes. The application activates when it detects environmental sounds above a certain threshold (e.g., sleep talking). It records for the duration of the sound and deactivates when the sound stops. A sensitivity threshold of "medium" was set by the participants. Moreover, we instructed participants to turn on the app when they went to bed and to place the recorder near the bed every night for one week. After their last awakening in the morning, participants turned off the app. Moreover, we trained participants to audio-record each recalled mental activity as accurately as possible, distinguishing when they recalled more than one dream report, and fill out sleep diaries [40] within a maximum of 15 min after awakening. Audio recordings of dream content ensure high compliance and more accurate reports [41]. We checked to see if there were other roommates or bed partners (this was investigated with a specific question in the questionnaire on case history: "Do you sleep alone or share your bed/room with another person? (a) By yourself; (b) With another person"). This revealed that 14 sleep talkers in the pre-pandemic group and 11 in the pandemic group shared a bed or room. To make sure that the recordings were carried out by the subjects themselves, we asked them to hold the phone close to the pillow on their side. This made it possible to distinguish any sounds made by other people due to proximity to the phone.

2.3. Measures

2.3.1. Sleep Diaries

The subjective assessment of night's sleep was obtained by sleep diaries (see [35]). The following variables were extracted:

- Sleep onset latency (SOL): the time (in minutes) it takes to fall asleep after turning off the light;
- Number of nocturnal awakenings (NOA);
- Intra-night wakefulness (ISW): subjective duration of wakefulness (in minutes) from falling asleep to the final awakening;
- Total sleep time in minutes (TST): the amount of time spent asleep;
- Total bed time (TBT): the amount of time from lights off to the final awakening;
- Sleep efficiency (SE = TST/TBT × 100).

Moreover, the sleep diaries included three items, rated on a five-point Likert scale, representing a direct judgment of sleep quality from the subjects:

- Sleep depth (1 = very light sleep, 5 = very deep sleep)
- Sleep quiet (1 = very disturbed sleep, 5 = very quiet sleep)
- Sleep restless (1 = very low-rest sleep, 5 = very high-rest sleep)

2.3.2. Dream Reports

Two independent judges (R.C. and M.B.) evaluated the dream reports of each subject. The judges transcribed and counted the number of dream recalls (DRs). They then performed the "pruning" (removal of all repetitions and subjects' inferences) to compute the total number of words (total word count—TWC). Then, they rated three qualitative variables of dreams on a Likert scale from one to six points [42]:

- Emotional intensity—EL: 1 = very low emotional intensity; 6 = very high emotional intensity;
- Vividness—VV: 1 = no image, just thinking about objects; 2 = very vague; 3 = less vague; 4 = moderately clear and vivid; 5 = clear and reasonably vivid; 6 = clear and vivid as normal vision;
- Bizarreness—B: 1 = scenes/thoughts closely related to everyday life, something that belongs to the individual's life or to current reality, is plausible; 2 = belongs to reality but not to the life of the individual, seems unusual, strange, or illogical; 3 = plot discontinuous, changes in setting, slightly inappropriate roles; 4 = plot highly discontinuous, unlikely elements and particular settings appear, inappropriate roles; 5 = abundance of unlikely elements, settings with imaginary elements, metamorphoses, imaginary characters; 6 = impossible settings, fantastic elements and characters (with unlikely characteristics), illogicality of the plot.

The two judges resolved any discordance by consensus, and the inter-rater reliability of each scale was substantial (Cohen's K > 70).

2.3.3. Sleep Talking Episodes

Two experimenters (R.C. and M.B.) assessed, transcribed, and counted ST episodes for each subject by listening to the audio records.

2.4. Statistical Analyses

The statistical procedures were carried out using the Statistical Package for Social Sciences (SPSS) version 25.0.

In order to verify the absence of differences between the two groups in MUPS, the variables were computed with a Chi-square test for independence and Fisher's exact test (FET) when appropriate by grouping the Likert scale scores into three categories: (I) absence/very low frequency (1–2), (II) low frequency (3–4), and (III) high frequency (5–7).

The PSQI global score was compared using Student's *t*-test for independent samples.

The normal distribution of the original data was checked and skewed data distributions (i.e., SOL, NOA, ISW, DR, TWC, EL, and STs) were transformed into normal distribution using $\log_{10}$ transformation.

The number of STs, the variables of sleep logs, and dream variables collected each day during weekly home monitoring were averaged for each participant.

To assess the effect of COVID-19 on ST episodes, we performed a one-way ANOVA with GROUP (PN, pre-PN) as the between factor and STs as dependent variable.

We also performed a one-way multivariate analysis of variance (MANOVA), with GROUP (PN, pre-PN) as a between factor and sleep measures and DR as dependent variables. The dependent variables considered were SOL, NOA, ISW, TST, TBT, SE, sleepiness, depth, quiet, restlessness, and DR. The MANOVA agreed to observe differences in sleep variables and DR frequency between the two groups.

Finally, we investigated whether the pandemic scenario, ST, or sleep fragmentation (ISW) could predict dream variables. Therefore, we performed multiple linear regressions considering GROUP, STs, and ISW as independent variables to assess the best explanatory variables of the dream measures (TWC, EL, VV, B). We checked for multicollinearity between the predictors by calculating variance inflation factors (VIF < 3) before performing the regression.

3. Results

3.1. Preliminary Questionnaire

Chi-square results show that PN and pre-PN did not differ in terms of the MUPS variables (Table 1). Similarly, the PSQI global score was not significantly different between the two groups, as shown in Table 1.

Table 1. Results of comparisons of measures derived from the Munich Parasomnia Questionnaire (MUPS) and the Pittsburgh Sleep Quality Index (PSQI).

	Preliminary Questionnaires							
MUPS	**Pandemic (%)**			**Pre-Pandemic (%)**			**Fisher's Exact**	*p*
	I	**II**	**III**	**I**	**II**	**III**		
Hypnic jerks	11.10	33.30	55.60	11.10	22.20	66.70	0.95	0.73
Rhythmic foot movements	59.30	18.50	22.20	40.70	14.80	44.40	3.02	0.28
Rhythmic movement disorders	92.60	7.40	0.00	81.50	11.10	7.40	2.08	0.50
Exploding head syndrome	74.10	14.80	11.10	85.20	7.40	7.40	1.13	0.62
Hypnagogic hallucinations	63.00	25.90	11.10	77.80	14.80	7.40	1.48	0.56
Periodic leg movements	81.50	11.10	7.40	70.40	18.50	11.10	0.99	0.65
Nocturnal leg cramps	55.60	37.00	7.40	59.30	29.60	11.10	0.54	0.85
Sleep-related bruxism	63.00	7.40	29.60	66.70	3.70	29.60	0.48	1.00
Abnormal swallowing	81.50	7.40	11.10	85.20	11.10	3.70	1.22	0.75
Groaning	48.10	14.80	37.00	55.60	18.50	25.90	0.83	0.76
Sleep enuresis	96.30	3.70	0.00	96.30	3.70	0.00		1.00
Nightmares	3.70	37.00	59.30	3.70	48.10	48,10	0.95	0.79
Sleep terrors	70.40	14.80	14.80	51.90	29.60	18.50	2.19	0.35
Nocturnal eating	100.00	0.00	0.00	88.90	7.40	3.70	2.74	0.24
Sleep-related eating	92.60	7.40	0.00	96.30	3.70	0.00		1.00
Confusing arousals	59.30	25.90	14.80	59.30	22.20	18.50	0.27	1.00
Sleep paralysis	85.20	7.40	7.40	88.90	7.40	3.70	0.56	1.00
Sleepwalking	88.90	11.10	0.00	77.80	14.80	7.40	2.02	0.39
Violent behavior	77.80	18.50	3.70	85.20	11.10	3.70	0.84	0.85
RBD	55.60	33.30	11.10	51.90	33.30	14.80	0.27	1.00
Others	88.90	7.40	3.70	66.70	18.50	14.80	3.71	0.18
PSQI **Global Score**	**Pandemic** 5.85 (0.55)			**Pre-Pandemic** 6.04 (0.52)			*t* −0.25	*p* 0.80

MUPS, Munich Parasomnia Questionnaire; RBD, REM Behavior Disorder; PSQI, Pittsburgh Sleep Quality Index.

3.2. Home Monitoring

Table 2 reports the descriptive statistics of sleep measures, dreams variables, and STs. About 54 participants, 48 sleep talkers (pandemic group, n = 26 (96.30%); pre-pandemic group, n = 22 (82.48%)) recalled at least one dream report.

Table 2. Sleep and dream measures in the two groups.

Variables		Pandemic Mean (sd)	Pre-Pandemic Mean (sd)
STs *		1.417 (1.06)	0.996 (1.51)
Sleep variables	SOL *	14.087 (10.22)	10.102 (4.09)
	NOA *	1.218 (0.79)	1.593 (1.37)
	ISW *	4.320 (3.23)	6.709 (7.74)
	TBT	493.330 (49.14)	503.367 (76.60)
	TST	441.661 (36.90)	457.102 (64.07)
	SE (%)	90.040 (4.38)	90.930 (5.19)
	Sleep depth	3.787 (0.58)	3.73 (0.58)
	Sleep quiet	3.371 (0.63)	3.312 (0.72)
	Sleep restless	3.453 (0.59)	3.279 (0.56)

Table 2. *Cont.*

Variables		Pandemic Mean (sd)	Pre-Pandemic Mean (sd)
	DR frequency *	0.701 (0.52)	0.599 (0.69)
	TWC *	73.059 (64.14)	124.819 (88.89)
Dream variables	EL *	2.429(1.12)	2.767 (0.88)
	B	1.903 (0.92)	2.750 (0.86)
	VV	2.972 (0.82)	3.582 (0.90)

* Values are reported as raw data. STs, sleep talking episodes; SOL, sleep onset latency; NOA, number of nocturnal awakenings; ISW, intra-sleep wakefulness; TBT, total bed time; TST, total sleep time; SE, sleep efficiency; DR, dream recall; TWC, total word count; EL, emotional intensity; B, bizarreness; VV, vividness.

PN showed a significantly higher weekly ST frequency (mean [SE] PN = 0.17 [±0.7] pre-PN = −0.27 [±0.9]) than the pre-PN group (F = 6.156; p = 0.016) (Figure 2).

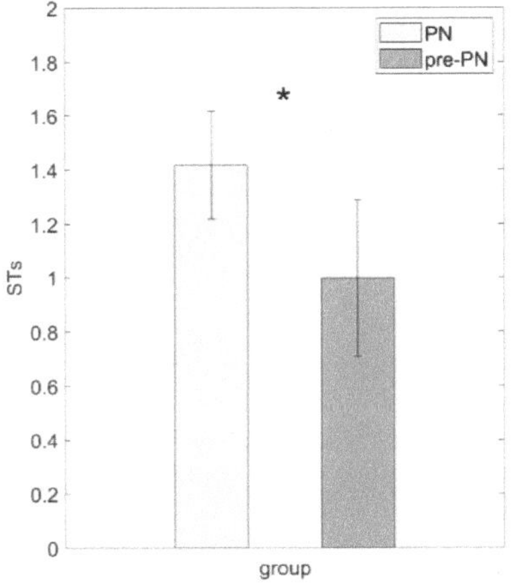

Figure 2. Results of the comparison between the pandemic group and pre-pandemic group. * indicates a statistical significant difference. Values are reported as raw frequencies.

The MANOVA performed on sleep log measures and dream frequency showed no significant differences between PN and pre-PN (Wilks' λ = 0.786, $F_{10,43}$ = 1.172, p = 0.335; η_p^2 = 0.214).

As shown in Table 3, the regression models were statistically significant for the EL and B dream variables (EL: adjusted R^2 = 0.129, p < 0.05; B: adjusted R^2 = 0.163, p < 0.05). Specifically, partial correlations show that higher frequency of STs (β = −0.393, t = −2.626, p = 0.012) predicted lower EL in dream reports, and the pre-pandemic group (β = −0.395, , p = 0.011) presented higher B in dream content.

Table 3. The multiple linear regression models.

Dependent Variables	Predictors	Standardized β Coefficients	t	p
TWC				
Adjusted R^2 = 0.087	Group	−0.289	−1.869	0.068
F = 2.488	STs	−0.166	−1.082	0.285
p = 0.073	ISW	0.019	0.128	0.899
EL				
Adjusted R^2 = 0.129	Group	−0.090	−0.599	0.553
F = 3.327	STs	−0.393	−2.626	0.012 *
p = 0.028	ISW	0.030	0.208	0.836

Table 3. *Cont.*

Dependent Variables	Predictors	Standardized β Coefficients	t	p
B				
Adjusted R^2 = 0.163	Group	−0.395	−2.673	0.011 *
F = 4.057	STs	−0.018	−0.124	0.902
p = 0.012	ISW	0.171	1.217	0.230
VV				
Adjusted R^2 = 0.056	Group	−0.336	−2.138	0.038
F = 1.928	STs	−0.006	−0.037	0.971
p = 0.139	ISW	0.015	0.097	0.923

STs, sleep talking episodes; ISW, intra-sleep wakefulness; TWC, total word count; EL, emotional intensity; B, bizarreness; VV, vividness. * indicates statistical significance

4. Discussion

Our study aimed to investigate the phenomenon of ST during the pandemic. The results revealed a significant increase in the occurrence of STs in a group of sleep talkers monitored during the pandemic compared to a group of sleep talkers monitored in a pre-pandemic period.

Consistent with our findings, studies on parasomnias during the pandemic have shown an increase in the number of parasomnia episodes. In particular, the general population showed higher episodes of nightmares, especially in women and people with sleep disturbances, symptoms of anxiety or depression, and stress associated with COVID-19 [21,22,24,28,43]. Similarly, the International COVID-19 Sleep Study observed a higher prevalence of dream-enacting behavior in the general population than previous epidemiologic studies [30]. DEB has been associated with some factors, such as lifestyle, post-traumatic stress disorder (PTSD), depression, and anxiety symptoms [30].

Previous studies have shown that some stressful factors and poor sleep hygiene could facilitate the onset of parasomnia episodes [29]. For instance, nightmares are often associated with PTSD [44–48] and are reactive to stressful life events [49–52]. Moreover, several studies have found that sleep paralysis may be facilitated by stress, PTSD, and irregular sleep–wake schedules [53–56]. Finally, episodes of NREM parasomnias appear to be triggered by sleep-disrupting factors, and many clinical observations support the hypothesis that stress and anxiety are related to the frequency of episodes, although evidence in the literature is sparse [57].

Along this vein, symptoms of anxiety, depression, and impaired sleep quality during the COVID-19 pandemic [3,9,58,59] may be associated with high levels of distress, which may facilitate parasomnia episodes [60], including ST. It has been observed that ST is associated with more disturbed sleep and altered macrostructure than healthy subjects [34,35]. It is hypothesized that there is a sensitivity in the sleep talker population when important changes occur that may alter sleep patterns and affect stress and mood, promoting a greater susceptibility to the onset of parasomnia.

The current study did not find worse sleep during the pandemic period. It must be considered that the pandemic group was recorded for an extended period after the second wave, during which the pandemic restrictions underwent several changes. Sleep quality in the pandemic group may have been improved compared to the first wave and partially restored, more closely resembling the pre-pandemic period [6,16,17].

Regarding dream activity, the regression model indicates that a higher frequency of ST episodes predicted lower emotional intensity. This result suggests that dreams of sleep talkers may be characterized by lower emotional intensity, independent of the pandemic state. As suggested by a previous study, low EL may reflect emotional regulation processes [35]. Dream-related emotional regulation attenuates the emotional intensity of waking activity [61,62]. Moreover, ST has been proposed as a dream-enactment behavior that may provide direct access to cognitive processes during sleep [33–35], and STs are often characterized by emotional content [33,63]. Therefore, STs may reflect the mechanism of dreams in the elaboration of emotional experiences.

An alternative explanation is that the low emotional intensity may represent an intrinsic feature of sleep talkers' dreams. Different parasomnias may manifest qualitative differences in dream activity [64–66]. Studies analyzing the qualitative characteristics of sleep mentation in RBD and disorders of arousals (DOAs) report that dream content associated with DEB episodes, collected in the morning after overnight laboratory monitoring, was less frequently recalled in RBD and DOAs [66]. Moreover, sleepwalkers recalled more vivid and complex plots than RBD patients, whereas bizarreness did not differ between the two groups [66].

In contrast, the finding on bizarreness is not consistent with the literature on dream activity during COVID-19. To date, the qualitative characteristics of dreams in ST are poorly investigated. Dream features of parasomnias could be due to the sleep stage from which the dream experience is collected (e.g., REM sleep in RBD, NREM sleep in somnambulism), as dream reports differ between REM-associated (dreamlike: more emotional, vivid, and bizarre mentation) and NREM-associated (thought-like: less emotional intensity and more realistic content) [67–70]. ST can occur during all sleep phases, although more frequently during NREM sleep periods [34,71,72], especially during N2 [34]. Moreover, the alteration in macrostructure in ST showed a lower amount of REM sleep and a higher amount of NREM sleep [34]. From this perspective, we speculatively suggest that a high frequency of ST could influence sleep architecture, which may have resulted in greater dream recall from NREM and lower bizarreness in the pandemic group. On the other hand, we should consider the possibility that the two groups may have basic and uncontrolled differences that explain this result.

Limitations

One of the most critical limitations of the study is its cross-sectional rather than longitudinal design. We cannot rule out the possibility that subjects in the two groups had prior differences in the frequency of episodes. The pandemic group was also recorded over a long period characterized by different phases of the pandemic trend, which may have influenced the results. In addition, the online survey did not investigate the presence or absence of previous COVID-19 infection.

Moreover, we can only indirectly hypothesize that stressful and emotional factors related to COVID-19 affected the production of STs because direct measures of daytime emotions were not collected.

Additional limitations include the small sample size, which does not allow generalization to the population of sleep talkers, and the lack of objective sleep measures. Instrumental measures (i.e., polysomnography) would have allowed for analysis of micro- and macrostructural sleep and EEG features of dreaming.

Finally, it should be considered that one of the advantages of the study is that sleep talkers were recorded in an ecological a organic environment. However, it is important to consider the intrinsic limitations of the voice-activated application. It may not have been able to capture all ST episodes, and it may not have been able to distinguish the sound produced by other individuals sharing the bed or room with the experimental subjects.

5. Conclusions

To the best of our knowledge, the present study was the first to investigate the phenomenon of sleep talking during the pandemic compared to a pre-pandemic period.

The results show an increase in ST episodes during the pandemic, which may have been due to the negative impact of COVID-19 on sleep and mood. We hypothesize a sensitivity of sleep talkers to high levels of stress or changes in mood tone, which would facilitate the occurrence of ST.

Furthermore, both the pandemic scenario and the frequency of STs affected the dream contents of sleep talkers.

Overall, understanding which factors may determine the frequency of parasomniac episodes may shed light on the etiology of the phenomenon. In addition, it would be possible to define treatment protocols when the phenomenon is disruptive.

Author Contributions: Conceptualization, M.C., S.S., M.G. and L.D.G.; data curation, M.C., S.S., V.A., M.G., R.C., M.D.B. and L.D.G.; formal analysis, M.C., S.S., V.A. and M.G.; funding acquisition, M.C. and L.D.G.; investigation, M.C., R.C., M.D.B. and A.M.; methodology, M.C., S.S., V.A., M.G., A.C. and L.D.G.; project administration, L.D.G.; resources, M.G., A.C. and L.D.G.; software, S.S., V.A. and A.C.; supervision, L.D.G.; validation, M.G. and L.D.G.; visualization, M.C. and S.S.; writing—original draft, M.C. and S.S.; writing—review and editing, M.C., S.S., V.A., M.G., A.C. and L.D.G.. All authors have read and agreed to the published version of the manuscript.

Funding: This research was funded by Sapienza University of Rome, "Avvio alla Ricerca 2020" (grant number: AR120172941614FC), to M.C.

Institutional Review Board Statement: The Institutional Review Board of the Department of Psychology of Sapienza University of Rome approved the present study (protocol number of 04 February 2020: 0000226). The study complied with the Declaration of Helsinki.

Informed Consent Statement: Informed consent was obtained from all subjects involved in the study.

Data Availability Statement: The data presented in this study are available on request to the corresponding author. The data are not publicly available due to legal and ethical reasons.

Acknowledgments: Thanks to Camilla Cacciapuoti, Dalila Francesca, and Virginia Casagrande for help with data collection.

Conflicts of Interest: The authors declare no conflict of interest.

References

1. Jahrami, H.; BaHammam, A.S.; Bragazzi, N.L.; Saif, Z.; Faris, M.; Vitiello, M.V. Sleep Problems during the COVID-19 Pandemic by Population: A Systematic Review and Meta-Analysis. *J. Clin. Sleep Med.* **2021**, *17*, 299–313. [CrossRef] [PubMed]
2. Rajkumar, R.P. COVID-19 and Mental Health: A Review of the Existing Literature. *Asian J. Psychiatr.* **2020**, *52*, 102066. [CrossRef] [PubMed]
3. Salfi, F.; Amicucci, G.; Cascioli, J.; Corigliano, D.; Viselli, L.; Tempesta, D.; Ferrara, M. The Impact of Home Confinement Due to COVID-19 Pandemic on Sleep Quality and Insomnia Symptoms among the Italian Population. 2020. Available online: https://www.researchgate.net/profile/Federico-Salfi/publication/344337922_The_impact_of_home_confinement_due_to_COVID-19_pandemic_on_sleep_quality_and_insomnia_symptoms_among_the_Italian_population/links/5f69c99392851c14bc8e03c4/The-impact-of-home-confinement-due-to-COVID-19-pandemic-on-sleep-quality-and-insomnia-symptoms-among-the-Italian-population.pdf (accessed on 15 September 2020).
4. Cao, W.; Fang, Z.; Hou, G.; Han, M.; Xu, X.; Dong, J.; Zheng, J. The Psychological Impact of the COVID-19 Epidemic on College Students in China. *Psychiatry Res.* **2020**, *287*, 112934. [CrossRef] [PubMed]
5. Casagrande, M.; Favieri, F.; Tambelli, R.; Forte, G. The Enemy Who Sealed the World: Effects Quarantine Due to the COVID-19 on Sleep Quality, Anxiety, and Psychological Distress in the Italian Population. *Sleep Med.* **2020**, *75*, 12–20. [CrossRef]
6. Alfonsi, V.; Gorgoni, M.; Scarpelli, S.; Zivi, P.; Sdoia, S.; Mari, E.; Fraschetti, A.; Ferlazzo, F.; Giannini, A.M.; De Gennaro, L. COVID-19 Lockdown and Poor Sleep Quality: Not the Whole Story. *J. Sleep Res.* **2021**, *30*, e13368. [CrossRef] [PubMed]
7. Mazza, C.; Ricci, E.; Colasanti, M.; Ferracuti, S.; Napoli, C.; Roma, P. A Nationwide Survey of Psychological Distress among Italian People during the COVID-19 Pandemic: Immediate Psychological Responses and Associated Factors. *Int. J. Environ. Res. Public Health* **2020**, *17*, 3165. [CrossRef]
8. Moccia, L.; Janiri, D.; Pepe, M.; Dattoli, L.; Molinaro, M.; De Martin, V.; Chieffo, D.; Janiri, L.; Fiorillo, A.; Sani, G.; et al. Affective Temperament, Attachment Style, and the Psychological Impact of the COVID-19 Outbreak: An Early Report on the Italian General Population. *Brain. Behav. Immun.* **2020**, *87*, 75–79. [CrossRef]
9. Salfi, F.; Lauriola, M.; Amicucci, G.; Corigliano, D.; Viselli, L.; Tempesta, D.; Ferrara, M. Gender-Related Time Course of Sleep Disturbances and Psychological Symptoms during the COVID-19 Lockdown: A Longitudinal Study on the Italian Population. *Neurobiol. Stress* **2020**, *13*, 100259. [CrossRef]
10. Ballesio, A.; Zagaria, A.; Musetti, A.; Lenzo, V.; Palagini, L.; Quattropani, M.C.; Vegni, E.; Bonazza, F.; Filosa, M.; Manari, T.; et al. Longitudinal Associations between Stress and Sleep Disturbances during COVID-19. *Stress Health* **2022**, *38*, 919–926. [CrossRef]
11. Holzinger, B.; Mayer, L.; Nierwetberg, F.; Klösch, G. COVID-19 Lockdown—Are Austrians Finally Able to Compensate Their Sleep Debt? *Sleep Med. X* **2021**, *3*, 100032. [CrossRef]
12. Cellini, N.; Canale, N.; Mioni, G.; Costa, S. Changes in Sleep Pattern, Sense of Time and Digital Media Use during COVID-19 Lockdown in Italy. *J. Sleep Res.* **2020**, *29*, e13074. [CrossRef] [PubMed]

13. Kocevska, D.; Blanken, T.F.; Van Someren, E.J.W.; Rösler, L. Sleep Quality during the COVID-19 Pandemic: Not One Size Fits All. *Sleep Med.* **2020**, *76*, 86–88. [CrossRef] [PubMed]

14. Wright, K.P.; Linton, S.K.; Withrow, D.; Casiraghi, L.; Lanza, S.M.; de la Iglesia, H.; Vetter, C.; Depner, C.M. Sleep in University Students Prior to and during COVID-19 Stay-at-Home Orders. *Curr. Biol.* **2020**, *30*, R797–R798. [CrossRef] [PubMed]

15. Salfi, F.; Amicucci, G.; Corigliano, D.; D'Atri, A.; Viselli, L.; Tempesta, D.; Ferrara, M. Changes of Evening Exposure to Electronic Devices during the COVID-19 Lockdown Affect the Time Course of Sleep Disturbances. *Sleep* **2021**, *44*, zsab080. [CrossRef]

16. Salfi, F.; D'Atri, A.; Tempesta, D.; Ferrara, M. Sleeping under the Waves: A Longitudinal Study across the Contagion Peaks of the COVID-19 Pandemic in Italy. *J. Sleep Res.* **2021**, *30*, e13313. [CrossRef] [PubMed]

17. Massar, S.A.A.; Ng, A.S.C.; Soon, C.S.; Ong, J.L.; Chua, X.Y.; Chee, N.I.Y.N.; Lee, T.S.; Chee, M.W.L. Reopening after Lockdown: The Influence of Working-from-Home and Digital Device Use on Sleep, Physical Activity, and Wellbeing Following COVID-19 Lockdown and Reopening. *Sleep* **2022**, *45*, zsab250. [CrossRef] [PubMed]

18. Gorgoni, M.; Scarpelli, S.; Alfonsi, V.; De Gennaro, L. Dreaming during the COVID-19 Pandemic: A Narrative Review. *Neurosci. Biobehav. Rev.* **2022**, *138*, 104710. [CrossRef] [PubMed]

19. Fränkl, E.; Scarpelli, S.; Nadorff, M.R.; Bjorvatn, B.; Bolstad, C.J.; Chan, N.Y.; Chung, F.; Dauvilliers, Y.; Espie, C.A.; Inoue, Y.; et al. How Our Dreams Changed during the COVID-19 Pandemic: Effects and Correlates of Dream Recall Frequency—A Multinational Study on 19,355 Adults. *Nat. Sci. Sleep* **2021**, *13*, 1573–1591. [CrossRef] [PubMed]

20. Schredl, M.; Bulkeley, K. Dreaming and the COVID-19 Pandemic: A Survey in a U.S. Sample. *Dreaming* **2020**, *30*, 189–198. [CrossRef]

21. Scarpelli, S.; Alfonsi, V.; Mangiaruga, A.; Musetti, A.; Quattropani, M.C.; Lenzo, V.; Freda, M.F.; Lemmo, D.; Vegni, E.; Borghi, L.; et al. Pandemic Nightmares: Effects on Dream Activity of the COVID-19 Lockdown in Italy. *J. Sleep Res.* **2021**, *30*, e13300. [CrossRef]

22. Solomonova, E.; Picard-Deland, C.; Rapoport, I.L.; Pennestri, M.H.; Saad, M.; Kendzerska, T.; Veissiere, S.P.L.; Godbout, R.; Edwards, J.D.; Quilty, L.; et al. Stuck in a Lockdown: Dreams, Bad Dreams, Nightmares, and Their Relationship to Stress, Depression and Anxiety during the COVID-19 Pandemic. *PLoS ONE* **2021**, *16*, e0259040. [CrossRef]

23. Scarpelli, S.; Nadorff, M.R.; Bjorvatn, B.; Chung, F.; Dauvilliers, Y.; Espie, C.A.; Inoue, Y.; Matsui, K.; Merikanto, I.; Morin, C.M.; et al. Nightmares in People with COVID-19: Did Coronavirus Infect Our Dreams? *Nat. Sci. Sleep* **2022**, *14*, 93–108. [CrossRef]

24. Gorgoni, M.; Scarpelli, S.; Alfonsi, V.; Annarumma, L.; Cordone, S.; Stravolo, S.; De Gennaro, L. Pandemic Dreams: Quantitative and Qualitative Features of the Oneiric Activity during the Lockdown Due to COVID-19 in Italy. *Sleep Med.* **2021**, *81*, 20–32. [CrossRef]

25. Scarpelli, S.; Alfonsi, V.; Gorgoni, M.; Musetti, A.; Filosa, M.; Quattropani, M.C.; Lenzo, V.; Vegni, E.; Borghi, L.; Margherita, G.; et al. Dreams and Nightmares during the First and Second Wave of the COVID-19 Infection: A Longitudinal Study. *Brain Sci.* **2021**, *11*, 1375. [CrossRef]

26. Scarpelli, S.; Alfonsi, V.; Camaioni, M.; Gorgoni, M.; Albano, A.; Musetti, A.; Quattropani, M.C.; Plazzi, G.; De Gennaro, L.; Franceschini, C. Longitudinal Findings on the Oneiric Activity Changes Across the Pandemic. *Nat. Sci. Sleep* **2023**, *15*, 435–447. [CrossRef]

27. Remedios, A.; Marin-Dragu, S.; Routledge, F.; Hamm, S.; Iyer, R.S.; Orr, M.; Meier, S.; Michael, S. Nightmare Frequency and Nightmare Distress during the COVID-19 Pandemic. *J. Clin. Sleep Med.* **2023**, *19*, 163–169. [CrossRef]

28. Pesonen, A.K.; Lipsanen, J.; Halonen, R.; Elovainio, M.; Sandman, N.; Mäkelä, J.M.; Antila, M.; Béchard, D.; Ollila, H.M.; Kuula, L. Pandemic Dreams: Network Analysis of Dream Content During the COVID-19 Lockdown. *Front. Psychol.* **2020**, *11*, 573961. [CrossRef]

29. Montplaisir, J.; Zadra, A.; Nielsen, T.; Petit, D. Basic Science, Technical Considerations and Clinical Aspects. In *Sleep Disorders Medicine*, 4th ed.; Springer: Berlin/Heidelberg, Germany, 2017; pp. 1–1269. [CrossRef]

30. Liu, Y.; Partinen, E.; Chan, N.Y.; Dauvilliers, Y.; Inoue, Y.; De Gennaro, L.; Plazzi, G.; Bolstad, C.J.; Nadorff, M.R.; Merikanto, I.; et al. Dream-Enactment Behaviours during the COVID-19 Pandemic: An International COVID-19 Sleep Study. *J. Sleep Res.* **2023**, *32*, e13613. [CrossRef]

31. Koulack, D.; Goodenough, D.R. Dream Recall and Dream Recall Failure: An Arousal-Retrieval Model. *Psychol. Bull.* **1976**, *83*, 975–984. [CrossRef]

32. van Wyk, M.; Solms, M.; Lipinska, G. Increased Awakenings from Non-Rapid Eye Movement Sleep Explain Differences in Dream Recall Frequency in Healthy Individuals. *Front. Hum. Neurosci.* **2019**, *13*, 370. [CrossRef]

33. Alfonsi, V.; D'Atri, A.; Scarpelli, S.; Mangiaruga, A.; De Gennaro, L. Sleep Talking: A Viable Access to Mental Processes during Sleep. *Sleep Med. Rev.* **2019**, *44*, 12–22. [CrossRef]

34. Mangiaruga, A.; D'Atri, A.; Scarpelli, S.; Alfonsi, V.; Camaioni, M.; Annarumma, L.; Gorgoni, M.; Pazzaglia, M.; De Gennaro, L. Sleep Talking versus Sleep Moaning: Electrophysiological Patterns Preceding Linguistic Vocalizations during Sleep. *Sleep* **2022**, *45*, zsab284. [CrossRef]

35. Camaioni, M.; Scarpelli, S.; Alfonsi, V.; Gorgoni, M.; De Bartolo, M.; Calzolari, R.; De Gennaro, L. The Influence of Sleep Talking on Nocturnal Sleep and Sleep-Dependent Cognitive Processes. *J. Clin. Med.* **2022**, *11*, 6489. [CrossRef]

36. Hublin, C.; Kaprio, J.; Partinen, M.; Koskenvuo, M. Parasomnias: Co-Occurrence and Genetics. *Psychiatr. Genet.* **2001**, *11*, 65–70. [CrossRef]

37. Klackenberg, G. Incidence of Parasomnias in Children in a General Population. In *Sleep and Its Disorders in Children*; Guilleminault, C., Ed.; Raven Press: New York, NY, USA, 1987.

38. Curcio, G.; Tempesta, D.; Scarlata, S.; Marzano, C.; Moroni, F.; Rossini, P.M.; Ferrara, M.; De Gennaro, L. Validity of the Italian Version of the Pittsburgh Sleep Quality Index (PSQI). *Neurol. Sci.* **2013**, *34*, 511–519. [CrossRef]

39. Fulda, S.; Hornyak, M.; Müller, K.; Cerny, L.; Beitinger, P.A.; Wetter, T.C. Development and Validation of the Munich Parasomnia Screening (MUPS): A Questionnaire for Parasomnias and Nocturnal Behaviors. *Somnologie* **2008**, *12*, 56–65. [CrossRef]

40. De Gennaro, L.; Ferrara, M.; Cristiani, R.; Curcio, G.; Martiradonna, V.; Bertini, M. Alexithymia and Dream Recall upon Spontaneous Morning Awakening. *Psychosom. Med.* **2003**, *65*, 301–306. [CrossRef]

41. Casagrande, M.; Cortini, P. Spoken and Written Dream Communication: Differences and Methodological Aspects. *Conscious. Cogn.* **2008**, *17*, 145–158. [CrossRef]

42. De Gennaro, L.; Cipolli, C.; Cherubini, A.; Assogna, F.; Cacciari, C.; Marzano, C.; Curcio, G.; Ferrara, M.; Caltagirone, C.; Spalletta, G. Amygdala and Hippocampus Volumetry and Diffusivity in Relation to Dreaming. *Hum. Brain Mapp.* **2010**, *32*, 1458–1470. [CrossRef]

43. Lehmann, S.; Skogen, J.C.; Haug, E.; Mæland, S.; Fadnes, L.T.; Sandal, G.M.; Hysing, M.; Bjørknes, R. Perceived Consequences and Worries among Youth in Norway during the COVID-19 Pandemic Lockdown. *Scand. J. Public Health* **2021**, *49*, 755–765. [CrossRef]

44. Ohayon, M.M.; Morselli, P.L.; Guilleminault, C. Prevalence of Nightmares and Their Relationship to Psychopathology and Daytime Functioning in Insomnia Subjects. *Sleep* **1997**, *20*, 340–348. [CrossRef]

45. Berlin, R.M.; Litovitz, G.L.; Diaz, M.A.; Ahmed, S.W. Sleep Disorders on a Psychiatric Consultation Service. *Am. J. Psychiatry* **1984**, *141*, 582–584. [CrossRef]

46. Cernovsky, Z.Z.; Paitich, D.; Crawford, G. MMPI and Nightmare Reports in Women Addicted to Alcohol and Other Drugs. *Percept. Mot. Skills* **1986**, *62*, 717–718. [CrossRef]

47. Munezawa, T.; Kaneita, Y.; Osaki, Y.; Kanda, H.; Ohtsu, T.; Suzuki, H.; Minowa, M.; Suzuki, K.; Higuchi, S.; Mori, J.; et al. Nightmare and Sleep Paralysis among Japanese Adolescents: A Nationwide Representative Survey. *Sleep Med.* **2011**, *12*, 56–64. [CrossRef]

48. Tanskanen, A.; Tuomilehto, J.; Viinamäki, H.; Vartiainen, E.; Lehtonen, J.; Puska, P. Nightmares as Predictors of Suicide. *Sleep* **2001**, *24*, 845–848. [CrossRef]

49. Kales, A.; Soldatos, C.R.; Caldwell, A.B.; Charney, D.S.; Kales, J.D.; Markel, D.; Cadieux, R. Nightmares: Clinical Characteristics and Personality Patterns. *Am. J. Psychiatry* **1980**, *137*, 1197–1201. [CrossRef]

50. Kramer, M.; Schoen, L.S.; Kinney, L. Psychological and Behavioral Features of Disturbed Dreamers. *Psychiatr. J. Univ. Ott.* **1984**, *3*, 102–106.

51. Berquier, A.; Ashton, R. Characteristics of the Frequent Nightmare Sufferer. *J. Abnorm. Psychol.* **1992**, *101*, 246–250. [CrossRef]

52. Husni, M.; Cernovsky, Z.Z.; Koye, N.; Haggarty, J. Nightmares of Refugees from Kurdistan. *J. Nerv. Ment. Dis.* **2001**, *189*, 557–559. [CrossRef]

53. Takeuchi, T.; Fukuda, K.; Sasaki, Y.; Inugami, M.; Murphy, T.I. Factors Related to the Occurrence of Isolated Sleep Paralysis Elicited during a Multi-Phasic Sleep-Wake Schedule. *Sleep* **2002**, *25*, 89–96. [CrossRef]

54. Hinton, D.E.; Pich, V.; Chhean, D.; Pollack, M.H.; McNally, R.J. Sleep Paralysis among Cambodian Refugees: Association with PTSD Diagnosis and Severity. *Depress. Anxiety* **2005**, *22*, 47–51. [CrossRef]

55. Ohayon, M.M.; Shapiro, C.M. Sleep Disturbances and Psychiatric Disorders Associated with Posttraumatic Stress Disorder in the General Population. *Compr. Psychiatry* **2000**, *7*, 628–637. [CrossRef]

56. Yeung, A.; Xu, Y.; Chang, D.F. Prevalence and Illness Beliefs of Sleep Paralysis among Chinese Psychiatric Patients in China and the United States. *Transcult. Psychiatry* **2005**, *42*, 135–145. [CrossRef]

57. Castelnovo, A.; Lopez, R.; Proserpio, P.; Nobili, L.; Dauvilliers, Y. NREM Sleep Parasomnias as Disorders of Sleep-State Dissociation. *Nat. Rev. Neurol.* **2018**, *14*, 470–481. [CrossRef]

58. Morin, C.M.; Bjorvatn, B.; Chung, F.; Holzinger, B.; Partinen, M.; Penzel, T.; Ivers, H.; Wing, Y.K.; Chan, N.Y.; Merikanto, I.; et al. Insomnia, Anxiety, and Depression during the COVID-19 Pandemic: An International Collaborative Study. *Sleep Med.* **2021**, *87*, 38–45. [CrossRef] [PubMed]

59. Salari, N.; Hosseinian-Far, A.; Jalali, R.; Vaisi-Raygani, A.; Rasoulpoor, S.; Mohammadi, M.; Rasoulpoor, S.; Khaledi-Paveh, B. Prevalence of Stress, Anxiety, Depression among the General Population during the COVID-19 Pandemic: A Systematic Review and Meta-Analysis. *Global. Health* **2020**, *16*, 57. [CrossRef] [PubMed]

60. Barone, D.A.; Henchcliffe, C. Rapid Eye Movement Sleep Behavior Disorder and the Link to Alpha-Synucleinopathies. *Clin. Neurophysiol.* **2018**, *129*, 1551–1564. [CrossRef]

61. Scarpelli, S.; Alfonsi, V.; Gorgoni, M.; De Gennaro, L. What about Dreams? State of the Art and Open Questions. *J. Sleep Res.* **2022**, *31*, e13609. [CrossRef] [PubMed]

62. Scarpelli, S.; Bartolacci, C.; D'Atri, A.; Gorgoni, M.; De Gennaro, L. The Functional Role of Dreaming in Emotional Processes. *Front. Psychol.* **2019**, *10*, 459. [CrossRef]

63. Rechtschaffen, A.; Goodenough, D.R.; Shapiro, A. Patterns of Sleep Talking. *Arch. Gen. Psychiatry* **1962**, *7*, 418–426. [CrossRef]

64. Castelnovo, A.; Loddo, G.; Provini, F.; Miano, S.; Manconi, M. Mental Activity during Episodes of Sleepwalking, Night Terrors or Confusional Arousals: Differences between Children and Adults. *Nat. Sci. Sleep* **2021**, *13*, 829–840. [CrossRef] [PubMed]

65. Fasiello, E.; Scarpelli, S.; Gorgoni, M.; Alfonsi, V.; Galbiati, A.; De Gennaro, L. A Systematic Review of Dreams and Nightmares Recall in Patients with Rapid Eye Movement Sleep Behaviour Disorder. *J. Sleep Res.* **2023**, *32*, e13768. [CrossRef] [PubMed]

66. Uguccioni, G.; Golmard, J.-L.; de Fontreaux, A.N.; Leu-Semenescu, S.; Brion, A.; Arnulf, I. Fight or Flight? Dream Content during Sleepwalking/Sleep Terrors vs. Rapid Eye Movement Sleep Behavior Disorder. *Sleep Med.* **2013**, *14*, 391–398. [CrossRef] [PubMed]

67. Antrobus, J. Dreaming: Cognitive Processes During Cortical Activation and High Afferent Thresholds. *Psychol. Rev.* **1991**, *98*, 96–121. [CrossRef] [PubMed]

68. Casagrande, M.; Violani, C.; Lucidi, F.; Buttinelli, E.; Bertini, M. Variations in Sleep Mentation as a Function of Time of Night. *Int. J. Neurosci.* **1996**, *85*, 19–30. [CrossRef] [PubMed]

69. Foulkes, D. Nonrapid Eye Movement Mentation. *Exp. Neurol.* **1967**, *19*, 28–38. [CrossRef] [PubMed]

70. Rechtschaffen, A.; Verdone, P.; Wheaton, J. Reports of mental activity during sleep. *Can. Psychiatr. Assoc. J.* **1963**, *8*, 409–414. [CrossRef] [PubMed]

71. MacNeilage, P.F. Motor Control of Serial Ordering of Speech. *Psychol. Rev.* **1970**, *77*, 182–196. [CrossRef]

72. Arkin, A.M.; Toth, F.; Baker, J.; Hastey, J.M. The Frequency of Sleep Talking in the Laboratory among Chronic Sleep Talkers and Good Dream Recallers. *J. Nerv. Ment. Dis.* **1970**, *151*, 369–374. [CrossRef]

Article

Frequent Lucid Dreaming Is Associated with Meditation Practice Styles, Meta-Awareness, and Trait Mindfulness

Elena Gerhardt [1] and Benjamin Baird [2,*]

[1] Institute of Psychology, Osnabrück University, 49076 Osnabrück, Germany; egerhardt@uni-osnabrueck.de
[2] Department of Psychology, The University of Texas at Austin, Austin, TX 78712, USA
* Correspondence: baird@utexas.edu

Abstract: Lucid dreaming involves becoming aware that one's current experience is a dream, which has similarities with the notion of mindfulness—becoming aware of moment-to-moment changes in experience. Additionally, meta-awareness, the ability to explicitly notice the current content of one's own mental state, has also been proposed to play an important role both in lucid dreaming and mindfulness meditation practices. However, research has shown conflicting strengths of associations between mindfulness, meditation, and lucid dreaming frequency, and the link between lucid dreaming and meta-awareness has not yet been empirically studied. This study evaluated the associations between lucid dreaming frequency and different meditation practice styles, mindfulness traits, and individual differences in meta-awareness through an online survey ($n = 635$). The results suggest that daily frequent meditators experience more lucid dreams than non-frequent meditators. However, weekly frequent meditators did not have a higher lucid dreaming frequency. A positive association was observed between open monitoring styles of meditation and lucid dreaming. The findings also indicate that meta-awareness is higher for meditators and weekly lucid dreamers. Furthermore, frequent lucid dreaming was commonly associated with a non-reactive stance and experiencing transcendence. Overall, the findings suggest a positive relationship between specific meditation practices and lucid dreaming as well as the importance of meta-awareness as a cognitive process linking meditation, mindfulness, and lucid dreaming.

Keywords: lucid dreaming frequency; mindfulness; meditation practices; meta-awareness

1. Introduction

Being explicitly aware of one's own mental state and maintaining present-centered awareness—paying attention to moment-to-moment changes in thoughts, emotions, and perceptions with a non-judgmental stance—are essential elements of the definition of mindfulness [1,2]. Becoming aware that one is dreaming while still being asleep defines the nocturnal state of lucid dreaming [3,4]. In both mindful awareness and lucid dreaming, there is an explicit awareness of one's current mental state that characterizes meta-awareness [5,6]. This awareness can be either propositional ("I am dreaming" or "I am mind-wandering") or non-propositional, sustaining a peripheral awareness of engagement with a chosen object or the ongoing realization of being in a dream [7]. Mindfulness can be cultivated through the practice of meditation, while the ability to induce lucid dreams can be trained through various methods, including cognitive practices, induction devices, or the use of substances [8–10].

Scientific research has focused primarily on meditation practices within Buddhism, which is divided into numerous lineages [11]. In Tibetan Buddhism, the achievement of continuous conscious awareness during all stages of sleep and dreaming is highly valued [12,13]. Tibetan "Dream and Sleep Yoga" teachings, a set of daytime and nighttime practices for gaining awareness during dreams and using lucid dreams as a platform for various meditation practices, were reserved for advanced practitioners [14,15]. Lucid

Citation: Gerhardt, E.; Baird, B. Frequent Lucid Dreaming Is Associated with Meditation Practice Styles, Meta-Awareness, and Trait Mindfulness. *Brain Sci.* **2024**, *14*, 496. https://doi.org/10.3390/brainsci14050496

Academic Editor: Maurizio Gorgoni

Received: 15 April 2024
Revised: 3 May 2024
Accepted: 8 May 2024
Published: 14 May 2024

dreaming and dreamless sleep awareness are seen as pathways to spiritual growth and enlightenment [14,15]. Meditation consists of deliberate actions such as observing, concentrating, letting go, generating, visualizing, and shifting attention from one mental object to another, all anchored in conscious awareness. Central to this practice is meta-awareness that integrates these various activities. Such a practice is diversely implemented across cultural traditions and secular settings [16].

Meditative practices may be differentiated depending on the direction and dynamics of attention. For instance, focused attention (FA) meditation involves practices that narrow the scope of attention and cultivate one-pointed concentration on a single object, such as the observation of the breath. Open monitoring (OM) meditation involves the non-reactive monitoring of the content of experience from moment to moment [17]. FA meditation requires the stability of attention on an object or activity, whereas OM meditation is based on the openness and expansiveness of awareness, monitoring changes over time [18]. Hence, meditation practices cultivate specific mental skills, including enhanced attentional stability and the monitoring of one's mental state (i.e., meta-awareness) [18]. Meta-awareness is an essential cognitive capacity discussed as the primary cognitive component for the state shift of non-lucid to lucid dreams [19]. Non-propositional meta-awareness is considered particularly relevant to lucid dreaming [7]. Meta-awareness becomes crucial at the moment of becoming aware of the bizarreness of the dream plot, recognizing the dream signs, or passively observing the ongoing dream [20–23]. Even in a stable lucid dream, one has a sustained awareness of the dream state while experiencing events in the dream [7,24].

Building on the idea that not just the waking state offers self-reflective thought and cognition but that self-reflective awareness can also be achieved in the dream state, a theoretical consideration of connecting lucidity with mindfulness influenced by meditation was made, placing waking experiences, dreams, and several stages of lucidity on a continuum of self-reflectiveness [25]. Taking this idea further, enhancing mindfulness during the day is thought to also increase mindfulness during the night [26]. This is derived from the continuity hypothesis that waking memories, dispositions, and habits are incorporated and transferred into the dream state [27].

Despite strong theoretical linkages, the degree of associations between meditation, mindfulness, and lucid dreaming have been conflicting in empirical studies. Overall, findings suggest an increase in the frequency of spontaneous lucid dreams in frequent or long-term meditators [28–34]. Moreover, evidence suggests that the extent of meditation experience can alter the strength of the association between mindfulness and lucid dreaming [33]. One recent investigation did not report an increase in lucid dreaming frequency due to meditation experience. However, it did find that certain facets of mindfulness were positively correlated with lucid dreaming frequency [35]. In a recent study by Geise and Smith [36], the Transcendence subscale of the Relaxation, Mindfulness, and Meditation Experience Tracker was found to be a significant predictor of lucid dreaming frequency. However, the total score on the Freiburg Mindfulness Inventory, both subscales, Presence and Acceptance, and measures of meditation frequency or experience did not show significant correlations with lucid dreaming frequency [36]. A positive association between an estimate of the number of lucid dreams during one year and the total number of years of meditation experience has also been observed [31].

Gackenbach et al. [29] found that intensive and frequent meditators rooted in Transcendental Meditation, on average, experience lucid dreams once or more per week. Despite the methodological issues regarding the potentially biased selection of participants for the study, the evidence for higher lucid dreaming frequencies in populations of long-term and frequent meditators should be considered. Previous findings from Baird et al. [28] indicated that long-term meditators have more spontaneous lucid dreams compared to inexperienced meditators. Furthermore, the associations between aspects of trait mindfulness measured by the Five Facet Mindfulness Questionnaire, such as Acting with Awareness and Observing, and the Decentering subscale of the Toronto Mindfulness Scale, were higher in long-term meditators with frequent lucid dreams. However, meditation novices did not show an increase in lucid dreaming frequency after 8-week mindfulness meditation training. A recent study validated an indirect lucid dream experience questionnaire in Spanish and examined, among other dream-related constructs, meditation practices, experience, and aspects of mindfulness [34]. The results suggested that time spent in OM meditation was positively correlated with higher scores on the lucid dream aspects of insight and control. The study did not examine lucid dreaming frequency, nor did it validate whether participants ever had a lucid dream.

These results are in line with the interpretation that meditation training improves metacognitive skills with the enhancement of dispositional mindfulness, which in turn could increase nighttime meta-awareness in order to promote the state shift of the onset of lucid dreaming. Research shows that different meditation techniques within various frameworks and traditions have different effects [37,38]. Furthermore, there has not been any study investigating which meditation technique is associated with higher lucid dreaming frequency. This study therefore explored the connection between meditation practices, meditation frequency, dispositional mindfulness, and lucid dreaming frequency. First, we sought to replicate the empirical findings that frequent meditators exhibit higher lucid dreaming frequencies. Second, we evaluated individual differences between frequent meditators and non-meditators on all mindfulness facets and dreaming variables to further explore the relationship between trait mindfulness and lucid dreaming. Third, we studied the relationship between lucid dreaming frequency and specific meditation techniques/practice styles. The main hypothesis was that particularly open monitoring (OM) meditation, and possibly focused attention (FA) meditation, would have a positive association with lucid dreaming frequency, as both practices emphasize the cultivation of meta-awareness and sustained attention monitoring. Moreover, it was expected that meta-awareness would be higher in frequent meditators compared to non-frequent meditators, but also in weekly lucid dreamers compared to non-weekly lucid dreamers. Lastly, we explored the role of meta-awareness in the association of meditation frequency and lucid dreaming frequency.

2. Methods

2.1. Participants

In total, 635 participants completed the online survey. Only persons who met the following criteria were asked to complete the survey: all participants (1) must be at least 18 years and no more than 75 years old, and (2) must be fluent in English. The upper limit of age was set, as lucid dream incidences and cognitive capacities have been shown to decline over age [39]. The convenience sample splits up into a German student population from Osnabrück University (Uos, $n = 72$) and students from the University of Texas at Austin (UT Austin, $n = 272$), as well as a general mixed international sample of 291 respondents. The study was approved by the Institutional Review Board of the University of Texas at Austin (STUDY00003582). Notably, all participants who claimed to have experienced at least one lucid dream had to verify their understanding of the lucid dream experience. Participants who did not pass the verification were not eligible and were thus excluded. All respondents were grouped depending on their meditation frequency [40,41]. Participants were either classified as non-frequent meditators (i.e., meditating less than once per week), as weekly frequent meditators (WFMs; meditating once or more per week), or as daily frequent meditators (DFMs; meditating at least twice per day or multiple times daily). Within the Osnabrück University sample, 66 participants were eligible (49 females, 16 males, and 1 non-binary; age = 22.94 ± 5.6 (M ± SD)). Based on meditation frequency, 2 students were classified as DFMs, 11 meditated weekly, and 53 were non-frequent meditators. Within the student population from UT Austin, 241 were classified as eligible (156 female, 81 male, and 4 non-binary; age = 19.43 ± 1.89 (M ± SD)). Based on meditation frequency, 1 student was classified as DFM, 29 meditated weekly, and 211 were non-frequent meditators. Within the general mixed population, out of 291 initial respondents, 270 were eligible, more male respondents completed the survey (112 females, 149 males, 4 non-binary, and 5 self-described, e.g., "Genderfluid"), and their ages ranged from 18 to 75, with an average of 37.74 ± 16.16 (M ± SD). Within the general mixed sample, 35 participants were DFMs, 117 were WFMs, and 118 were classified as non-frequent meditators.

2.2. Procedure

The survey was internationally distributed, with a focus on Europe and the United States. Recruitment started in January 2023 and ended in July 2023. Data collection and the entire recruitment process were conducted online. All study materials were provided in English and implemented using Qualtrics software (https://www.qualtrics.com) (accessed on 1 January 2023). The survey completion took, on average, 54.8 (Mdn) minutes in the general sample, while the student populations needed between 20.9 (UT Austin; Mdn) and 35.7 (Uos; Mdn) minutes. The platforms for distributing the survey varied. The link and an invitation protocol were sent out to several lucid dreaming experts, institutes, and other researchers. Additionally, placement on several social media platforms, starting with Reddit, Facebook, Instagram, Forums, and other websites like Dream Views, YouTube, and LinkedIn, achieved a wide reach. Furthermore, meditation and wellness centers facilitated study participant recruitment by distributing materials through email lists.

Students from UT Austin were reached via the SONA system. Psychology and Cognitive Science students from Osnabrück University were contacted via mailing lists. All student populations were compensated with one credit, which corresponds to an hour of participation, for their research participation sheets. Respondents in the mixed general sample did not receive any compensation. All participants were provided with the informed consent document and a short introduction. After the agreement, all participants received the following sections in the same order: Demographics, Dream Survey, and Meditation Experience Questionnaire. For these instruments, branching allowed the researchers to efficiently present participants with in-depth questions based on their previous experience. The following instruments were presented to all in a randomized order: Multidimensional Awareness Scale, Toronto-Mindfulness Scale, Relaxation, Meditation, and Mindfulness Experiences Questionnaire, and Five Facet Mindfulness Questionnaire, along with two other scales which were part of another research scope (Mysticism Scale and Indirect Realism Scale).

2.3. Measures

In order to achieve a comprehensive assessment while maintaining time efficiency, shortened versions of many instruments were implemented. For internal consistency of the scales, McDonald's omega total was preferred over Cronbach's alpha. Since it can be assumed that not all items contribute equally to their score, McDonald's omega is a more accurate reliability estimate, especially for multidimensional or ordinal scales [42–44]. Omega can be described as the proportion of variance in observed scores that can be attributed to a single underlying factor or to the common variance among the items on a scale [45,46]. As with Cronbach's alpha, larger values indicate a higher reliability [47].

Dream Recall and Lucid Dreaming Experience. Lucid and ordinary dream experiences were recorded with an adapted dream survey. The original questionnaire developed by Baird et al. [48] was modified to fit the specific aims of this study. All participants reported dream recall frequency and lucid dreaming frequency on a 16-point Likert scale, extending the established scales by Schredl and Erlacher [49]: *0 = never; 1 = less than 1 (lucid) dream per year; 2 = 1 (lucid) dream per year; 3 = 2 (lucid) dreams per year; 4 = 3–5 (lucid) dreams per year; 5 = 6–8 (lucid) dreams per year; 6 = 9–11 (lucid) dreams per year; 7 = 1 (lucid) dream per month; 8 = 2 (lucid) dreams per month; 9 = 3 (lucid) dreams per month; 10 = 1 (lucid) dream per week; 11 = 2 (lucid) dreams per week; 12 = 3–4 (lucid) dreams per week; 13 = 5–6 (lucid) dreams per week; 14 = 1 (lucid) dream per night; 15 = more than 1 (lucid) dream per night.* Based on the methodology of Stumbrys, Erlacher, and Malinowski [33], class means transformed the ordinal scores into metric frequencies either as units per month (0 → 0, 1 → 0.0714, 2 → 0.0833, 3 → 0.1667, 4 → 0.3333, 5 → 0.5833, 6 → 0.8333, 7 → 1, 8 → 2, 9 → 3, 10 → 4,

11 → 8, 12 → 13.5, 13 → 23.5, 14 → 30, 15 → 33) or units per week (0 → 0, 1 → 0.0185, 2 → 0.0192, 3 → 0.0385, 4 → 0.0769, 5 → 0.1346, 6 → 0.1923, 7 → 0.25, 8 → 0.50, 9 → 0.75, 10 → 1, 11 → 2, 12 → 3.5, 13 → 5.5, 14 → 7, 15 → 9). The same class means-recoded 16-point scale was given for lucid dream induction frequency per month. Participants received a written definition along with the scales: "Lucid dreaming is a special sort of dream in which you know that you are dreaming while still in the dream. Typically, you tell yourself "I'm dreaming!" or "This is a dream!". In some cases, you may also control the content of the dream and alter the dream events as well as your actions voluntarily".

Respondents who had previously experienced lucid dreams were asked detailed questions regarding their lucid dream experiences, their ability to control lucid dreams, and their training in lucid dream induction techniques. In addition to the monthly lucid dreaming frequency, the number of lucid dreams in the previous six-month period was assessed, which is a summative measure (i.e., an overall measurement taken after a period of time has passed) of lucid dreaming frequency as opposed to a formative approach (i.e., a measurement at shorter time intervals for each week or month). All items were presented either as an open text field or as a Likert-type format. The following single items were used: success of the lucid dream induction: "If you decide to have a lucid dream on a given night, how likely will you succeed?" (0 = *very unlikely*; 4 = *very likely*); wake-initiated lucid dream occurrences (0 = *never*; 4 = *always*); and how often one experiences a detached observer stance in the lucid dream (0 = *never*; 4 = *always*). To verify participants' understanding of the lucid dream state, they were required to provide a brief report of one of their lucid dreams, detailing how they realized that they were dreaming. As all scales were adapted or created for this study, traditional reliability measures were not applicable. Nonetheless, a strong correlation was found between the frequency of lucid dreams per month and the number of lucid dreams in the previous six-month period (r_{sp} = 0.93, *p* < 0.0001). Participants who reported experiencing lucid dreams at least once per month were categorized as monthly frequent lucid dreamers (MFLDs), while those who reported experiencing lucid dreams at least once or more per week were classified as weekly frequent lucid dreamers (WFLDs), extending the standard classification convention [9,29,50,51].

Meditation Experience and Frequency. A revised version of the Meditation Experience Questionnaire [28] was utilized to assess the quantitative experience of meditation practices. To cover various meditation frameworks, three options of Buddhism (Theravadan, Tibetan, or Mixed) were extended to 18 different meditation frameworks: 9 religious/spiritual-oriented traditions (Theravadan, Tibetan, and Zen Buddhism; Daoism; Yoga; Sufism; Judaism; Christianity; and Shamanism) and 9 secular-oriented frameworks were included (app-guided, online-based, Vipassana, self-guided, Yoga, Thai Chi/Qigong, MBSR-based, non-dual meditation, and Transcendental Meditation), plus the option to specify an individual framework and tradition. Two items assessed previous meditation experience (*yes*/*no*), and meditation frequency. Meditation frequency was measured with a 16-point scale (0 = *never*, 15 = *more than 1 meditation per day*). Class means transformed the ordinal scale into metric units per week (0 → 0, 1 → 0.0185, 2 → 0.0192, 3 → 0.0385, 4 → 0.0769, 5 → 0.1346, 6 → 0.1923, 7 → 0.25, 8 → 0.50, 9 → 0.75, 10 → 1, 11 → 2, 12 → 3.5, 13 → 5.5, 14 → 7, 15 → 9). If participants meditated at least once per week, they reported how many different techniques they used in their meditation practice on a regular basis. In addition to questions about the quantitative meditation routine, they reported on the styles practiced while meditating, as well as meditation training and retreat experience. For each regularly practiced meditation framework, respondents provided an estimate of the length of an average meditation session in minutes, the frequency per day, the number of days per week, and the number of years of practice within the meditation framework. Based on this account, the total number of hours spent in meditation for each framework per week was calculated: $\frac{Minutes\ per\ Session\ *\ Freqency\ per\ Day\ *\ Days\ with\ Practice}{60}$. Moreover, the participants indicated the percentage of the total time they dedicated to each meditation framework in which a specific quality was facilitated: "Please indicate what percentage of your average meditation time you spend on a specific meditation technique". Respondents were presented with six options: open monitoring (OM) meditation, focused attention (FA) meditation, loving-kindness/emotionally toned (LK) meditation, meditation to recognize the nature of the mind, non-dual meditation, and one option for an individually specified technique.

Meta-Awareness. The Meta-Awareness subscale of the Multidimensional Awareness Scale (MAS) captured the cognitive ability to recognize one's current mental state based on self-assessment [52]. The item "I am aware of my thoughts and feelings as I experience them" reveals the direct aim of measuring the trait aspects of the cognitive process, as the instruction asked participants to indicate the extent to which the given statements represent the typical experience of their thoughts or feelings. The MAS-MA subscale consists of 7 items selected from the original 25-item MAS scale. The items were rated on a 7-point Likert scale (1 = *strongly disagree*, 7 = *strongly agree*). Reliability was found to be good, $\omega_t = 0.89$.

Trait Mindfulness. The measurement of trait mindfulness was performed using multiple instruments: the Toronto-Mindfulness Scale (TMS) [53,54], a short version of the Five Facet Mindfulness Questionnaire (FFMQ) [55,56], and the Relaxation, Meditation, and Mindfulness Experience Questionnaire (RMMtm) [36,57].

The TMS is a 13-item mindfulness inventory constructed as a 5-point Likert-scale (0 = *not at all*; 4 = *very much*). The original scale measured state mindfulness and the survey incorporated the trait version of the TMS, asking for experiences during the previous 7 days [53]. The scale was derived from a two-component definition of mindfulness: Curiosity and Decentering [1]. Curiosity includes 6 items and refers to the openness to explore one's internal states as they occur. Decentering, measured by 7 items, refers to the ability to maintain a stance of detachment from one's thoughts and emotions, with the capacity not to be carried away by thoughts and emotions [54]. The scales were found to be reliable, with $\omega_t = 0.91$ for Curiosity and $\omega_t = 0.85$ for Decentering.

Drawing from various definitions of mindfulness, a psychometrically validated trait mindfulness measure, the FFMQ, was developed. It combines five mindfulness instruments, and factor analysis on a large sample revealed five factors [58]. Its condensed form has 20 items rated on a 5-point Likert scale (1 = *never or rarely true*, 5 = *very often or always true*) [56]. The brief version measures five distinct but related components of mindfulness, with 4 items for each scale: (1) Observing: the ability to notice and attend to internal and external experiences, $\omega_t = 0.76$; (2) Describing: the capacity to articulate one's experiences in words, $\omega_t = 0.77$; (3) Acting with awareness: as opposed to "being on autopilot", the degree of presence and awareness while engaged in activities, $\omega_t = 0.88$; (4) Nonjudgment of inner experience: the ability to refrain from judging inner experiences as neither good nor bad, $\omega_t = 0.88$; (5) Nonreactivity to inner experience: the attitude of allowing thoughts and feelings to arise and pass without being caught up in or swept away by them, allowing these experiences to exist without interference, $\omega_t = 0.77$. The psychometric properties of the short version of the FFMQ by Tran et al. [56] demonstrated moderate to strong reliability.

The Relaxation, Meditation, and Mindfulness Tracker (RMMtm) by Smith [59] comprises a self-reported broad-spectrum inventory for assessing dispositional mindfulness. The 7-point Likert scale with 32 items captures trait mindfulness (1 = *never*, 2 = *not this month, but once or twice this year*; 3 = *about once this month*; 4 = *about once a week*; 5 = *about 2 or 3 times a week*; 6 = *about every day*; 7 = *several times daily*). The RMMtm includes various experiences associated with the practice of mindfulness meditation techniques that manifest as characteristics over time. The instrument was derived from third-wave mindfulness theory to capture all essential phenomenological states of mindfulness, represented on a continuum with varying levels of mindfulness [60]: (1) Mindful Relaxation, which assesses initial mindful relaxation experiences; (2) Mindful Quiet Focus, which captures the stillness and observational nature of meditation; (3) Mindful Engagement, which measures the ability to remain present and compassionate during activities; and (4) Mindful Transcendence, the deepest level observed in long-term practitioners in spiritual or non-secular contexts [61]. Due to expected variations in levels of mindfulness among long-term and frequent meditators compared to other groups, separate component analysis for each population sample is recommended [36,57]. A principal component analysis identified 2 distinct factors. Items 1–24 loaded on the first factor; this dimension is interpreted as Mindful Relaxation and Focus, $\omega_t = 0.97$. The first dimension includes aspects of physical relaxation but also cognitive and emotional aspects of mindfulness: "I was living in the present moment, not past or future concerns" or "I felt selfless/caring/compassion". Items 25–32 loaded on the second factor, compromising Mindful Transcendence, $\omega_t = 0.93$. Self-transcendence, in general, is defined as the capacity to expand self-boundaries [62]. The subscale refers to transpersonal self-transcendence: "I had a sense of what is timeless, boundless, infinite". Due to the sample-specific component analysis, reliability measures were not comparable to previous research. However, reliability analysis indicated satisfying internal consistencies.

2.4. Statistical Analysis

Statistical analyses were performed using R, version 4.3.1, and SPSS, version 29. For the primary data analysis, only the general mixed sample was utilized, following manual validation of each participant's lucid dream report. The student populations were not included in the data analysis for the research questions addressed in this study, since there was an insufficient number of participants per group. Data management was based on the functions of the tidyverse package in R [63]. A total of 270 participants out of 291 were considered eligible for data analysis. The more liberal Shapiro–Wilk tests revealed significant deviations from normal distribution for most of the measures, e.g., lucid dreaming frequency (Shapiro–Wilk's $W = 0.590$, $p < 10^{-14}$), dream recall frequency (Shapiro–Wilk's $W = 0.855$, $p < 10^{-15}$), lucid dream induction frequency (Shapiro–Wilk's $W = 0.540$, $p < 10^{-22}$), and meditation frequency (Shapiro–Wilk's $W = 0.813$, $p < 10^{-16}$). Therefore, non-parametric independent two-sample permutation tests were utilized for assessing overall group differences. For each group, a set of multiple tests with adjusted p-values based on Benjamini and Hochberg and the False Discovery Rate (FDR) correction were reported [64].

Monte Carlo permutation tests with $R = 10{,}000$ permutations were implemented for group comparisons [65]. For most group comparisons, the Monte Carlo permutation test of the mean was calculated. All variables besides the ordinal scaled WILD frequency and the ordinal variable related to the detached observer stance were continuous variables. When comparing the total hours of meditation per week and the largest number of years for a meditation practice, a Monte Carlo permutation test of the median was used. In addition to that, Spearman's correlation was preferred over the Pearson correlation due to the influence of the largest values in the weekly hours of meditation for each meditation practice style.

Prior to this, an exploratory principal component analysis found the sample-specific RMMtm scales [36]. Bartlett's test of sphericity was significant ($X^2 = 8357.33$, df = 496, $p < 0.001$), indicating that the variables were sufficiently intercorrelated to proceed with principal component analysis (PCA). A PCA with Kaiser normalization and an Oblimin rotation method yielded a two-factor solution in the general population with tools from the psych package [66]. The Kaiser–Meyer–Olkin Measure was 0.953, which indicates good sampling adequacy. Factor extraction was based on the scree plot, indicating two factors, while parallel analysis yielded two factors and Kaiser–Gutman criteria indicated four factors. The two-factor solution accounted for 62% of the cumulative variance, compared to 67% for the three-factor and 70% for the four-factor solution. The tools from the MBESS package calculated omega total (ω_t) estimates instead of the psych tools for a more conservative reliability measure [43,47].

Multiple regression analysis for the monthly lucid dreaming frequency, predicted by the RMMtm Mindful Transcendence subscale, the TMS Decentering and Curiosity subscale, the MAS-MA subscale, weekly meditation frequency, and age, was implemented with the Boot and LessR package in R (i.e., Table S1, Figures S1 and S2) [67,68]. Due to violations of heteroscedasticity and normal distribution, examined by means of the visual plotting of the predicted values against the standardized residuals, the coefficients were tested based on

10,000 bootstrapped bias-corrected confidence intervals. Assumptions were investigated via the LessR package: no VIF values exceeded 5, and none of Cook's distance indexes was larger than 1, with the highest VIF being 3.634 and the largest Cook's distance index being 0.12.

The exploratory analysis of the relationship between lucid dreaming, meditation, and meta-awareness was performed with model 4 of the PROCESS macro for R [69]. In this model, meta-awareness was used as a mediator, meditation frequency as the independent variable, and the number of lucid dreams in the previous six months was used as the outcome variable. The model controlled for covariates including lucid dream induction frequency, dream recall frequency, and age. It must be stated that for this analysis, 25 participants who did not have prior experience with lucid dreams were excluded, due to missing data. Bias-corrected accelerated (BCa) confidence intervals for the coefficients were bootstrapped with R = 10,000 replicates, as was the confidence interval for the indirect effect. Standard errors were computed using heteroscedasticity-consistent estimates due to the heteroscedasticity of the residuals and the nonnormality of the dependent variable (Shapiro–Wilk's $W = 0.381$, $p < 10^{-22}$). The model parameters were standardized, and the random seed was fixed to 9999 to ensure the reproducibility of the results.

3. Results

In the general sample spanning all groups, participants reported an average of 5.92 ± 3.03 [M $\pm$ SD, $n = 270$] dreams recalled weekly and 4.26 ± 7.65 [M $\pm$ SD, $n = 270$] remembered lucid dreams per month. A crucial number of participants had at least one lucid dream (90.74%), induced 3.74 ± 7.72 [M $\pm$ SD, $n = 245$] lucid dreams each month, and indicated an average of 27.80 lucid dreams over the previous half year (SD = 71.90, Range = 0–720). A total of 143 participants (52.96%) had at least one lucid dream per month, with an average of 7.80 ± 9.16 [M $\pm$ SD] lucid dreams per month, defined as monthly frequent lucid dreamers, whereas 73 experienced at least one lucid dream per week (27.04%), with an average of 13.70 ± 9.67 [M $\pm$ SD] monthly lucid dreams, and thus were classified as WFLDs; see Table 1.

Table 1. Number of participants in groups based on lucid dreaming and meditation frequency.

	Frequent Meditators		Non-Frequent Meditators		
	DFMs	**WFMs**	**Infrequent Meditators**	**Never Have Meditated**	
NWLDs	22	85	72	18	
	107		90		$n = 197$
WFLDs	13	32	19	9	
	45		28		$n = 73$
	35	117	91	27	
	$n = 152$		$n = 118$		$n = 270$

Note: WFLDs = weekly lucid dreamers; NWLDs = non-weekly lucid dreamers; DFMs = daily frequent meditators; WFMs = weekly frequent meditators.

Across the whole general sample, bivariate correlations between dream variables yielded a significant positive correlation between weekly dream recall and monthly lucid dreaming frequency ($r_{sp} = 0.44$, $p < 0.001$) as well as lucid dreams in the previous six months ($r_{sp} = 0.36$, $p < 0.001$). In addition, monthly lucid dream induction frequency correlated with the monthly lucid dreaming frequency ($r_{sp} = 0.46$, $p < 0.001$) and the estimated lucid dreams in the previous six months ($r_{sp} = 0.47$, $p < 0.001$). Furthermore, self-reported meta-awareness correlated positively with monthly lucid dreams ($r_{sp} = 0.23$, $p < 0.001$), while less strongly with the summative measure of lucid dreams in the previous six months ($r_{sp} = 0.14$, $p = 0.012$). Weekly meditation frequency and age were not significantly associated with either the monthly lucid dreaming frequency or the total number of lucid dreams in the most recent six-month period (all $p > 0.05$, Table 2). However, weekly meditation frequency was significantly associated with age ($r_{sp} = 0.38$, $p < 0.001$), meta-awareness ($r_{sp} = 0.25$, $p < 0.001$), and the frequency of wake-initiated lucid dreams ($r_{sp} = 0.15$, $p < 0.001$).

Table 2. Descriptive statistics and bivariate intercorrelations of important variables.

	M (SD)	DRF	M LDF	6M LDF	LDIF	Induce S.	WILD	META	Medi
Weekly Dream Recall	5.92 (3.03)								
Monthly LDF	4.26 (7.65)	0.47 **							
6-Month LDF [a]	27.79 (71.89)	0.35 **	0.93 **						
LD Induction F [a]	3.74 (7.72)	0.45 **	0.47 **	0.47 **					
Induction Success [a]	1.5 (1.12)	0.32 **	0.66 **	0.62 **	0.42 **				
WILD Freq [a]	1.31 (1.15)	0.08	0.53 **	0.51 **	0.28 **	0.57 **			
MAS-MA	5.60 (0.97)	0.19 *	0.23 **	0.15 **	0.09	0.28 **	0.22 **		
Weekly Meditation Freq	3.24 (3.36)	0.14 *	0.06	−0.02	0.06	0.09	0.15 **	0.25 **	
Age	37.74 (16.16)	−0.07	−0.01	−0.06	−0.07	0.02	0.13	0.25 **	0.38 **

Note. ** = $p < 0.001$; * = $p < 0.05$; $n = 270$; [a] = 25 participants excluded; Correlations based on Spearman's correlation coefficient; Monthly LDF (M LDF) = Monthly lucid dreaming frequency; 6-Month LDF [a] (6M LDF) = Total number of lucid dreams during the most recent six-month period; LD Induction F [a] (Induce S.) = Monthly lucid dream induction frequency; Induction Success [a] (0 = *very unlikely*; 4 = *very likely*); WILD Freq [a] (WILD) = Frequency of wake-initiated lucid dreams (0 = *never*; 4 = *always*); DRF = Weekly dream recall frequency; MAS-MA (META) = Meta-Awareness subscale; Weekly Meditation Freq (Medi) = Weekly meditation frequency.

Out of 270 respondents, 243 (90.01%) meditated at least once, of which 91 (33.70%) meditated infrequently, with 0.22 ± 0.26 [M ± SD] weekly meditations. A total of 152 respondents (56.30%) reported meditating at least once a week; the average weekly meditation frequency was 5.58 (SD = 2.72, Range = 1–9). Among the frequent meditators, 35 practiced up to multiple times daily (12.96%), termed DFMs. A total of 13 WFMs and 4 DFMs violated the instructions of the survey, leading to an unrealistic meditation practice time (i.e., 0 minutes of meditation or more than 50 hours per week). Hence, they were excluded from specific analyses. A total of 135 weekly meditators stated a median experience of 7.1 years (SD = 13.7, Range = 0.5–55) in their most consistently practiced meditation framework and a median of 4.67 practiced meditation hours weekly (SD = 7.3, Range = 0.133–45). When looking at the specific meditation styles and techniques, the average total hours per week were divided into six different meditation techniques. The highest practiced minutes averaged over frameworks and traditions was non-dual meditation with 47.44 ± 107.27 [M ± SD], and then OM meditation with 47.01 ± 96.22 [M ± SD], closely followed by FA meditation with 42.07 ± 80.14 [M ± SD], meditation to recognize the true nature of the mind with 17.82 ± 72.32 [M ± SD], LK meditation to cultivate emotional capacities with 20.5 ± 50.3 [M ± SD], and other contemplative techniques with 8.34 ± 25.75 [M ± SD].

Meditation frameworks diverged, falling either within traditions with spiritual and religious backgrounds or within secular frameworks. On average, participants engaged in 2.65 ± 1.96 [M ± SD, $n = 141$, Range = 1–11] traditions or frameworks for their meditation practices. Most participants meditated in a self-guided (69) and an online-based (40) setting, followed by meditation included in secular Yoga practices (29), app-guided meditation (27), Vipassana (24), Transcendental Meditation (15), non-dual meditation (15), meditation included in secular Thai Chi/ Qigong practices (12) and mindfulness-based stress reduction (MBSR) meditation (8). Within the non-secular traditions, most practitioners were rooted in Tibetan Buddhism (34), Zen Buddhism (29), and religious or spiritual Yoga practices (15), as well as Theravadan Buddhism (14), followed by Shamanistic (9) and Christian (5) meditation techniques, concluding with meditation rooted in Daoism-based (2) and Sufi traditions (2). Notably, Judaistic meditation practices were not represented. Participants also had the flexibility to specify an additional tradition (10) or secular approach (14).

3.1. Lucid Dreaming in Meditators

All frequent meditators together (collapsing DFMs and WFMs) had an insignificantly higher monthly lucid dreaming frequency compared to non-frequent meditators. In addition, the numbers of lucid dreams in the previous six-month period revealed insignificantly higher numbers of lucid dreams in frequent meditators compared to non-frequent meditators (see Table 3, right panel), whereas the dream recall frequency was significantly higher in frequent meditators than in non-frequent meditators. All frequent meditators had significantly more wake-initiated lucid dreams compared to the WILD occurrences in non-frequent meditators, while also being more often in the role of a detached observer in the lucid dream compared to non-frequent meditators. Having control in the lucid dream, as is possible in waking life, showed no significant difference between frequent meditators and the rest of the participants ($p > 0.05$; Table 3, right panel).

Table 3. Group comparisons for lucid dream experience variables between different meditation groups.

| | DFMs > Non-Frequent Medi. ($n = 35$) | | | | Frequent Medi. > Non-Frequent Medi. ($n = 152$) | | | |
| | | | ($n = 118$) | | | | ($n = 118$) | |
	M (SD)	M (SD)	M Diff	p [a]	M (SD)	M (SD)	M Diff	p [a]
Frequency								
LDF 6M	56.7 (156.6)	20.4 (41.9)	33.1	0.03 *	33.3 (87.7)	20.4 (41.9)	12.84	0.11
LDF monthly	6.44 (9.1)	3.55 (7.0)	2.89	0.03 *	4.80 (8.1)	3.55 (7.0)	1.25	0.11
DRF weekly	7.04 (2.5)	5.40 (3.1)	1.65	0.01 *	6.33 (2.9)	5.40 (3.1)	0.92	0.02 *
WILD	1.64 (1.4)	1.14 (1.1)	0.50	0.03 *	1.44 (1.2)	1.14 (1.1)	0.29	0.04 *
Control								
Control PWL	2.03 (1.1)	2.57 (1.1)	−0.54	0.03 *	2.38 (1.2)	2.57 (1.1)	−0.19	0.11
Experience								
Detached observer	1.52 (1.0)	1.10 (1.0)	0.42	0.03 *	1.56 (1.0)	1.10 (1.1)	0.46	0.0006 *

Note: M Diff = Mean difference between observed mean and mean constructed under the null hypothesis with R = 10,000 permutations; p [a] = p-value corrected with false discovery error rate, * = $p < 0.05$; DFMs = daily frequent meditators; Frequent-Medi = Frequent meditators; LDF 6M = Number of lucid dreams during the previous six-month period, LDF monthly = Monthly lucid dreaming frequency, DRF weekly = Weekly dream recall frequency; WILD = Frequency of wake-initiated lucid dreams; Control PWL = Control aspects of the lucid dream possible during waking life.

Among the weekly meditators, 45 practitioners reported experiencing lucid dreams once or more per week. When correcting for the hours spent in meditation, 38 WFLD frequent meditators dedicated significantly more time to meditation each week (Mdn = 5.29, SD = 9.29) in comparison to their NWLD frequent meditating counterparts (Mdn = 3.67, SD = 6.32, n = 97) [p = 0.039; Monte Carlo permutation test]. The number of years of meditation experience for WFLD frequent meditators (Mdn = 7.00, *SD* = 13.84) did not differ significantly from NWLD frequent meditators (Mdn = 6.00, SD = 13.50) [p = 0.400; Monte Carlo permutation test].

When looking at the group with more intensive meditation practice, DFMs had significantly more lucid dreams per month compared to non-frequent meditators; see Table 3, left panel. In addition, there were significantly higher numbers of lucid dreams in the previous six-month period in DFMs compared to non-frequent meditators. Dream recall frequency differed significantly from non-frequent meditators. DFMs were less often controlling in the way that is possible during waking life compared to non-frequent meditators. Moreover, DFMs were significantly more often in the detached observer stance in a lucid dream than non-frequent meditators. Furthermore, DFMs experienced significantly more compared to non-frequent meditators; see Table 3, left panel.

After excluding the non-valid cases, DFMs reached significance with a median of 9.33 ± 7.52 [Mdn ± SD, n = 31] hours of weekly meditation compared to 3.50 ± 6.80 [Mdn ± SD, n = 104] practiced hours in WFMs [$p < 0.0001$; Monte Carlo permutation test].

There was no significant difference in meditation experience in years ($p > 0.05$). A total of 12 DFMs who were classified as WFLDs engaged in 3.58 ± 2.35 [M ± SD] meditation frameworks or traditions: self-guided meditation (5), Transcendental Meditation/TM (5), meditation from Shamanistic traditions (4), Vipassana meditation (4), Tibetan (4), Zen (4), and Theravadan (2) Buddhism, non-dual meditation (3), and meditation in secular Thai Chi/ Qigong practices (4). There was no significant difference in the total hours of meditation per week between WFLDs and NWLDs in daily frequent meditators ($p > 0.05$). DFMs who lucid-dream weekly had significantly more meditation experience, with a median of 25 ± 16.85 years [Mdn ± SD, $n = 11$], compared to non-weekly lucid dreaming DFMs, with a median of 5 ± 14.66 years (Mdn ± SD, $n = 19$) [$p = 0.029$; Monte Carlo permutation test].

3.2. Lucid Dreaming and Meditation Practices

The assessment of the varied qualities cultivated during meditation, based on specific meditative practices, was accomplished by distributing the total weekly practice hours across the percentage of the techniques each participant practiced, and then averaging the practice time for these qualities/techniques across all traditions for each participant. As hypothesized, the open monitoring meditation practice exhibited a significant positive bivariate correlation with the number of lucid dreams in the previous six-month period ($r_{sp} = 0.16$, $p = 0.037$). Hence, more weekly practiced hours of OM meditation were associated with more lucid dreams per month. Other techniques did not show any significant relationship: FA meditation ($r_{sp} = 0.08$, $p > 0.05$), meditation related to the nature of the mind ($r_{sp} = 0.08$, $p > 0.05$), nondual meditation ($r_{sp} = 0.06$, $p > 0.05$), and LK meditation ($r_{sp} = 0.03$, $p > 0.05$) showed insignificant bivariate correlations with the number of lucid dreams during the previous six months. Furthermore, when looking at the association between meditation practices and monthly lucid dreaming frequency, there was no significant bivariate association (all $p > 0.05$).

3.3. Lucid Dreaming and Mindfulness Instruments

Weekly frequent lucid dreamers descriptively scored more highly in all mindfulness measurements, except for the Describing facet of the FFMQ. If collapsing participants across all groups (DFMs, WFMs, and non-frequent meditators): WFLD (M = 3.55, SD = 0.79) exceeded NWLD (M = 3.23, SD = 0.77) in the Nonreactivity subscale of the FFMQ [$p = 0.018$; Monte Carlo permutation test with FDR correction]. Furthermore, the sample-specific Mindful Transcendence subscale was also higher in WFLDs (M = 3.59, SD = 1.69) compared to NWLDs (M = 3.05, SD = 1.56) [$p = 0.027$; Monte Carlo permutation test with FDR correction]. Within non-frequent meditators, no differences reached significance after controlling for multiple testing when comparing WFLDs and NWLDs (all $p > 0.05$; Table 4, right panel).

Table 4. Group comparisons for all instruments between frequent and non-frequent lucid dreamers within meditation groups.

	DFMs				Frequent Meditators				Non-Frequent Meditators			
	WFLDs	NWLDs			WFLDs	NWLDs			WFLDs	NWLDs		
	M (SD)	M (SD)	M Diff [a]	*p* [a]	M (SD)	M (SD)	M Diff [a]	*p* [a]	M (SD)	M (SD)	M Diff [a]	*p* [a]
FFMQ												
Observing	4.50 (0.07)	3.94 (0.6)	0.56	0.02 *	4.16 (0.8)	3.96 (0.7)	0.20	0.12	3.9 (0.7)	3.8 (0.8)	0.10	0.25
Describing	4.31 (0.8)	3.80 (0.6)	0.51	0.04 *	3.60 (1.04)	3.61 (0.71)	−0.01	0.47	3.38 (0.8)	3.11 (0.9)	0.27	0.23
Actaware	3.40 (0.9)	3.38 (0.9)	0.02	0.46	3.22 (0.9)	3.12 (0.8)	0.10	0.33	2.91 (0.9)	2.67 (0.8)	0.24	0.23
Nonjudge	4.27 (1.1)	3.74 (0.9)	0.53	0.14	3.93 (0.8)	3.89 (0.8)	0.04	0.46	3.54 (1.0)	3.35 (1.0)	0.19	0.24
Nonreact	4.02 (0.7)	3.47 (0.7)	0.55	0.04 *	3.6 (0.7)	3.42 (0.7)	0.26	0.14	3.32 (0.9)	3.00 (0.8)	0.32	0.23
TMS												
Decentering	2.59 (1.0)	2.53 (0.7)	0.07	0.46	2.54 (0.8)	2.32 (0.8)	0.22	0.14	1.73 (1.0)	1.60 (0.8)	0.14	0.25
Curiosity	2.41 (1.1)	2.27 (0.7)	0.14	0.42	2.30 (1.0)	2.15 (0.8)	0.14	0.26	1.86 (0.7)	1.65 (1.0)	0.15	0.24

Table 4. *Cont.*

	DFMs				Frequent Meditators				Non-Frequent Meditators			
	WFLDs	**NWLDs**			**WFLDs**	**NWLDs**			**WFLDs**	**NWLDs**		
	M (SD)	**M (SD)**	**M Diff** [a]	*p* [a]	**M (SD)**	**M (SD)**	**M Diff** [a]	*p* [a]	**M (SD)**	**M (SD)**	**M Diff** [a]	*p* [a]
RMMtm Transcend	4.52 (1.2)	3.80 (1.8)	0.72	0.18	4.01 (1.7)	3.56 (1.6)	0.55	0.14	2.78 (1.4)	2.45 (1.3)	0.33	0.23

Note: M Diff [a] = Mean difference between observed mean and mean constructed under the null hypothesis and R = 10,000 permutations; *p* [a] = *p*-value corrected with false discovery rate for each set of tests; * = $p < 0.05$; DFMs = daily frequent meditators; WFLDs = weekly lucid dreamers; NWLDs = non-weekly lucid dreamers; TMS = Toronto Mindfulness Scale; FFMQ = Five Facet Mindfulness Questionnaire; Actaware = Acting with Awareness; Nonjudge = Nonjudgment; Nonreact = Nonreactivity; RMMtm = Relaxation, Mindfulness and Meditation Tracker.

Within the frequent meditators, WFLDs scored descriptively higher in all mindfulness measurements but none of the differences reached statistical significance after controlling for multiple testing (all $p > 0.05$; Table 4, middle panel). In contrast, within the DFMs, WFLDs surpassed NWLDs in the FMMQ mindfulness aspects of Nonreactivity, Describing, and Observing (Table 4, left panel).

3.4. Individual Differences in Meta-Awareness

Meta-awareness differed significantly between meditators: DFMs scored highest at 6.06 ± 0.68 [M ± SD] compared to WFMs at 5.7 ± 0.91 [M ± SD] ($p = 0.0224$) and non-frequent meditators at 5.34 ± 1.04 [M ± SD] ($p = 0.0007$). In addition, meta-awareness scores between WFMs and non-frequent meditators reached significance [$p = 0.0035$; Monte Carlo permutation test with FDR correction]. Separately, across the general sample for the lucid dreaming groups, meta-awareness was highest in WFLDs, scoring 5.90 ± 0.98 [M ± SD], compared to monthly frequent lucid dreamers, who scored 5.56 ± 0.90 [M ± SD] ($p = 0.0247$), and non-frequent lucid dreamers, who scored 5.44 ± 0.98 [M ± SD] ($p = 0.0032$) [Monte Carlo permutation tests with FDR correction]. However, monthly frequent lucid dreamers could not reach significantly higher scores in meta-awareness compared to non-frequent lucid dreamers [$p = 0.2256$; Monte Carlo permutation test with FDR correction]. Within the non-frequent meditators, WFLDs also showed higher values with 5.82 ± 0.91 [M ± SD] in meta-awareness compared to NWLDs with 5.20 ± 1.03 [M ± SD] [$p = 0.0047$; Monte Carlo permutation test with FDR correction]. In contrast, WFLDs within the weekly frequent meditators could not reach significantly higher values, 5.73 ± 1.13 [M ± SD], than NWLDs who meditate once a week, 5.71 ± 0.83 [M ± SD] ($p = 0.4346$). However, DFMs who are weekly frequent lucid dreamers scored significantly higher, 6.44 ± 0.48 [M ± SD], compared to non-weekly lucid DFMs, 5.83 ± 0.69 [M ± SD] ($p = 0.0079$) [Monte Carlo permutation test with FDR correction].

3.5. Meditation Frequency, Mindfulness, and Meta-Awareness for Lucid Dreaming

Mindful Transcendence, measured using the RMMtm, and the two dimensions, Decentering and Curiosity, of the TMS, as well as the MAS subscale measuring meta-awareness, in addition to weekly meditation frequency, were utilized to predict monthly lucid dreaming frequency. Therefore, a multiple linear regression model with BCa bootstrapped confidence intervals was performed, given the violation of the assumptions of normality and homoscedasticity of the residual. All predictors accounted for $R^2 = 0.098$, $F_{(6, 263)} = 4.780$, $p < 0.001$. Transcendence [b = 1.228, BCa 95% CI (0.423, 1.219)], in addition to the MAS-MA subscale [b = 1.420, BCa 95% CI (0.633, 1.405)], had significant predictive regression coefficients for the monthly frequency of lucid dreams (Table 5). Age was statistically controlled for and did not have a significant coefficient. Therefore, higher scores on the Transcendence subscale, as well as higher scores on the MAS-MA meta-awareness subscale, are associated with a higher frequency of lucid dreams per month.

Table 5. Multiple regression model for the relationship between lucid dreaming frequency and aspects of mindfulness, transcendence, meditation frequency, and meta-awareness, controlled for age.

Model	b	Bias [a]	SE [a]	95% BCa CI [a] Lower	Upper
Constant	−4.136	0.058	2.552	−9.346	0.727
Mindful Transcendence	1.228 *	−0.007	0.423	0.474	2.146
Decentering TMS	−0.303	−0.009	0.835	−1.921	1.367
Curiosity TMS	−0.441	0.004	0.792	−2.025	1.059
Meta-Awareness MAS	1.420 *	−0.006	0.633	0.227	2.684
Weekly Meditation Freq	0.029	0.002	0.171	−0.305	0.369
Age	−0.055	−0.0003	0.029	−0.113	0.0003

Note. [a] = Bootstrapped with R = 10,000 replicates; * = significant coefficient; SE [a] = standard error of the unstandardized coefficient; Model summary: R^2 = 0.098; Residual standard error: 7.350; $F_{(6, 263)}$ = 4.780; p = 0.0001; n = 270; Weekly Meditation Freq = Weekly meditation frequency.

Two regression models and one mediation analysis were conducted to further elucidate the relevance of meta-awareness for the association between meditation frequency and lucid dreaming frequency. Specifically, the relationship between weekly meditation frequency and the number of lucid dreams in the previous six-month period (Lucid Dreams 6M) was tested with meta-awareness as the mediator. Weekly dream recall frequency, age, and monthly lucid dream induction frequency were included as covariates. Considering the violations of normality, homoscedasticity, and concerns with leverage and outliers, the coefficients' confidence intervals from both regression models, as well as the test of the indirect path, were bootstrapped with R = 10,000 runs, and a heteroscedasticity-consistent standard error was used.

The first model, with meta-awareness as the outcome variable, explained a total of R^2 = 0.124 variance, $F_{(4, 240)}$ = 9.836, $p < 0.0001$. Age, dream recall, and meditation frequency significantly predicted meta-awareness (Table 6).

Table 6. Regression models for complete mediation of the relationship of lucid dreaming occurrences in the previous 6-month period and weekly meditation frequency mediated by meta-awareness.

Outcome Variable	b	SE [a]	95% BCa CI [a] Lower	Upper	R^2	F^{hc}
META-AWARENESS					0.124	9.836 ***
Intercept	4.661 *	0.221	4.215	5.085		
Meditation Freq	0.047 *	0.020	0.009	0.086		
Induction Freq	0.002	0.009	−0.017	0.018		
Dream Recall Freq	0.048 *	0.022	0.004	0.091		
Age	0.014 *	0.004	0.006	0.021		
LUCID DREAMS 6M					0.305	4.778 ***
Intercept	−17.849	20.459	−62.946	9.033		
Meditation Freq	0.702	1.366	−1.610	3.904		
Meta-Awareness	6.481 *	3.626	0.122	14.650		
Induction Freq	4.913 *	1.485	2.358	8.296		
Dream Recall Freq	0.305	0.939	−1.605	2.125		
Age	−0.357 *	0.201	−0.857	−0.037		

Note. [a] = Bootstrapped with R = 10,000 replicates; SE [a] = standard error of the unstandardized coefficient; * = significant coefficient; *** = $p < 0.001$; F^{hc} = robust estimate for standard error due to heteroscedasticity; n = 245; Lucid Dreams 6M = number of lucid dreams in the previous six-month period; Meditation Freq = Weekly meditation frequency; Induction Freq = Monthly frequency of induction of lucid dreams; Dream Recall Freq = Weekly dream recall frequency.

More frequent meditation was associated with higher meta-awareness (b = 0.048, BCa 95% CI [0.009, 0.086]). Dream recall frequency (b = 0.047, BCa 95% CI [0.004, 0.091]) and age (b = 0.013, BCa 95% CI [0.004, 0.091]) were both significantly positively correlated with meta-awareness. Predicting lucid dreaming frequency (Lucid Dreams 6M) explained $R^2 = 0.305$ variance, F (5, 239) = 4.778, $p < 0.0001$. The mediator analysis of the coefficient product of the indirect path, the coefficient of predicting meta-awareness based on meditation frequency, and Lucid Dreams 6M based on meta-awareness resulted in a significant positive indirect effect (b = 0.303, BCa 95% CI [0.013, 1.009]; see Figure 1. The direct effect of meditation frequency on Lucid Dreams 6M revealed an insignificant positive effect (b = 0.703, 95% CI [−2.030, 3.435]; $p = 0.613$). Two covariates, higher lucid dream induction frequency (b = 4.913, BCa 95% CI [2.357, 8.296]) and younger age (b = −0.357, BCa 95% CI [−0.857, −0.037]), were significantly associated with higher numbers of lucid dreams in the previous six months.

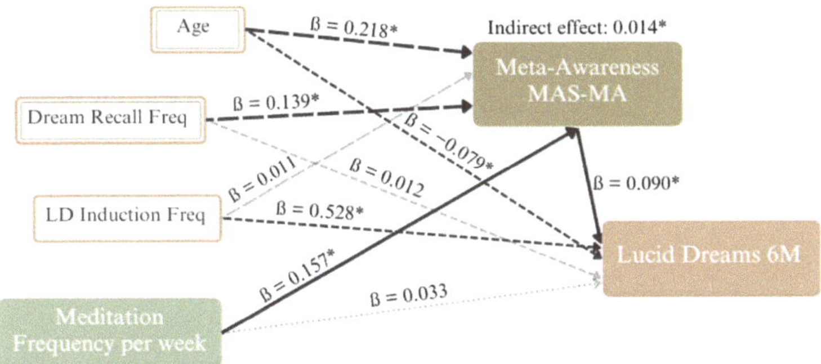

Figure 1. Complete mediation model for the relationship between meditation frequency and the total number of lucid dreams in the previous six-month period with age, weekly dream recall, and monthly lucid dream induction frequencies as covariates. Note: $n = 245$. Bootstrapped BCa with R = 10,000. The model includes only standardized coefficients and effects; * = significant effect; grey dotted line = the insignificant direct effect; large dashed grey line = the insignificant effect of the covariate on the mediator; small grey dashed line = the effect of the covariate on the dependent variable; small dashed black lines = the significant effects of the covariates on the dependent variable; large dashed black lines = the significant effects of the covariates on the mediator; solid black lines = the indirect path.

4. Discussion

The present study explored the link between lucid dreaming, dispositional mindfulness, and meditation practices, emphasizing the role of meta-awareness. The results from this study indicate that individuals who meditate more than once per day have more lucid dreams compared to infrequent meditators. Therefore, the hypothesis that frequent meditation is associated with more lucid dreams was supported, adding to the already existing body of work [28,33]. The results also partially confirmed the hypothesis that open monitoring meditation was positively associated with increased lucid dreaming. In addition, meta-awareness was found to be highest in daily frequent meditators and was also elevated in weekly lucid dreamers without meditation experience. Together, these findings indicate a link between lucid dreaming, meta-awareness, and OM meditation. As OM meditation is known to enhance sustained open awareness, this supports the idea that non-propositional and sustained meta-awareness could be a key capacity for lucid dreaming [7]. Exploratory analyses highlighted the mediating role of meta-awareness in the relationship between weekly meditation frequency and lucid dream occurrences during the previous six months. Furthermore, the evidence that aspects of trait mindfulness are associated with frequent lucid dreaming points towards the continuity of mindful awareness from waking consciousness to sleeping consciousness [70,71]. In addition to meta-awareness, transcendence also emerged as a positive predictor of monthly lucid dreaming frequency across all groups, replicating previous research [36].

Several conceptual links have been made between lucid dreaming and meditation practices, postulated to be influenced by regulating attention and meta-awareness [19,28,33]. A meta-analysis looked at meditation practices across various traditions and backgrounds and extracted the various effects of different meditation styles and their influence on cognitive capacities [72]. Several meditation practices influence the dynamics and direction of attention: in FA meditation, attention is directed towards one object, e.g., focusing on the breath and keeping the concentration on the same object up to the whole session of meditation [17]. This process involves meta-awareness to recognize the wandering mind. Once the inner focus is distracted, e.g., by thoughts about the future, meta-awareness can detect the distraction and thus return to the initial object of focus [8,18]. In some practice styles, OM meditation is postulated to evolve from FA meditation [17]. As individuals progress in their meditation practice, they cultivate monitoring skills that become the crucial point to the practice of OM meditation. The practitioner attempts to remain in a state of pure observation, vigilantly attending and monitoring each moment-to-moment event in awareness without anchoring the focus on any particular object [17].

Existing studies have not examined the specific meditative practices that might lead to more frequent lucid dreams. The purpose of this study was to study the link between specific meditation practices and lucid dreams by investigating the time spent on five different meditation styles per week. The results partially confirmed the main hypothesis that FA meditation and OM meditation would correlate with an increased number of monthly lucid dreams, with a positive association observed only between the duration of averaged weekly OM practice and the number of lucid dreams over the previous six months.

However, it should be noted that there was a discrepancy between the formative and summative measurement of lucid dreaming frequency. Specifically, in contrast to the lucid dreams in the previous 6 months, there was no significant association between the average time spent in OM meditation and average monthly lucid dreaming frequency among meditators. One reason for this discrepancy may be that the 6-month measure is a more accurate assessment of lucid dreaming frequency as it is more precise (it requires that participants report a specific number of lucid dreams) and over a proximal, well-defined time interval (the previous 6 months). Another possibility is that meditation experience and practice could influence lucid dreaming, but possibly only in individuals who have already had lucid dreams. This is consistent with research where meditation has been successfully integrated as a complementary technique in combination with cognitive and substance-enhanced lucid dream induction methods [73].

Only a handful of studies have examined the relationship between meditation experience and lucid dreaming frequency [26,28,29,31,33–36,74]. Baird and colleagues [28] observed a higher frequency of spontaneous lucid dreams per month among long-term meditators compared to those with less meditation experience. The current study supports these findings, with daily frequent meditators reporting a greater number of lucid dreams per month than infrequent meditators. Consequently, this study supports the notion that regular and intensive meditation is associated with an increased incidence of lucid dreams [28]. In particular, it was found that meditating several times a day was associated with a weekly occurrence of lucid dreams, a finding that is consistent with results observed in practitioners of Transcendental Meditation [29].

On a cognitive, psychological, and neuropsychological level, researchers have suggested that the link between lucid dreaming and meditation could be an increase in self-reflectiveness and meta-awareness [28,31,33,75]. Given that no earlier research has examined individual differences in self-reported meta-awareness within and between lucid dreamers and meditators, this study analyzed individual differences in meta-awareness using the MAS-MA scale. Daily frequent meditators reported the highest self-rated meta-awareness on this scale within the general sample. Furthermore, weekly lucid dreamers showed higher meta-awareness scores compared to both monthly and infrequent lucid dreamers. Monthly frequent lucid dreamers showed only a marginal increase in meta-awareness compared to infrequent lucid dreamers across the sample. Notably, even among non-meditators, weekly lucid dreamers expressed higher meta-awareness than non-weekly lucid dreamers. Only within the DFM subgroup was weekly lucid dreaming associated with higher meta-awareness compared to non-weekly lucid dreamers. These results implicate an effect of meditation on meta-awareness, as well as an effect of meta-awareness on lucid dreaming, pointing towards a mediating effect of meta-awareness on lucid dreaming frequency.

Indeed, the results of an exploratory mediation analysis supported the hypothesis that meta-awareness fully mediates the positive association between meditation frequency per week and the number of lucid dreams in the previous six-month period. However, it has to be noted that a mediation model with this study design cannot imply a causal relationship since the variables were assessed simultaneously, and there might be other variables and cognitive mechanisms involved that were not captured. Nonetheless, the findings are consistent with the hypothesis that higher meta-awareness is associated with more lucid dreams [7,19].

Consistent with previous findings by Stumbrys et al. [33] and Baird et al. [28], the results suggest that lucid dreaming is associated with specific aspects of mindfulness that differ depending on whether an individual has experience with meditation. WFLDs reported higher scores on the Non-reactivity subscale of the Five Facet Mindfulness Questionnaire (FFMQ) than did non-weekly lucid dreamers. Nonreactivity involves the ability to experience thoughts and feelings without becoming caught up in them [1]. Higher scores in this area for WFLDs may indicate a more developed ability to observe experiences without immediate reaction, which may be advantageous for recognizing the dream state and becoming lucid, without being caught up in the event and thereby losing lucidity. Further analysis showed that WFLDs also had significantly higher scores on the RMMtm Mindful Transcendence subscale compared to NWLDs. This facet reflects a heightened awareness of the present experience and a more transcendent perspective that may facilitate the detachment necessary for dream lucidity. Mindful Transcendence measured in this study is comparable to the transcendence subscale found in the sample of students in the study by Geise and Smith [36]. We replicated that feeling more self-transcendent, like "I felt connected. I felt at one with everything and humanity. I felt in harmony with the world. I felt a sense of belonging. A part of something larger", positively predicted lucid dreaming frequency per month [36]. In contrast, there were no significant differences in mindfulness facets found between WFLDs and NWLDs within the group of non-frequent meditators and all frequent meditators, DFMs and WFMs together, after adjustment for multiple comparisons. This is in contrast with the study of long-term meditators, where non-frequent meditators scored higher on the Describing scale of the FFMQ [28].

There has been no other study to date that investigated aspects of lucid dream experiences in meditators. Here, we found that daily frequent meditators reported having more wake-initiated lucid dreams, were more often in the observing stance in a lucid dream, and were less often actively exerting ordinary forms of control of the dream compared to infrequent meditators. Overall, these results support the notion that consistent, intensive meditation practice may enhance one's ability to maintain a nonreactive, observing stance, and, by extension, enhance the state shift from ordinary to lucid dreams. The results indicate that the frequency and depth of mindfulness practices are linked to the experience of lucid dreaming and underscore the importance of considering individual differences in meditation practice and experience for lucid dreaming.

5. Limitations and Future Directions

It is crucial to highlight that the general mixed sample included a substantial proportion of frequent meditators, 56% of respondents, as well as an extremely high number of monthly frequent (53%) and weekly frequent (27%) lucid dreamers. This distribution does not reflect the general population, a discrepancy that is highlighted when compared to student populations or representative cohort studies [49,51]. Meta-analytic research suggests a 23% prevalence of monthly lucid dreams and a 55% likelihood of having at least one lucid dream [76]. The overrepresentation in this study is likely due to self-selection bias, disproportionately including individuals interested in lucid dreaming, dreaming, and meditation [77].

The introduction of a 16-point Likert scale marked a major advance in the assessment of lucid dreaming frequency, providing more variability but making direct comparisons with responses from previous studies more difficult. Many studies operationalized lucid dreaming frequency based on the scale developed by Schredl and Erlacher [78] as a formative measure (giving an estimate of how many lucid dreams one has per month), but few studies also incorporated a summative approach to measuring lucid dreams [31]. The correlation analysis related to the hours spent in OM meditation showed a discrepancy between the two measures of lucid dreaming frequency, although the association between the two frequency measures was high. The discrepancy needs to be considered when interpreting the relationship between lucid dreams and meditation practices.

The cross-sectional nature of the study with a single data collection point limits conclusions to correlational and associative implications. Biases such as social desirability and performance bias, particularly prevalent in a convenience sample likely to be interested in meditation and (lucid) dreaming, may bias self-reported measures of meta-awareness and mindfulness [79]. This potential response bias and selective sample underscore the need for caution when interpreting the results. Recall bias presents a major challenge in retrospective assessments, potentially distorting the accuracy of participant reports of their experiences of lucid dreaming and meditation practices [80]. This bias can be particularly problematic when relying on individuals to estimate the frequency of events over extended periods of time. Along this line, dream recall is usually underestimated in retrospective measurements compared to daily log-books [81]. Ecological Momentary Assessment (EMA) could mitigate this problem by collecting data in real time, thereby providing more accurate and immediate reports of experiences as they occur [82]. Implementing EMA in future studies would not only increase the reliability of self-reported data but also provide greater insight into the dynamic interplay between mindfulness practices and lucid dreaming. Future investigations should include a wide scale for lucid dreaming frequency, encompassing the higher end of variability in lucid dreamers. Also, it could be fruitful to compare different meditation practices within different frameworks and traditions (e.g., app-based meditation vs. Transcendental Meditation).

6. Conclusions

The findings of this study validate the association between frequent meditation, specifically open monitoring (OM) meditation, and increased lucid dreaming frequency, and support a role of meta-awareness in enhancing lucid dream experiences. These results suggest that OM meditation enhances sustained meta-awareness, which is essential for recognizing and maintaining lucidity in dreams. It would be worthwhile in future research to test this hypothesis through a random-assignment meditation intervention pre-post design study. In addition, the experiences of expert meditators and those practicing dream Yoga should be explored to further understand consciousness in different sleep states. Longitudinal studies and intensive retreats may prove valuable in assessing the effects of meditation on lucid dreaming. Complementary methods, such as sleep diaries and EMA, could allow for the detailed tracking of lucid dreams and meditation practices. Investigating neurophysiological changes of expert meditators during sleep might also shed light on the neural underpinnings of meditation-related changes in consciousness.

Supplementary Materials: The following supporting information can be downloaded at: https://www.mdpi.com/article/10.3390/brainsci14050496/s1, Table S1: Factor loading matrix of RMMtm using PCA and Oblimin rotation; Figure S1: Scree plot of PCA result; Figure S2: Scree plot of parallel analysis of PCA result.

Author Contributions: Conceptualization, E.G., B.B.; methodology, E.G., B.B.; software, E.G., B.B.; formal analysis, E.G., B.B.; investigation, E.G., B.B.; data curation, E.G., B.B.; writing—original draft preparation, E.G.; writing—review and editing, E.G., B.B., supervision, B.B.; visualization, E.G; project administration, E.G., B.B. All authors have read and agreed to the published version of the manuscript.

Funding: This study did not receive specific funding from public, private, or nonprofit sources.

Institutional Review Board Statement: The study was conducted according to the guidelines of the Declaration of Helsinki and approved by the IRB Review Board of The University of Texas at Austin (STUDY00003582, 11 July 2022).

Informed Consent Statement: Informed consent was obtained from all subjects involved in the study.

Data Availability Statement: The raw data supporting the conclusions of this article will be made available by the authors on request. The data are not publicly available due to privacy restrictions.

Acknowledgments: We thank Lukasz Stasielowicz for his help with data collection and advice. We also thank our colleagues who reviewed and commented on the manuscript. Finally, we thank Nino Rekhviashvili for her supportive comments on aspects related to meditation. The acknowledgments do not imply that the individuals mentioned endorse the views expressed in this manuscript. The authors are solely responsible for the views expressed in this article.

Conflicts of Interest: The authors declare no conflicts of interest.

References

1. Bishop, S.R.; Lau, M.; Shapiro, S.; Carlson, L.; Anderson, N.D.; Carmody, J.; Segal, Z.V.; Abbey, S.; Speca, M.; Velting, D. Mindfulness: A Proposed Operational Definition. *Clin. Psychol. Sci. Pract.* **2004**, *11*, 230. [CrossRef]
2. Kabat-Zinn, J. An Outpatient Program in Behavioral Medicine for Chronic Pain Patients Based on the Practice of Mindfulness Meditation: Theoretical Considerations and Preliminary Results. *Gen. Hosp. Psychiatry* **1982**, *4*, 33–47. [CrossRef] [PubMed]
3. van Eeden, F. A Study of Dreams. *Proc. Soc. Psych. Res.* **1913**, *26*, 431–461.
4. LaBerge, S.P.; Nagel, L.E.; Dement, W.C.; Zarcone, V.P. Lucid Dreaming Verified by Volitional Communication during Rem Sleep. *Percept. Mot. Skills* **1981**, *52*, 727–732. [CrossRef] [PubMed]
5. Baird, B.; Erlacher, D.; Czisch, M.; Spoormaker, V.I.; Dresler, M. Consciousness and Meta-Consciousness during Sleep. In *Handbook of Behavioral Neuroscience*; Elsevier: Amsterdam, The Netherlands, 2019; Volume 30, pp. 283–295; ISBN 1569-7339.
6. Schooler, J.W.; Smallwood, J.; Christoff, K.; Handy, T.C.; Reichle, E.D.; Sayette, M.A. Meta-Awareness, Perceptual Decoupling and the Wandering Mind. *Trends Cogn. Sci.* **2011**, *15*, 319–326. [CrossRef] [PubMed]
7. Dunne, J.D.; Thompson, E.; Schooler, J. Mindful Meta-Awareness: Sustained and Non-Propositional. *Curr. Opin. Psychol.* **2019**, *28*, 307–311. [CrossRef] [PubMed]
8. Dahl, C.J.; Lutz, A.; Davidson, R.J. Reconstructing and Deconstructing the Self: Cognitive Mechanisms in Meditation Practice. *Trends Cogn. Sci.* **2015**, *19*, 515–523. [CrossRef] [PubMed]
9. Stumbrys, T.; Erlacher, D.; Schädlich, M.; Schredl, M. Induction of Lucid Dreams: A Systematic Review of Evidence. *Conscious. Cogn.* **2012**, *21*, 1456–1475. [CrossRef]
10. Tan, S.; Fan, J. A Systematic Review of New Empirical Data on Lucid Dream Induction Techniques. *J. Sleep Res.* **2023**, *32*, e13786. [CrossRef]
11. Harvey, P. *An Introduction to Buddhism: Teachings, History and Practices*; Cambridge University Press: Cambridge, UK, 2012; ISBN 978-1-139-85126-8.
12. Gillespie, G. Lucid Dreams in Tibetan Buddhism. In *Conscious Mind, Sleeping Brain: Perspectives on Lucid Dreaming*; Springer: Berlin/Heidelberg, Germany, 1988; pp. 27–35.
13. Gillespie, G. Dreams and Dreamless Sleep. *Dreaming* **2002**, *12*, 199–207. [CrossRef]
14. Norbu, N.; Katz, M. *Dream Yoga and the Practice of Natural Light*; Snow Lion Pub.: Ithaca, NY, USA, 2002; ISBN 1-55939-161-8.
15. Wangyal, T. *The Tibetan Yogas of Dream and Sleep*; Shambhala Publications: Boulder, CO, USA, 1998; ISBN 978-1-55939-101-6.
16. Sparby, T.; Sacchet, M.D. Defining Meditation: Foundations for an Activity-Based Phenomenological Classification System. *Front. Psychol.* **2022**, *12*, 795077. [CrossRef] [PubMed]
17. Lutz, A.; Slagter, H.A.; Dunne, J.D.; Davidson, R.J. Attention Regulation and Monitoring in Meditation. *Trends Cogn. Sci.* **2008**, *12*, 163–169. [CrossRef]
18. Lutz, A.; Jha, A.P.; Dunne, J.D.; Saron, C.D. Investigating the Phenomenological Matrix of Mindfulness-Related Practices from a Neurocognitive Perspective. *Am. Psychol.* **2015**, *70*, 632–658. [CrossRef] [PubMed]
19. Baird, B.; Mota-Rolim, S.A.; Dresler, M. The Cognitive Neuroscience of Lucid Dreaming. *Neurosci. Biobehav. Rev.* **2019**, *100*, 305–323. [CrossRef] [PubMed]
20. LaBerge, S.P. The Psychophysiology of Lucid Dreaming. In *Conscious Mind, Sleeping Brain: Perspectives on Lucid Dreaming*; Gackenbach, J., LaBerge, S., Eds.; Springer New York: Boston, MA, USA, 1988; pp. 135–153; ISBN 978-1-4757-0423-5.

21. LaBerge, S.P. *Lucid Dreaming: An Exploratory Study of Consciousness during Sleep*; Stanford University: Stanford, CA, USA, 1980; ISBN 9798660816505.
22. LaBerge, S.P.; Rheingold, H. *Exploring the World of Lucid Dreaming*; Ballantine Books: New York, NY, USA, 1990; ISBN 0-345-42012-8.
23. Hearne, K.M.T. Lucid Dreams: An Elecro-Physiological and Psychological Study. Ph.D. Thesis, University of Liverpool, Liverpool, UK, 1978.
24. Maleeh, R.; Konjedi, S. Meta-Awareness, Mind Wandering and Negative Mood in the Context of the Continuity Hypothesis of Dreaming. *Phenom. Cogn. Sci.* **2022**, *23*, 105–131. [CrossRef]
25. Moffitt, A.; Hoffmann, R.; Mullington, J.; Purcell, S.; Pigeau, R.; Wells, R. Dream Psychology: Operating in the Dark. *Conscious Mind Sleep. Brain Perspect. Lucid Dreaming* **1988**, 429–439. [CrossRef]
26. Rider, R.L. Exploring the Relationship between Mindfulness in Waking and Lucidity in Dreams A Thesis Submitted to the Faculty of. Ph.D. Thesis, Drexel University, Philadelphia, PA, USA, 2012.
27. Bell, A.P.; Hall, C.S. *The Personality of a Child Molester: An Analysis of Dreams*; Transaction Publishers: Piscataway, NJ, USA, 2011; ISBN 1-4128-1847-8.
28. Baird, B.; Riedner, B.A.; Boly, M.; Davidson, R.J.; Tononi, G. Increased Lucid Dream Frequency in Long-Term Meditators but Not Following Mindfulness-Based Stress Reduction Training. *Psychol. Conscious. Theory Res. Pract.* **2019**, *6*, 40. [CrossRef] [PubMed]
29. Gackenbach, J.; Cranson, R.; Alexander, C. Lucid Dreaming, Witnessing Dreaming, and the Transcendental Meditation Technique: A Developmental Relationship. *Lucidity Lett.* **1986**, *5*.
30. Hunt, H.T. Lucid Dreaming as a Meditative State: Some Evidence from Long-Term Meditators in Relation to the CognitivePsychological Bases of Transpersonal Phenomena*. In *Dream Images*; Routledge: New York, NY, USA, 1991; ISBN 978-1-315-22470-1.
31. Hunt, H.T.; Ogilvie, R.D. Lucid Dreams in Their Natural Series: Phenomenological and Psychophysiological Findings in Relation to Meditative States. In *Conscious Mind, Sleeping Brain: Perspectives on Lucid Dreaming*; Springer: Berlin/Heidelberg, Germany, 1988; pp. 389–417.
32. McLeod, B.; Hunt, H. Meditation and Lucid Dreams. *Lucidity Lett.* **1983**, *2*.
33. Stumbrys, T.; Erlacher, D.; Malinowski, P. Meta-Awareness during Day and Night: The Relationship between Mindfulness and Lucid Dreaming. *Imagin. Cogn. Personal.* **2015**, *34*, 415–433. [CrossRef]
34. García-Campayo, J.; Moyano, N.; Modrego-Alarcón, M.; Herrera-Mercadal, P.; Puebla-Guedea, M.; Campos, D.; Gascón, S. Validation of the Spanish Version of the Lucidity and Consciousness in Dreams Scale. *Front. Psychol.* **2021**, *12*, 742438. [CrossRef]
35. Tzioridou, S.; Dresler, M.; Sandberg, K.; Mueller, E.M. The Role of Mindful Acceptance and Lucid Dreaming in Nightmare Frequency and Distress. *Sci. Rep* **2022**, *12*, 15737. [CrossRef]
36. Geise, C.; Smith, J.C. Mindfulness, Meditation, and Lucid Dreaming: A Narrow vs. Broad-Spectrum Mindfulness Study. *Imagin. Cogn. Personal.* **2023**, *42*, 263–280. [CrossRef]
37. Fucci, E.; Abdoun, O.; Caclin, A.; Francis, A.; Dunne, J.D.; Ricard, M.; Davidson, R.J.; Lutz, A. Differential Effects of Non-Dual and Focused Attention Meditations on the Formation of Automatic Perceptual Habits in Expert Practitioners. *Neuropsychologia* **2018**, *119*, 92–100. [CrossRef]
38. Hasenkamp, W.; Barsalou, L. Effects of Meditation Experience on Functional Connectivity of Distributed Brain Networks. *Front. Hum. Neurosci.* **2012**, *6*, 38. [CrossRef]
39. Schredl, M.; Göritz, A.S. Changes in Dream Recall Frequency, Nightmare Frequency, and Lucid Dream Frequency over a 3-Year Period. *Dreaming* **2015**, *25*, 81–87. [CrossRef]
40. Lykins, E.L.B.; Baer, R.A. Psychological Functioning in a Sample of Long-Term Practitioners of Mindfulness Meditation. *J. Cogn. Psychother.* **2009**, *23*, 226–241. [CrossRef]
41. Riordan, K.; Simonsson, O.; Frye, C.; Vack, N.; Sachs, J.; Fitch, D.; Goldman, R.I.; Chiang, E.; Dahl, C.; Davidson, R. How Often Should I Meditate? A Randomized Trial Examining the Role of Meditation Frequency When Total Amount of Meditation Is Held Constant. *J. Couns. Psychol.* **2023**, *71*, 104. [CrossRef]
42. Dunn, T.J.; Baguley, T.; Brunsden, V. From Alpha to Omega: A Practical Solution to the Pervasive Problem of Internal Consistency Estimation. *Br. J. Psychol.* **2014**, *105*, 399–412. [CrossRef]
43. Hayes, A.F.; Coutts, J.J. Use Omega Rather than Cronbach's Alpha for Estimating Reliability. But…. *Commun. Methods Meas.* **2020**, *14*, 1–24. [CrossRef]
44. McDonald, R.P. *Test Theory: A Unified Treatment*; Lawrence Erlbaum Associates Publishers: Mahwah, NJ, USA, 1999; pp. xi, 485; ISBN 978-0-8058-3075-0.
45. McNeish, D. Thanks Coefficient Alpha, We'll Take It from Here. *Psychol. Methods* **2018**, *23*, 412–433. [CrossRef] [PubMed]
46. Viladrich, C.; Angulo-Brunet, A.; Doval, E. A Journey around Alpha and Omega to Estimate Internal Consistency Reliability. *An. Psicol.* **2017**, *33*, 755–782. [CrossRef]
47. Malkewitz, C.P.; Schwall, P.; Meesters, C.; Hardt, J. Estimating Reliability: A Comparison of Cronbach's α, McDonald's Ωt and the Greatest Lower Bound. *Soc. Sci. Humanit. Open* **2023**, *7*, 100368. [CrossRef]
48. Baird, B.; Castelnovo, A.; Gosseries, O.; Tononi, G. Frequent Lucid Dreaming Associated with Increased Functional Connectivity between Frontopolar Cortex and Temporoparietal Association Areas. *Sci. Rep.* **2018**, *8*, 17798. [CrossRef] [PubMed]
49. Schredl, M.; Erlacher, D. Lucid Dreaming Frequency and Personality. *Personal. Individ. Differ.* **2004**, *37*, 1463–1473. [CrossRef]
50. Dalai Lama; Varela, F. *Sleeping, Dreaming, and Dying: An Exploration of Consciousness*; Simon and Schuster: New York, NY, USA, 1997; ISBN 0-86171-123-8.

51. Snyder, T.J.; Gackenbach, J. Individual Differences Associated with Lucid Dreaming. In *Conscious Mind, Sleeping Brain: Perspectives on Lucid Dreaming*; Gackenbach, J., LaBerge, S.P., Eds.; Springer New York: Boston, MA, USA, 1988; pp. 221–259; ISBN 978-1-4757-0423-5.

52. DeMarree, K.G.; Naragon-Gainey, K. Individual Differences in the Contents and Form of Present-Moment Awareness: The Multidimensional Awareness Scale. *Assessment* **2022**, *29*, 583–602. [CrossRef] [PubMed]

53. Davis, K.M.; Lau, M.A.; Cairns, D.R. Development and Preliminary Validation of a Trait Version of the Toronto Mindfulness Scale. *J. Cogn. Psychother.* **2009**, *23*, 185–197. [CrossRef]

54. Lau, M.A.; Bishop, S.R.; Segal, Z.V.; Buis, T.; Anderson, N.D.; Carlson, L.; Shapiro, S.; Carmody, J.; Abbey, S.; Devins, G. The Toronto Mindfulness Scale: Development and Validation. *J. Clin. Psychol.* **2006**, *62*, 1445–1467. [CrossRef] [PubMed]

55. Baer, R.; Smith, G.T.; Hopkins, J.; Krietemeyer, J.; Toney, L. Using Self-Report Assessment Methods to Explore Facets of Mindfulness. *Assessment* **2006**, *13*, 27–45. [CrossRef]

56. Tran, U.S.; Glück, T.M.; Nader, I.W. Investigating the Five Facet Mindfulness Questionnaire (FFMQ): Construction of a Short Form and Evidence of a Two-Factor Higher Order Structure of Mindfulness. *J. Clin. Psychol.* **2013**, *69*, 951–965. [CrossRef]

57. Malakoutikhah, A.; Zakeri, M.A.; Dehghan, M. A comparison between the relaxation/meditation/mindfulness tracker inventory and the Freiburg Mindfulness Inventory for predicting general health, anxiety, and anger in adult general population. *Front. Psychol.* **2022**, *13*, 810383. [CrossRef] [PubMed]

58. Baer, R.; Gu, J.; Strauss, C. Five Facet Mindfulness Questionnaire (FFMQ). In *Handbook of Assessment in Mindfulness Research*; Medvedev, O.N., Krägeloh, C.U., Siegert, R.J., Singh, N.N., Eds.; Springer International Publishing: Cham, Switzerland, 2022; pp. 1–23; ISBN 978-3-030-77644-2.

59. Smith, J.C. RMM 3.1a. Manual. Available online: https://drive.google.com/file/u/0/d/1-9MrBVVsOEwzHiu2HyplXvjLVJNJfeZm/view?usp=embed_facebook (accessed on 15 January 2024).

60. Smith, J.C. The psychology of relaxation, meditation, and mindfulness: An introduction to RMM theory, practice, and assessment. In *Principles and Practice of Stress Management*, 4th ed.; Lehrer, P.M., Woolfolk, R.L., Eds.; The Guilford Press: New York, NY, USA, 2021; pp. 39–57.

61. Smith, J.C. *Mindfulness Reimagined and the Image Priming Method*; Kendall Hunt: Dubuque, IA, USA, 2022.

62. Reed, P.G. Theory of Self-Transcendence. *Middle Range Theory Nurs.* **2008**, *3*, 105–129.

63. Wickham, H.; Averick, M.; Bryan, J.; Chang, W.; McGowan, L.D.; François, R.; Grolemund, G.; Hayes, A.; Henry, L.; Hester, J.; et al. Welcome to the Tidyverse. *J. Open Source Softw.* **2019**, *4*, 1686. [CrossRef]

64. Benjamini, Y.; Hochberg, Y. Controlling the False Discovery Rate: A Practical and Powerful Approach to Multiple Testing. *J. R. Stat. Soc. Ser. B (Methodol.)* **1995**, *57*, 289–300. [CrossRef]

65. Manly, B.F.J.M.; Bryan, F.J. *Randomization, Bootstrap and Monte Carlo Methods in Biology*, 3rd ed.; Chapman and Hall/CRC: New York, NY, USA, 2017; ISBN 978-1-315-27307-5.

66. Revelle, W. *Psych: Procedures for Psychological, Psychometric, and Personality Research*; Northwestern University: Evanston, IL, USA, 2023.

67. Canty, A.; Ripley, B.; Brazzale, A.R. Package 'Boot'. Version 1.3-30, 2024. Available online: https://cran.r-project.org/web/packages/boot/boot.pdf (accessed on 20 February 2024).

68. Gerbing, D.; Gerbing, M.D.W.; KernSmooth, S. Package 'LessR'. Version 4.3.3, 2024. Available online: https://cran.opencpu.org/web/packages/lessR/lessR.pdf (accessed on 10 March 2024).

69. Hayes, A.F. *Introduction to Mediation, Moderation, and Conditional Process Analysis: A Regression-Based Approach*; Guilford publications: New York, NY, USA, 2017; ISBN 1-4625-3466-X.

70. Domhoff, G.W. The Invasion of the Concept Snatchers: The Origins, Distortions, and Future of the Continuity Hypothesis. *Dreaming* **2017**, *27*, 14–39. [CrossRef]

71. Lee, M.-N.; Kuiken, D. Continuity of Reflective Awareness Across Waking and Dreaming States. *Dreaming* **2015**, *25*, 141–159. [CrossRef]

72. Fox, K.C.R.; Dixon, M.L.; Nijeboer, S.; Girn, M.; Floman, J.L.; Lifshitz, M.; Ellamil, M.; Sedlmeier, P.; Christoff, K. Functional Neuroanatomy of Meditation: A Review and Meta-Analysis of 78 Functional Neuroimaging Investigations. *Neurosci. Biobehav. Rev.* **2016**, *65*, 208–228. [CrossRef] [PubMed]

73. Sparrow, G.; Hurd, R.; Carlson, R.; Molina, A. Exploring the Effects of Galantamine Paired with Meditation and Dream Reliving on Recalled Dreams: Toward an Integrated Protocol for Lucid Dream Induction and Nightmare Resolution. *Conscious. Cogn.* **2018**, *63*, 74–88. [CrossRef] [PubMed]

74. Mota-Rolim, S.A.; Targino, Z.; Souza, B.; Blanco, W.; Araujo, J.; Ribeiro, S. Dream Characteristics in a Brazilian Sample: An Online Survey Focusing on Lucid Dreaming. *Front. Hum. Neurosci.* **2013**, *7*, 836. [CrossRef]

75. Gackenbach, J. Frameworks for Understanding Lucid Dreaming: A Review. *Dreaming* **1991**, *1*, 109. [CrossRef]

76. Saunders, D.T.; Roe, C.A.; Smith, G.; Clegg, H. Lucid Dreaming Incidence: A Quality Effects Meta-Analysis of 50 Years of Research. *Conscious. Cogn.* **2016**, *43*, 197–215. [CrossRef] [PubMed]

77. Bethlehem, J. Selection Bias in Web Surveys. *Int. Stat. Rev.* **2010**, *78*, 161–188. [CrossRef]

78. Schredl, M. Reliability and Stability of a Dream Recall Frequency Scale. *Percept. Mot. Skills* **2004**, *98*, 1422–1426. [CrossRef] [PubMed]

79. Krumpal, I. Determinants of Social Desirability Bias in Sensitive Surveys: A Literature Review. *Qual. Quant.* **2013**, *47*, 2025–2047. [CrossRef]
80. Colombo, D.; Suso-Ribera, C.; Fernández-Álvarez, J.; Cipresso, P.; Garcia-Palacios, A.; Riva, G.; Botella, C. Affect Recall Bias: Being Resilient by Distorting Reality. *Cogn. Ther. Res.* **2020**, *44*, 906–918. [CrossRef]
81. Aspy, D.J. Is Dream Recall Underestimated by Retrospective Measures and Enhanced by Keeping a Logbook? An Empirical Investigation. *Conscious. Cogn.* **2016**, *42*, 181–203. [CrossRef]
82. Raffaelli, Q.; Andrews, E.S.; Cegavske, C.C.; Abraham, F.F.; Edgin, J.O.; Andrews-Hanna, J.R. Dreams Share Phenomenological Similarities with Task-Unrelated Thoughts and Relate to Variation in Trait Rumination and COVID-19 Concern. *Sci. Rep.* **2023**, *13*, 7102. [CrossRef]

Review

Neuropsychopharmacological Induction of (Lucid) Dreams: A Narrative Review

Abel A. Oldoni [1], **André D. Bacchi** [2], **Fúlvio R. Mendes** [3], **Paula A. Tiba** [1] and **Sérgio Mota-Rolim** [4,*]

1. Center for Mathematics, Computing and Cognition, Federal University of ABC, São Bernardo do Campo 09606-045, Brazil; abel.o@aluno.ufabc.edu.br (A.A.O.); paula.tiba@ufabc.edu.br (P.A.T.)
2. Faculty of Health Sciences, Federal University of Rondonópolis, Rondonópolis 78736-900, Brazil; bacchi@ufr.edu.br
3. Center for Natural and Human Sciences, Federal University of ABC, São Bernardo do Campo 09606-045, Brazil; fulvio.mendes@ufabc.edu.br
4. Brain Institute, Federal University of Rio Grande do Norte, Natal 59078-970, Brazil
* Correspondence: sergioarthuro@neuro.ufrn.br

Abstract: Lucid dreaming (LD) is a physiological state of consciousness that occurs when dreamers become aware that they are dreaming, and may also control the oneiric content. In the general population, LD is spontaneously rare; thus, there is great interest in its induction. Here, we aim to review the literature on neuropsychopharmacological induction of LD. First, we describe the circadian and homeostatic processes of sleep regulation and the mechanisms that control REM sleep with a focus on neurotransmission systems. We then discuss the neurophysiology and phenomenology of LD to understand the main cortical oscillations and brain areas involved in the emergence of lucidity during REM sleep. Finally, we review possible exogenous substances—including natural plants and artificial drugs—that increase metacognition, REM sleep, and/or dream recall, thus with the potential to induce LD. We found that the main candidates are substances that increase cholinergic and/or dopaminergic transmission, such as galantamine. However, the main limitation of this technique is the complexity of these neurotransmitter systems, which challenges interpreting results in a simple way. We conclude that, despite these promising substances, more research is necessary to find a reliable way to pharmacologically induce LD.

Keywords: dream recall; lucid dreaming; REM sleep; metacognition; self-consciousness; dopamine; acetylcholine; galantamine; acetylcholinesterase inhibitors; sesquiterpene lactones

Citation: Oldoni, A.A.; Bacchi, A.D.; Mendes, F.R.; Tiba, P.A.; Mota-Rolim, S. Neuropsychopharmacological Induction of (Lucid) Dreams: A Narrative Review. *Brain Sci.* **2024**, *14*, 426. https://doi.org/10.3390/brainsci14050426

Academic Editor: Giuseppe Lanza

Received: 23 March 2024
Revised: 19 April 2024
Accepted: 22 April 2024
Published: 25 April 2024

1. Introduction

Dreams are characterized by perceptions, emotions, and cognitions that happen in any sleep stage. Lucid dreaming (LD) is a spontaneously rare type of dream, mainly associated with REM sleep, in which subjects recognize that they are dreaming. In this article, we review studies on substances derived from plants and drugs used to intensify oneiric activity and induce LD. We first describe the processes that regulate sleep and the main neurotransmitters that control REM sleep (see Supplementary Material). Next, we discuss the neurophysiology and phenomenology of LD to finally review substances that increase metacognition, REM sleep duration, and dream recall frequency, thus with the potential to induce LD.

2. Lucid Dreams: Definition and Neurophysiology

LD occurs when dreamers know they are dreaming; that is, they comprehend that the reality they are experiencing is not the waking one, but the sleeping one. Although LD is spontaneously rare in the general population, it is a learnable skill. LaBerge [1] underwent self-training for three years using cognitive tasks such as self-suggestion with a focus on his motivation and intention to stay lucid in the next dream he would have. Refining this training, he developed the technique mnemonic induction of lucid dreaming (MILD), which consists of remembering to perform future actions, using mental visualization or verbalization about the act that one would like to perform (e.g., "I will remember to recognize that I am in a dream while dreaming").

The objective verification of LD can occur in a real-time experimental setting, through communication by eye movements (which do not present complete muscle atonia as other body muscles) during rapid-eye-movement (REM) sleep [2]. By combining these movements before falling asleep, participants, when becoming lucid in the dream, can perform pre-agreed eye movements (e.g., left and right), which can be detected in the electrooculogram [3].

Although LD happens predominantly during REM sleep, it can also occur during sleep onset (N1) and light sleep (N2) stages [4,5]. Eye-signaling reports of LD in the deep sleep (N3) stage have not yet been found. However, practitioners of Transcendental Meditation report being able to maintain self-awareness during this deep sleep stage [6,7]. In addition, Yoga Nidra practitioners report the possibility of cultivating self-awareness throughout the entirety of sleep [8,9]. Physiological studies of LD substantially focus on the occurrence of lucid REM dreams, which are more easily verified. For convenience and based on the available literature, the focus of the description of LD physiology here will be exclusively on the phenomenon during REM sleep.

Using a combination of electroencephalography and functional magnetic resonance imaging, Dresler et al. [10] observed activation in neocortical and cortical networks during lucid REM sleep, mainly in the precuneus, occipitotemporal cortex, frontopolar cortex, and right dorsolateral prefrontal cortex. Among possible neurophysiological and phenomeno-logical mechanisms related to LD, it is suggested that the increased activity in the right dorsolateral prefrontal cortex may be related to self-centered metacognitive evaluation and, together with the parietal lobules, corresponds to working memory activation. Frontopolar areas are involved in the processing of internal states, emotions, and thoughts, which have been widely reported in LD. The higher difference in activation between lucid and non-lucid REM sleep was in the precuneus, which is related to self-referential processing, i.e., first-person perspective. The exceptional clarity of scenery (vividness) in LD [11] coincides with the activation of occipitotemporal cortices, which are ventral parts of visual processing and are involved in various aspects of conscious visual perception [10].

Low metacognition in non-lucid dreams (exemplified by a lack of volitional capacity, impaired critical thinking, and a lack of self-reflection on one's state) can be explained by the deactivation of the dorsolateral prefrontal cortex and the frontopolar cortex during REM sleep [12]. To test whether these areas are active during LD, Filevich et al. [13] conducted a study with thought monitoring tasks, metacognition questionnaires, and neuroimaging. When separating participants into high- and low-lucidity groups, the former showed increased gray matter volume in regions BA9, especially in the dorsolateral prefrontal cortex, and BA10, specifically in the frontopolar cortex, following the authors' hypothesis. The BA10 area has been associated with visual and metacognitive monitoring abilities. Additionally, greater activity was measured in these areas during metacognitive tasks compared to a control condition that did not require these abilities [13].

Another study [14] evaluated the neuroanatomy of frequent lucid dreamers' traits. The authors showed significantly increased functional connectivity at rest between the left anterior prefrontal cortex and bilateral angular gyrus, bilateral middle temporal gyrus, and right inferior frontal gyrus. In contrast to the findings of Filevich and colleagues, there was no difference in gray matter density between the two groups. Among the areas differentiated between LD and normal dreams, there was an increase in functional connectivity between the frontopolar cortex and temporoparietal associative areas, with overlap of the frontoparietal network. The authors conclude that the changes observed in the functional connectivity of the anterior prefrontal cortex may underline metacognitive judgments and functions that are presented in LD (Figure 1). They also suggest future studies involving cholinergic modulation, since pro-cholinergic drugs tend to increase frontoparietal activity, as will be detailed in the next section.

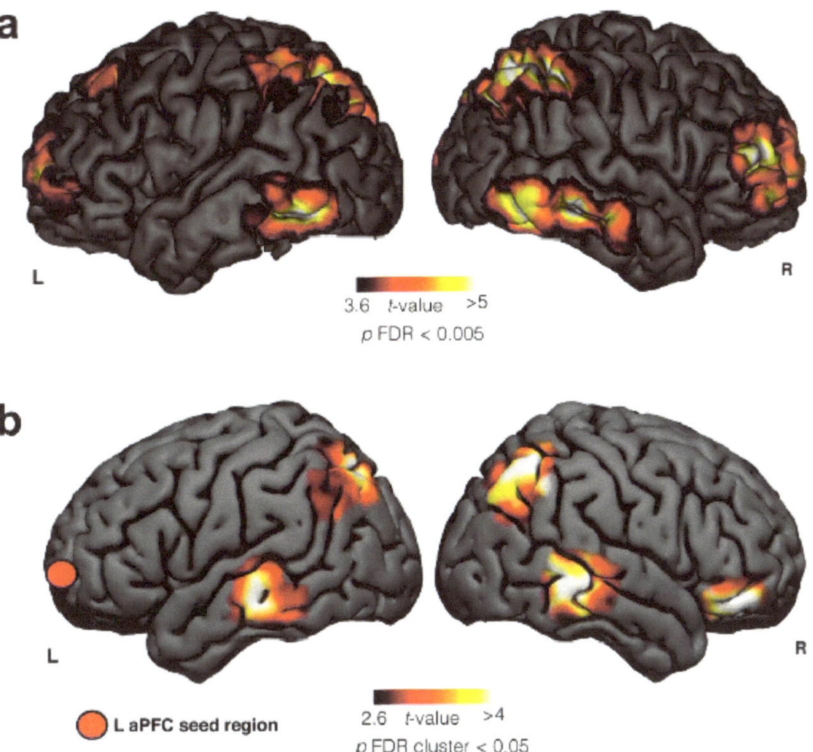

Figure 1. Frontal, parietal, and temporal brain areas are involved in LD, which is evidenced by two different methods of neuroimaging: (**a**) Blood-oxygen-level-dependent (BOLD) activity in an fMRI case report of LD. (**b**) Seed-based resting-state (SBRS) functional connectivity differences between frequent lucid dreamers and non-frequent lucid dreamers (control group). Adapted from [15] with permission from the authors. R = right side and L = left side. aPFC = anterior prefrontal cortex.

3. Pharmacological Induction of Lucid Dreams

Since spontaneous LD is rare, an effective method for its induction is currently desired, both for facilitating experimental studies and for understanding the neural basis of metacognition and dream self-awareness [16]. In this section, we first describe cholinergic and dopaminergic substances, which are the main candidates to induce LD. Then, we review some case reports of less-studied substances. Finally, potential natural candidates that increase dream vividness and/or recall, such as plants and herbs, are discussed.

3.1. Cholinergic Substances

Acetylcholinesterase inhibitors (AChEIs) constrain the enzyme that hydrolyzes acetylcholine (ACh) and increases its levels in the synapse. Many plants have been used in traditional medicine to treat cognitive deficits, including neurodegenerative disorders. These plants are a rich source of compounds with antioxidant activity and acetylcholinesterase inhibition. Most studies have focused on anticholinesterase alkaloids, but other major classes of compounds reported to have such activity are terpenoids, glycosides, flavonoids, and coumarins [17–20]. AChEIs, such as donepezil, rivastigmine, and galantamine, despite sharing this central mechanism, bear specific properties that can potentially engender subtle disparities in their effect profiles [21]. Donepezil is a reversible, long-acting, and selective AChEI with no inhibitory effect on butyrylcholinesterase (BuChE). The selectivity of acetylcholinesterase (AChE) over BuChE could hypothetically mitigate side effects associated with BuChE inhibition. Its reversible nature and long-term action may also contribute to its sustained effectiveness throughout the day at a single daily dose [22]. In contrast, rivastigmine, a "pseudo-irreversible" inhibitor of both AChE and BuChE, reversed this inhibitory effect over time. This property may allow for superior control over brain ACh levels, potentially minimizing ACh toxicity. Furthermore, rivastigmine might show selectivity for AChE in the cortex and hippocampus, as opposed to AChE in other brain regions, which could potentially enhance cognitive effectiveness with fewer peripheral side effects [23].

Galantamine is a natural tertiary alkaloid from the Amarylidacea family, originally isolated from the bulbs of Snowdrop and Narcissus species [24] (Figure 2). In contrast to donepezil and rivastigmine, galantamine is not only an AChEI, but also a positive allosteric modulator of nicotinic ACh receptors. This property allows galantamine to potentiate the effects of ACh on these receptors, possibly enhancing its effectiveness in improving cognitive function [25]. This dual mechanism of action could potentially elevate the efficacy of galantamine over other AChEIs that are devoid of this additional action on nicotinic receptors.

Figure 2. Galanthus flower, also known as Snowdrop [26].

There is ample evidence that REM sleep is modulated by ACh (see Supplementary Materials) and AChEIs are involved in the phasic activation and stabilization of REM sleep, which increases the chance of inducing LD [27]. LaBerge's patent [28] shows that AChEIs, such as galantamine, donepezil, rivastigmine, and huperzine A, significantly increase the clarity of cognition, lucidity, self-reflection, recall, control, bizarreness, and visual vividness, with few side effects. With donepezil, 8 out of 10 subjects experienced LD, and only 1 out of 10 reported LD on placebo night. However, it is important to note that this medication has some undesired side effects, such as nausea. Huperzine A, from *Huperzia serrata*, a potent plant-derived AChEI alkaloid, presents similar effects, but with a much higher dose compared to donepezil. Rivastigmine and galantamine had the same effects on lucidity, with fewer side effects. LaBerge proposes that in addition to AChEIs, other cholinergic agonists, such as muscarinic receptor agonists, or presynaptic receptor antagonists might induce LD.

More recently, LaBerge et al. [29] investigated whether galantamine would induce LD. In a final sample of 121 individuals, 3 doses of galantamine were distributed to be used over 3 days: 0 mg (control condition), 4 mg, and 8 mg. Galantamine was found to be a dose-dependent inducer of LD, with few or no side effects. There was also an increase in dream recall, sensory vividness, bizarreness, and dream complexity, as well as a decrease in negative emotions. Participants with previous experience with LD were more likely to experience LD, and of the 10 individuals who had never experienced LD, 4 of them reported one lucid dream episode with the higher dose.

Another study combined galantamine with two cognitive strategies: meditation and dream reliving (MDR) and wake back to bed (WBTB) [30]. MDR is a technique for reliving dreams with critical self-reflection; while the events are happening in mental imagery, the person tries potential ways to become lucid, like scenarios of nightmares. The meditation used involves practicing self-awareness and being non-reactive to stressful oneiric content. WBTB is an LD induction technique in which the subject wakes up late at night when REM sleep and vivid dreams are more prone to happen. Then, the person goes back to sleep, potentially experiencing the same dream, or having vivid dreams, augmenting the probability of achieving lucidity. The study used 35 participants, and 8 nights of dreaming were recorded. Galantamine outperformed placebo with both MDR and WBTB techniques, which did not differ in terms of lucidity. MDR increased reflexivity, as well as fear and threat, compared to WBTB. As the number of lucid dreams achieved was not mentioned in the article, Baird et al. [15] contacted the authors, and the results showed that 9% of participants reported LD in the WBTB + placebo condition and 11% in the MDR + placebo condition, while 40% of participants reported LD in the WBTB + galantamine condition, and 34% in the MDR + galantamine condition. These results highlight the potential drug candidate, galantamine, to induce LD. The authors suggest that MDR may bring exposure to trauma or conflict resolution, depending on the dreamer's intention.

Although AChEIs increase cholinergic activity, it seems that cholinergic agonists are not always effective in LD induction, as is the case of alpha-GPC. Alpha-GPC is a precursor of ACh, which, in contrast to ACh itself, can cross the blood–brain barrier. The procedures adopted in a study with alpha-GPC involved three non-consecutive nights: one with a placebo, one with the supplement, and one wash-out night [31]. Participants' reports were recorded, and the LuCiD scale was used to measure the level of consciousness in dreams. No inducing effect of this substance was observed; of the 33 participants, six experienced LD: two of the advanced practitioners, two of the novices with alpha-GPC, and two novices on placebo nights. One possible reason is that no mental training was used to achieve lucidity, and cognitive training is widely used in experimental studies [15].

Considering the pharmacological action of galantamine as an AChEI and a positive allosteric modulator of nicotinic ACh receptors, it is plausible to suggest that this unique combination of mechanisms may account for the superior efficacy observed in LD induction. In contrast, despite being AChEIs, donepezil and rivastigmine lack the same action on nicotinic receptors, potentially making them less effective than galantamine at inducing LD. In parallel, alpha-GPC, a cholinergic precursor that can increase available brain ACh levels, lacks the positive allosteric modulation of nicotinic ACh receptors. Furthermore, galantamine prolonged the action of ACh by inhibiting its degradation. In contrast, a-GPC enhanced the availability of choline, an ACh precursor, but did not directly influence ACh degradation. Thus, galantamine's ability to prolong ACh action may promote LD, and might be in accordance with the high modulation of lucid REM sleep by ACh [27]. In addition, ACh seems to modulate activity on the dorsolateral prefrontal cortex, an executive area that is activated during LD. However, these hypotheses remain rooted in the current understanding of the mechanisms of action of these drugs and the LD phenomenon. Further research is necessary to validate these propositions and better understand the relationship between cholinergic pharmacology and LD.

3.2. Dopaminergic Substances

Data obtained from experiments with rats [32] suggest that galantamine increases dopamine (DA) cell firing in the ventral tegmental area, through allosteric potentiation of nicotinic (not muscarinic) receptors. Specifically, it increases extracellular DA levels in the prefrontal cortex, especially the medial prefrontal cortex. The authors note that the cognitive mechanisms behind this effect are still not fully understood. AChEIs might play a role in the dopaminergic system through reward inhibition. Even though the mechanisms are still not fully understood, and different doses might affect this type of behavior, repression of irrelevant stimuli, such as seeking drugs, in both humans and rats seems to be prevalent. Suppressing irrelevant stimuli while enhancing others, as in the case documented by Kjaer et al. [33] during Yoga Nidra, might be a way to become lucid in dreams.

Yoga Nidra is a meditation technique systematized by Swami Satyananda Saraswati, who took the ancient scriptures contained in the Vedantic literature. In his book "Yoga Nidra", many definitions are made, such as "a technique in which you learn to relax consciously" (p. 4); "the consciousness is in a state between waking and sleep, but it is subject to neither [34]". This meditation has been gaining attention in recent years, having many benefits such as an increased subjective quality of sleep, as well as treating insomnia; decreasing stress, anxiety, and depression; and increasing psychological well-being [35].

An intriguing study demonstrated that during Yoga Nidra meditation, [11C]-Raclopride, a selective antagonist of D2 and D3 receptors, showed less binding in the ventral striatum compared to the control group when the practitioners of the meditation were at rest [33]. The radioligand had 7.9% less biding, which can be interpreted as endogenous DA in the ventral striatum being augmented by 65%. In addition, participants reported less readiness to act and enhanced sensory imagery. Parker et al. [8] argue that the former results only show a preliminary practice of Yoga Nidra, not the practice per se, since there are definitions of Yoga Nidra that take into account that there should be increased delta waves in the brain, with the subject simultaneously conscious. Indeed, a recent study has shown electrophysiological evidence of meditation-naïve participants in a 2-week intervention of Yoga Nidra, showing local sleep—that is, increased synchronization evidenced by delta waves—in the central area and reduction in the prefrontal area. This result was accompanied by enhanced subjective sleep quality and efficiency [36]. Even though delta waves might indicate loss of consciousness, Yoga Nidra shows the contrary. A similar phenomenon is the paradoxical pharmacological dissociation, defined as drug-induced states that enhance delta oscillations but preserve consciousness [37], which will be further commented on in the general discussion (see Section 5).

Yoga Nidra, as well as Dream Yoga (from Tibet), regard LD just as a step that could be facilitated by entering consciously into sleep. With this in mind, more neuroimaging studies regarding self-consciousness during sleep should be carried out to make solid conclusions on the subject, since being lucid in sleep might not be exclusively associated with executive control. Nevertheless, it can be hypothesized from the studies above that DA is involved in self-consciousness. Indeed, DA might play a crucial role in self-awareness. A causal role of the medial prefrontal cortex and medial parietal regions might be speculated. These regions have connections to the so-called default mode network (Figure 3), such as the angular gyrus, insula, striatum, and thalamus, where DA interacts with GABA-induced synchronized gamma oscillations [38].

A study found that synchronous transcranial alternating-current stimulation in theta frequency applied to the frontoparietal network improved working memory performance when cognitive demands were high. The frontoparietal network, and working memory ability, might be crucially involved in LD [14,15].

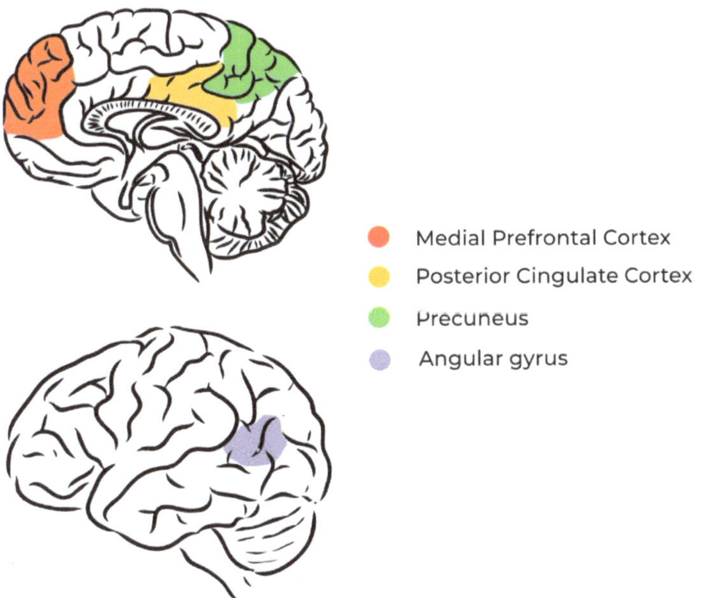

Figure 3. The main regions that comprise the default mode network (DMN). (**up**) Medial view: (brown) medial prefrontal cortex; (yellow) posterior cingulate cortex; (green) precuneus; (**down**) lateral view: (gray) angular gyrus. Adapted from [39] with permission from the authors.

In line with the evidence that DA is involved in metacognition, becoming lucid in dreams may involve conscious self-monitoring, and DA could play a crucial role in this process. Joensson et al. [40] aimed to investigate whether dopaminergic stimulation could enhance self-awareness, metacognition, and self-monitoring by administering L-dopa, a dopamine precursor, to healthy individuals. The results demonstrate that self-referential processing improved, leading to increased self-awareness and self-monitoring. In addition, individuals who took L-dopa exhibited improved performance compared with the control group. Thus, the subjective experience of consciousness became a more accurate predictor of performance. To further examine the neural basis of these improvements, the authors used magnetoencephalography to analyze brain regions implicated in self-monitoring. They found that dopaminergic stimulation increased activity in the medial prefrontal cortex, a region that may exhibit differential dopaminergic modulation during LD due to its association with conscious self-monitoring.

L-dopa is the natural amino acid precursor of DA [41] and is used in the treatment of Parkinson's disease, but it can have behavioral and cognitive side effects. Preliminary results show that this dopaminergic medication has positive effects on metacognition in Parkinson's disease patients [42], but it seems that the drug can cause vivid dreams in these patients [43–46]. REM sleep behavior disorder might contribute to the vivid dreams attributed to Parkinson's disease patients [39]. Occurring in approximately one-third of patients with Parkinson's disease, REM sleep behavior disorder is characterized by a loss of muscle atonia during REM, and is associated with aggressive and vivid dream content in Parkinson's disease patients [47]. Excluding Parkinson's disease patients with REM sleep behavior disorder, it was found that DA agonist dosage in the treatment of Parkinson's disease made dreams less vivid and less emotional [48]. It was also observed that visual vividness was related to amygdala volume and the thickness of the medial prefrontal cortex. Other drugs that can treat Parkinson's disease such as pramipexole, a non-ergot dopamine D2/D3 receptor agonist, seem to produce vivid dreams as side effects [49]. It is worth mentioning that, in normal individuals, it inhibits DA synthesis and release, but when

DA neurons are impaired—as in the case of Parkinson's disease—the drug acts as a potent postsynaptic dopamine D2 receptor agonist.

Taking these findings together, it is plausible to infer a causal role of DA in the generation of dreams, having an impact on oneiric emotional content and vividness. Both factors, when enhanced, increase the possibility of the dreamer to understand the narrative and become lucid. Knowing that DA is also related to metacognition, self-awareness, and suppressing irrelevant stimuli, future studies should address the role of DA in inducing vivid dreams and helping one become self-conscious when suppressing stimuli that are barriers to understanding the nature of the oneiric reality the dreamer is immersed in.

3.3. Case Reports with Other Substances

Haas et al. [50] report a case of a 65-year-old woman diagnosed with multiple myeloma, with a history of chronic pain. On the second day after her admission to the hospital, she complained of LD, which were terrifying (see lucid nightmare in the general discussion). The authors concluded that the cause was pregabalin, since upon discontinuing the medication, LD also ceased. Pregabalin is used to treat neuropathic pain; however, it has various side effects. It is a 3-isobutyl derivative of GABA, and the mechanisms of action of pregabalin are associated with its strong binding to alpha-2-delta sites, which are a subunit of voltage-dependent calcium channels in central nervous system tissues that regulate calcium influx into nerve terminals. This reduces the release of excitatory neurotransmitters including NA, glutamate, DA, 5-HT, and substance P.

In another case report, Mousailidis et al. [51] describe a patient who experienced visual hallucinations and agitation associated with an increase in the dose of pregabalin, which completely resolved after discontinuation of the drug. The manufacturer reports abnormal dreams as an infrequent characteristic of the drug's use.

In a recent study, an online survey was conducted on food and supplements that could affect nocturnal consciousness (e.g., dream recall, LD, and hypnagogic hallucinations) [52]. Partial correlations showed that taking vitamin supplements increases either dream recall or LD frequency. In addition, vitamin intake had a positive correlation with hypnagogic recall. The surprising result was that the consumption of fish was positively correlated with LD. The author speculates that this result might be explained by the omega-3 fatty acids that have already been suggested to underlie some cognitive abilities. However, these results should be taken with caution, since the individuals could be training LD or compensating for a deficiency in omega-3. Moreover, taking antidepressants positively correlated with LD, which might be linked to the REM sleep rebound caused by substances that affect the serotonin (5-HT) system. Another interesting result was that eating chili was correlated with hypnagogic recall, and this could be due to the capsaicin substance found in chili, which has already been linked to cognitive functioning, as the author mentions. These results could be promising for the intake of substances to induce LD. Although the main limitation of this research is that it is an online survey, and hence prone to subjectivity, the author emphasizes that field research could be fruitful in accessing variables related to sleep, dreaming, and LD.

Sergio [53] conducted a self-case study and began testing substances that stimulate the brain. 2-Dimethylaminoethanol (DMAE) proved to be positive in his personal experiences to induce LD. It is also described as a powerful stimulant of brain reticular formation, causing excitement and consciousness during REM sleep, and shortening the amount of sleep needed each night, promoting a sense of mental clarity upon waking. As a result, DMAE helps a person stay at a higher level of consciousness while dreaming, which facilitates the realization that one is dreaming. DMAE is a relatively non-toxic compound with vitamin-like properties similar to choline and is converted to choline in the body. They differ in terms of membrane synthesis because choline cannot be transported across cell membranes to the sites where it is needed, while DMAE can easily pass through cell membranes. Sergio advises to use 50 mg per day for 2 weeks to achieve peak effects. In combination, he would wake up 1 h before his usual wake-up time and engage in visualization techniques to become lucid, allowing him to enter directly into LD.

DMAE (Deanol) is a classical nootropic [54] found in many nutritional supplements, such as salt of tartaric acid. Interestingly, it can also be absorbed through eating fish, such as salmon and shellfish, which supports the study of Biehl, cited above, where eating fish was related to LD, and also with the self-experiment cited above where DMAE was shown to induce LD [53]. Delving into its properties as a cognitive booster, there is evidence that DMAE increased choline and ACh extracellular levels in rats, thus improving spatial memory. It has also been shown that, in human studies, DMAE combined with supplements (such as vitamins) improved alertness and attention [54]. Taking these results together, future studies should address whether frequent consumption of fish is actually related to LD.

Richter's [55] patent presents a nutritional supplement for improving sleep and increasing LD, containing *Calea ternifolia*, L-5-hydroxytryptophan (L-5-HTP), and vinpocetine. The supplement may include secondary ingredients, such as melatonin, Mugwort extract, DMAE, passionflower extract, green tea extract, and wild lettuce extract. It may also contain vitamins B, D, and C, zinc, magnesium, and calcium. Among the reasons why the supplement could enhance LD, the author reports that (1) *Calea ternifolia* is a known dream herb used by the Chontal Natives of Mexico (see the next session for more details); (2) green tea (*Camellia sinensis*) is a well-known beverage that is rich in antioxidants; (3) Mugwort (*Artemisia vulgaris*) extract has been used for medicinal purposes for centuries and is known to cause a dreamlike state of consciousness (it will be mentioned again in the next session); (4) passionflower (*Passiflora* spp.) has mild sedative properties that can be used to treat insomnia and anxiety; (5) wild lettuce is an ingredient in some sleep tonics and has historically been considered a mild sedative; (6) L-5-HTP is known to increase serotonin levels, which, in turn, improves sleep quality; (7) DMAE may increase acetylcholine levels; (8) Vinpocetine (from *Vinca minor*) is known to help memory and improve mental functioning; (9) melatonin has properties that are detailed below; (10) B vitamins are known to regulate the body's energy processes and promote good health; (11) vitamin C is also a necessary ingredient for maintaining good health; (12) zinc is used by hundreds of enzymes that regulate many of the body's functions; and (13) vitamin D, calcium, and magnesium are additional elements that are utilized by the body. The author presents two brief case reports of individuals who became lucid and controlled their dreams when taking this supplement.

Melatonin (N-acetil-5-metoxitriptamina) is a hormone produced by the pineal gland that informs an organism that it is night, which for diurnal animals (such as us, humans), represents rest and sleep. In this way, melatonin has been used as a sleep-promoting agent [56,57]. Besides this hypnotic function, it seems that melatonin can also increase dream recall frequency, vividness, and bizarreness [58]. This can be partially explained by better coordination of the sleep stages, especially REM sleep. Despite these results, as far as we know, there are no studies that investigate whether melatonin can be used to induce LD.

Interestingly, a pilot study demonstrated that vitamin B6 (pyridoxine) supplements taken before sleep can increase dream vividness and dream recall [59]. Another study showed that vitamin B6 significantly increased the amount of dream content that participants remembered but did not significantly affect dream vividness or bizarreness, nor did it significantly affect other sleep-related variables. However, another group of the experiment that took the B complex had worse sleep quality [60]. These two studies speculate that the likely cause of B6's impact on dreams is its role as a co-factor in the conversion of L-tryptophan to 5-HTP and the conversion of 5-HTP to 5-HT. In sleep studies, it is hypothesized that high levels of 5-HT before sleep suppress REM sleep in the early cycles, and thus induce REM sleep rebound at the end of the night, increasing and intensifying dreams. In addition, pyridoxine disulfide (called pyritinol, a derivative of B6) can be used as a nootropic. Increased levels of concentration of choline in cholinergic neurons have been observed in rats, and another study found learning and memory improvements with its administration [54]. Therefore, despite controversies with Ritcher's patent, where vitamin B seems to be used in general, B6 might be a way to help dreamers become lucid, although more studies on the B complex related to sleep are needed. Moreover, the derivative pyritinol might be a potential candidate to induce LD, but further studies should provide more evidence of its mechanism of action.

4. Plants with the Potential to Intensify Oneiric Activity

Humans have always used natural resources for the maintenance of life. Plants and herbs are used as traditional medicine through the extraction of specific molecules with therapeutic potential. On this topic, we first present plants documented culturally as intensifiers of lucid and non-lucid dreams, as well as cognition. Then, we review Ayurvedic medicine and nootropic substances that modulate oneiric activity.

4.1. Papaver somniferum

Sleep-inducing drugs were empirically administered in ancient times for different psychiatric diseases recognized as paranoia, and drugs with hypnotic effects were classified as temperature reducers [61]. One of these sleeping-inducing drugs was poppy (*Papaver somniferum*, Papaveraceae), the plant used in the production of opium. There are several lines of evidence of opium and poppy use in the ancient era. The use of opium poppy as a treatment for children's insomnia is described in the Egyptian Papyrus of Ebers dated to the 16th century BCE [62] and was later cited in books reporting the medicines used by Paracelsus, Galen, and Avicenna, among other important physicians and naturalists [62,63]. Greco-Roman physicians knew the dangers and benefits of the opium poppy and that greater dosages were required to obtain its pleasure effects, including sleep accompanied by alluring dreams and visions [64,65].

Several poppy-derived products have been used over time, such as raw opium, laudanum, laudanum tinctures, and poppy tea, among others. There are dozens of alkaloids in opium, including morphine, the most important narcotic substance of poppy. Morphine was first isolated in 1806 by Friedrich Sertürner, which demonstrated that it was the sleep-inducing narcotic substance of opium [63]. Sertürner first referred to the substance as the "*principium somniferum*" and later named it "morphium" after Morpheus, the Greek god of dreams. Later, the alkaloid was renamed "morphine" by Joseph Louis Gay-Lussac [63]. There are several representations of the narcotic properties of morphine and poppies in mythology. According to Greek mythology, Hypnos and Thanatos are the personifications of sleep and death and are usually associated with opium poppies [63]. A marble sculpture of Bertel Thorvaldsen of 1815 represents the Day and Night angels; Night is represented holding two children, sleep and death, and Night has her hair decorated with poppy branches [65] (Figure 4). Despite being a natural alkaloid, it is accepted nowadays that morphine is synthesized by mammalian cells from DA, although the function of endogenous morphine in the body is still a matter of debate [66].

Figure 4. Day and Night angels, by Bertel Thorvaldsen (1815). (**left**) The Day angel. (**right**) The Night angel, with poppy branches in her hair, holds two children: Hypnos, who represents sleep, and Thanatos, who represents death.

4.2. Calea ternifolia

Calea ternifolia (also known as *Calea zacatechichi*), pertaining to the Asteraceae family and known as "the dream herb", is an endemic species from Central America, traditionally used by the Chontal people from Mexico for divination due to its properties that increase dreaming. Mayagoitia et al. [67] reported some of its characteristics. The plant was collected and extracts were made with the supervision of the Chontal folk. Both extracts increased light sleep (N2 stage), the spontaneity of waking up during a nap period, and vividly hypnagogic imagery. The spontaneity of waking imagination also suggests hypnopompic imagery, and both of these states are used for inducing LD. Less deep sleep (N3 stage) and REM sleep was also observed, the latter contrasting with the literature that dreams are more associated with REM sleep, making way for dreams in other sleep stages, since dream recall was also increased by the extracts. It was also observed that methanol extract was associated with more recalled dreams, less content, and more vivid dreams. This larger effect of the extract suggests that the active compounds might be present in the polar fraction of the substance. *Calea* induced a discrete increase in all sensorial perceptions, discontinuity in thoughts, a rapid flux of ideas with difficulty in their retrieval, and statistically significant slowness of reaction time, which might have also induced a light hypnotic state [67,68].

One of the mechanisms of action of *Calea* is through its sesquiterpene lactones. A recent review found 37 sesquiterpene lactones in this plant [69]. There are not many reports of extracts of plants that have sesquiterpenes with AChEI activity compared to alkaloids, such as galantamine and huperzine A, but data indicate this type of activity mentioned in some plants and their respective sesquiterpenes. Recently, sesquiterpenes have been proposed for the treatment of Alzheimer's disease due to their AChEIs properties with much therapeutic potential, mainly lactone ones, which might inhibit AChE more than galantamine [70]. In this manner, *Calea* might be able to induce LD, possibly due to its AChEI properties.

4.3. Celastrus paniculatus

Another sesquiterpene mentioned by Arya et al. [70] is Dihidro-β-agarofurano found in the Celastraceae family, like in *Celastrus paniculatus*, which when administered in C. elegans had rejuvenating properties [71]. This plant has been used for more than a thousand years in Indian Ayurvedic medicine, commonly called "the intellect tree", from Hindu, Malkangani. *Celastrus paniculatus* has therapeutic potential, from oil and methanol extract to seeds [72,73]. Its neuroprotective potential has already been tested for schizophrenia induced by ketamine in rats, treating the disorder combined with clozapine or alone. Two weeks of *Celastrus paniculatus* treatment restored dendritic atrophy in the hippocampus and synaptic plasticity, also reducing AChE in the frontal cortex, hippocampus, and hypothalamus in stressed rats. Additionally, memory and cognition were augmented by *Celastrus*, possibly due to AChEI activity. In vivo studies showed restoration of working memory and spatial learning [72]. *Celastrus* was also observed to be nontoxic in the doses used, demonstrating its possible path for clinical studies in humans. Hence, it could be used for the induction of LD.

4.4. Silene capensis

Used by the Xhosa people from South Africa, *Silene capensis* is called the "African dream root". Hirst [74] presents Xhosa dream reports and discusses a medical category of plants that enhance dreams (ubulawu) through his ethnographic study. A beverage is prepared by the root being ground and shaken in a beaker of water until it produces a thick white foam, which novice diviners consume on an empty stomach to increase dreams. The goal is to fill the stomach with foam until regurgitation, which indicates that a sufficient amount has been consumed. Novices consume and wash themselves with the residues during a period of three days of the full moon. The root is also chewed. This plant is used for communication with ancestors through dreams, people with trouble remembering dreams, and to cause more vivid and memorable dreams. Evidence points out that the ubulawu root has principal effects on lucid and prophetic dreams through its triterpenoid saponins [74], which are known to have AChEIs properties [75]. Other plants from the same genus have been reported with these properties [76–78]. In addition, the family that these plants compose, Carophyllaceae, are known to produce foam, from saponins. These plants are used for the preparation of the psychoactive drink ubulawu, traditionally used in Southern Africa, in spiritual healing processes [79]. It is feasible that the plant has potential for LD induction, also by inhibiting AChE.

4.5. Artemisia vulgaris

Another common herb used for intensifying dreams is Mugwort (*Artemisia vulgaris*), like *Calea zacatechichi*, also from the Asteraceae family. By inhibiting the degradation of serotonin by the monoamine oxidase (MAO) enzyme, the herb has already been reported to cause vivid dreams, where the REM sleep rebound at the end of the night may contribute to this experience. It is also observed that some lactone sesquiterpenes in this plant might also have MAO inhibition properties [80]. It is noted that the herb is also an antioxidant—like many of the herbs reported here as well as galantamine—which supports the idea that Artemisia might help to trigger lucidity during dreams.

4.6. Withania somnifera

Traditionally used in Ayurvedic medicine, the herb *Withania somnifera*, commonly known as Ashwagandha, is classified in the Solanaceae family. Major bioactive compounds are steroidal lactones, called withanolides, which are potent antioxidants together with alkaloids. Restoring spatial memory, motor learning, muscarinic receptor activation, and oxidative stress are cited in Bashir and colleague's recent review [81]. They also cite anti-Alzheimer properties—inhibition of AChE and enhanced choline acetyltransferase level in rats—which are an effect of the active compound whitaferin-A (withanolide). Increased cholinergic transmission in the basal ganglia and cerebral cortex might explain the cognition improvements induced by these substances [81]. The ethanolic root extract improved gripping ability and motor movements, as well as increasing striatum DA in rats. Another ethanolic root extract had neuroprotective effects on nigrostriatal dopaminergic neurons [81].

In vitro screening of the extract was found to protect against acrolein-induced toxicity and, henceforth, could have a neuroprotective effect on patients with Alzheimer's disease, since acrolein is found to be significantly increased in these patients [82]. A mouse model of Parkinson's disease treated with *Withania somnifera* was found to inhibit the oxidative stress and apoptotic pathways of dopaminergic neurons (which could be related to pathways in basal ganglia, especially the substantia nigra). Improvement in motor deficits and enhanced quality of walking was observed [83]. Taking these results, AChEI activity, as already cited in many cases during this review, might be indicative that the plant can be used to induce LD.

A recent review of the effects of this herb on sleep included five randomized controlled trials, with healthy volunteers, stressed adults, or those with insomnia. In all studies, Ashwagandha was found to be beneficial and induced significant improvement in overall sleep compared to placebo groups. In addition, mental alertness was better, and reduced anxiety was observed [84]. Ashwagandha root extract was found to improve recall memory and focus, lower serum cortisol, improve sleep quality, and lower stress in a final sample of 125 individuals who took the extract (or placebo) for 90 consecutive days. Importantly, in this double-blind, randomized, parallel-group, two-arm, placebo-controlled trial, Ashwagandha was well tolerated [85]. Mild cognitive impairment has been ameliorated by Ashwagandha extract in a prospective, double-blind, placebo-controlled clinical study [86]. Task results can be summarized as improved memory, such as for faces and family pictures, as well as general and logical memory. Executive functions, including working memory, attention, and information processing speed, were improved compared to the placebo group.

The aforementioned results might favor *Withania somnifera* extract as a potential substance to induce LD. Importantly, it seems to be well tolerated in the general population. Unfortunately, we could not find any scientific studies of the plant being used to induce LD or to have oneiric properties. Moreover, sleep stages were not mentioned to have been affected in a recent review cited previously [84]. Even with no scientific reports, the molecular targets, as well as enhancing cognition, support the notion that this plant could be used for potentiating the probability of achieving lucidity in dreams.

4.7. Nootropics and Ayurvedic Medicine

Nootropics are a group of substances that might enhance cognitive function and can be further added to the list of candidates to induce LD. Some nootropics are found in the Ayurvedic system of knowledge too. Their mechanism of action, in general, is through improving glucose and oxygen brain supply [54]; henceforth, they might have antioxidant effects—such as many of the substances cited during this review. Thus, they might be efficient in improving cognition, such as combating oxidative stress in Alzheimer's disease. Malík and Tlustoš [54] cite other substances that can be regarded not just as cognitive enhancers during the day, but also during night sleeping. Meclofexatone is a combination of DMAE (Deanol) and synthetic auxin, and it seems to be twice as effective as DMAE regarding its effects on choline and ACh levels in the brain. This substance normalizes oxidative stress in rats, and increased mental alertness was observed in humans.

Ginkgo biloba is a plant with antioxidant properties due to triterpene lactone compounds, and it also has cholinergic mechanisms. It seems that, for the treatment of Alzheimer's disease, *Ginkgo* combined with donepezil, in contrast to donepezil alone, is superior in terms of safety and efficacy [87]. Human studies have shown improved working memory and information processing speed. Asiatic Pennywort (*Centella asiatica*), used in traditional medicine, has anti-Alzheimer's disease properties, and its main active compounds are triterpenoids, especially triterpene saponins such as those that might be contained in *Silene capensis*, which might have properties for increasing ACh synthesis, improving memory and cognition. In rats, learning and memory were improved through modulating DA, 5HT, and noradrenaline (NA) systems with an aqueous leaf extract [88]. The eclipta species (from the same family as *Artemisia* and *Calea*—Asteraceae) has its major compounds as alkaloids, triterpenoid saponins, volatile oil, sterols, and flavonoids. Studies have shown improved learning and memory abilities in rats. Butanol fractions increase ACh and reduce oxidative stress. Another interesting plant, Water Hyssop (*Bacopa monnieri*), which can be used in cooking, has effects on enhancing attention, cognition, and learning, restoring cognitive dysfunction possibly by increasing ACh in mice brains. *Convolvulus pluricaulis*, used in Ayurveda, seems to improve memory and cognition, as well as having AChEI properties [54,88].

Overall, these cited substances seem to have key properties found in many drugs that induce LD—enhancing memory, learning, cognition, working memory, and attention, as well as antioxidative and ACh-increasing properties—especially those with AChEI activity. Indian herbal formulations that have been studied in Alzheimer's disease might be promising in potentiating the possibility of achieving lucidity in dreams. These drugs have many types of plants in their constitution; for simplification, we will only cite the plants that are presented in this review, and for further investigation, we suggest seeing the review by Mehla et al. [88]. Improvements in learning, memory, and cognition, as well as AChEI activity, have been found in these compounds. Such compounds, even though they are not fully understood on a molecular level, might have properties for inducing LD.

5. General Discussion of the Neuropsychopharmacological Induction of Lucid Dreams

This review collected a wide diversity of articles that enabled a better understanding of the neurophysiological and phenomenological aspects of LD, thus promoting a comprehension of the use of exogenous substances to induce LD. For a summary of the studies that report the use of substances to induce oneiric activity, see Table 1 below.

Table 1. Studies that report the use of substances to induce lucid or state-like LD.

Authors (Year)	Study Design	Substances/Techniques	Mechanism of Action	Results	Commentary
LaBerge (2004) [28]	Patent	Donepezil, Rivastigmine, Galantamine, and Huperzine	Acetylcholinesterase inhibitors (AChEIs)	Donepezil: 90% of subjects induced LD. Similar dose-dependent results were encountered for rivastigmine and galantamine, with fewer side effects. The same for huperzine, but the dosage was inconclusive.	Reports a series of methods utilizing memory enhancing drugs such as those used for Alzheimer's disease—AChEIs. Other possible mechanisms cited: cholinergic agonists; muscarinic receptor agonists; allosteric modulators of ACh and nicotinic receptors.
LaBerge et al. (2018) [29]	Experimental study	Galantamine, WBTB, MILD, lectures about LD for recognizing dream cues	AChEI	62% reported LD: 14% on placebo; 27% with 4 mg; and 42% with 8 mg.	Galantamine induces LD in a dose-dependent manner. The integrated protocol seems to be effective in induction. Authors comment that galantamine may impact REM sleep, reducing latency and increasing phasic activity. Galantamine enhances dopaminergic neurotransmission, which might be involved in metacognition and conscious self-monitoring.
Sparrow et al. (2018) [30]	Experimental study	Galantamine, MDR, WBTB	AChEI	LD was reported by 40% of participants on WBTB + galantamine; 34% reported LD on MDR + galantamine.	MDR condition might expose traumas and conflicts, possibly for resolution.
Kern et al. (2017) [31]	Experimental study	L-alpha glycerylphosphorylcholine (α-GPC)	ACh precursor	No significant results.	No cognitive techniques were explicitly used in combination, which might impact results.
Haas et al. (2022) [50]	Case study	Pregabalin	Binding to alpha-2-delta sites, reducing excitatory neurotransmission. Anti-glutamatergic reports	Patient with multiple myeloma and history of chronic pain. Drug was administered causing lucid nightmares; after discontinuation, it ceased.	Abnormal dreams and visual hallucinations are uncommon symptoms of pregabalin.
Biehl (2022) [52]	Survey study	Vitamin intake, fish, fruit, and chili consumption, and antidepressants	Omega-3 (fish, which might contain DMAE), capsaicin (chili), antidepressants (availability of serotonin and receptors)	Significant correlations of the substances with LD.	Many of the substances affected dreaming, nightmares, LD, and hypnagogic state.
Sergio (1988) [53]	Self-case report	DMAE	Converted to choline by the body, stimulating reticular formation	Combining WBTB and visualization techniques, the author has suggested that it has great effects on inducing LD.	The lactate salt is the most effective, followed by the p-acetamidobenzoate salt, and the tartrate salt is the least effective.
Richter (2007) [55]	Patent	Primary ingredients of the supplement: Calea ternifolia, L-5-HTP, and vinpocetine	Calea ternifolia: AchEI L-5-HTP: increases serotonin Vinpocetine: anti-oxidant	The author gives two examples of people with recurrent nightmares. When taking the supplement, subjects started resolving conflicts through enhanced vividness and lucidity during dreams.	Secondary ingredient: melatonin; tertiary ingredients: wild lettuce extract, Mugwort extract (Artemisia vulgaris), DMAE, passionflower extract, and green tea extract.
Ebben et al. (2002) [59]	Experimental study	Pyridoxine (B6)	Co-factor for production of 5-HT	Increased recall and vividness in dreams.	High levels of serotonin before sleep suppress REM, causing rebound at the end of the night.

Table 1. *Cont.*

Authors (Year)	Study Design	Substances/Techniques	Mechanism of Action	Results	Commentary
Aspy et al. (2018) [60]	Experimental study	Pyridoxine and B Complex	Co-factor for production of 5-HT (B6)	Increased oneiric content recalled for B6. B complex worsened sleep quality.	Besides serotonin, nootropics derivatives from B6, like pyritinol, might combine effects for LD induction.
Mayagoitia et al. (1986) [67]	Experimental study	Calea zacatechichi	AChEI (sesquiterpene lactones)	Individuals reported increased light sleep; vividly hypnagogic imagery; less REM and deep sleep; more recalled and vivid dreams; and less content.	Many sesquiterpene lactones have been found in this plant. The Asteraceae family, including Calea, Artemisia, and Eclipta plants, have a potential role in the modulation of the cholinergic system to induce LD.
Hirst (2005) [74]	Ethnography study	Silene capensis	Saponins—AChEI	Main effects reported by diviners and novices were lucid and prophetic dreams.	Ubulawu drinks containing medicinal plants, which some have triterpenoid saponins for AChEIs, are used in rituals for divination for medicinal purposes.

Among the promising results, the AChEIs stand out, which are generously reported to induce LD. In addition to their scientific validation, they have already been used in various cultures. Galantamine and its influence on the dopaminergic system through the modulation of ACh can jointly generate self-monitoring phenomena, attention, cognitive clarity, recovery of memory about aspects of the self, working memory, and self-awareness. This is mainly due to the effects seen in the frontal lobe (especially the prefrontal cortex), hippocampus, and precuneus, highlighting the working memory modulated by both the cholinergic and dopaminergic systems. Based on this medicament, hybrid molecules can be created that are less toxic and inhibit AChE more than galantamine itself [24]. Therefore, it is a possibility to synthesize AChEIs for the induction of LD.

Although there are no reports of its potential related to dreams, *Celastrus paniculatus* is a strong candidate for LD induction, with AChEI mechanisms and modulating aspects similar to galantamine. An interesting mechanism is its restoration of dendritic atrophies in the hippocampus and prefrontal cortex related to depletion caused by AChE, which may be one of the mechanisms that give it the name of the "intellect tree" in the East, helping in memory and learning. Its mechanisms of action related to ACh and DA may be opportunistic in inducing LD, and studies on its therapeutic potential are well-received. *Artemisia vulgaris* and *Calea zacatechichi* could also have their studies deepened due to their AChEI potential, especially *Artemisia* due to the MAO inhibition property. The three plants mentioned in this paragraph contain sesquiterpene lactones, which should be a target in understanding the pharmacological induction of LD.

LD can be used to treat recurrent nightmares; however, in some cases, LD itself may be the symptom that needs to be treated. This is defined as a lucid nightmare, where the individual has little or no control over what happens in the dream and is consciously trapped and subjected to their fears and traumas. This might be one of the main limitations of LD therapy to recurrent nightmares [89]. From another perspective, it is observed that lucid nightmares are closely related to strong memories and traumas that have consolidated throughout an individual's life; thus, these experiences tend to manifest more frequently in dreams. It would be interesting to know if pleasurable life experiences also tend to manifest more in dreams and help to reach the threshold that makes the dream lucid, as seems to be the case with terrifying lucid nightmares. These impactful mechanisms are mainly related to DA, possibly by activating limbic and mesolimbic areas, as well as reality self-monitoring.

Another interesting point to note is the similarities between dreams, psychoses, and psychedelic substances. These states can serve as models for specific aspects of LD [90–92].

The internally generated hallucinations in dreams together with the knowledge that it is a perception generated within oneself and the hallucinations induced by psychedelics may share neural substrates related to the neurotransmission of 5-HT, where the functionalities of mental and sensory images of this neurotransmitter may play an important role. In addition, substances that help improve metacognition in psychotic patients, keeping them in a state of self-awareness without fully immersive delusions, should increase the likelihood of LD. An example of this is that AChEI drugs might be useful for the treatment of psychotic symptoms. Thus, the similarities between these states can help in understanding lucidity in a more general way, not just in dreams but in understanding the state of consciousness one is in. Paradoxical pharmacological dissociations are an interesting phenomenon since drugs inducing delta activity, a state normally associated with loss of consciousness, could provide biomarkers for the neural substrates of awareness, where the subject remains conscious. Delta-enhancing drugs such as N,N-dimethyltryptamine [60] do not necessarily induce loss of consciousness, and might even enhance it in some cases, such as augmented visual imagery in some episodes of LD.

Since LD is extremely sensitive to suggestion, for future studies, pharmacological techniques could be combined with behavioral/cognitive techniques. The latter includes a variety of types of induction techniques, which can be divided into dream-initiated LD (including MILD, reality testing, and Tholey's combined techniques) and wake-initiated LD (sense-initiated LD) (for review, see [93]). This division makes a distinction between those LD episodes that start during the dream and those that start as the subject enters consciously into sleep. Other ways to enhance the probability of becoming lucid include a dream diary and mindfulness interventions. Studies have already found a possible role of mindfulness practices and meditations in inducing LD [6,94–96], besides ancient techniques that cultivate self-consciousness during sleep. Further research should explore the AChEI properties of the various substances mentioned, integrating pharmacological and cognitive techniques for reliable LD induction.

6. Limitations, Perspectives, and Conclusions

Some limitations must be considered. Firstly, an important point to recognize in neurotransmission is the diversity and complexity of neurotransmitters, their different production sites and projection areas, and the great diversity of receptors. Each neurotransmitter can have distinct or similar functions to others, depending on these points. In addition, they modulate other neurotransmitter systems. Considering this extremely complex brain network that functions in circadian and homeostatic ways and produces the experience of human reality perception, what has been described here is a mere attempt at understanding how lucidity is achieved on a neurotransmitter level. Moreover, no studies have yet been conducted on eye movement verification related to the pharmacological induction of LD. Therefore, it would be interesting to both verify the eye movements during LD induced pharmacologically and test whether galantamine and its potential AChEIs activate areas related to self-awareness, self-monitoring, and metacognition, with emphasis on dopaminergic modulation. Furthermore, it is possible to utilize the interaction of plants with specific properties to facilitate the induction of LD. However, a comprehensive understanding is required to discern if there exists a synergy between these plant-based compounds, the potential risks associated with such synergies, and the cognitive–behavioral effects induced by these interactions. It is noteworthy that there is a scarcity of studies focusing on plants at present, which could potentially contribute to discussions advocating for streamlined regulatory processes for human trials involving plant-derived substances. This scarcity of research serves as an observable limitation in exploring the efficacy and safety of plant-based interventions for inducing LD.

This review integrated numerous studies to deepen the comprehension of LD. In light of this, we propose a research agenda to steer forthcoming investigations towards the pharmacological induction of LD. Future studies are urged to concentrate on elucidating the impact of various drugs and plants on sleep and dreaming, an area significantly underexplored in the current literature. Further details of this comprehensive research agenda are outlined below, delineating pathways for interdisciplinary collaboration and innovative methodological approaches.

Research agenda:

1. Future studies should focus on how the aforementioned drugs and plants have an impact on sleep and dreaming. In addition, there is a potential influence of the chronobiological effects of time on the administration of these drugs; some of them (e.g., nootropics) mostly affect the waking state, but also might impact nocturnal consciousness. The literature on this type of information is scarce.
2. The pharmacodynamics of plants and drugs should also be researched, mostly those that have similar pharmacological, cognitive, behavioral, and neurophysiological aspects to LD.
3. Sesquiterpenes from known plants, which possibly already have approval for human testing (to reduce bureaucracy), should be researched using Ellman's method. This allows for the elaboration of an AChEI's concentration and efficacy, which can be compared to medications such as galantamine and donepezil.
4. Better understanding how sesquiterpenes influence DA and ACh concentrations in the brain would be a great step, allowing for inferring its action in specific brain areas.
5. Pharmacological protocols for the understanding of the synergism between potential drugs/plants to induce LD is crucial. In this way, a combination can be used to enhance the probability of LD and to know whether there are side effects
6. Depending on the focus of future studies, many methodologies can be created. A mixture of induction techniques, including behavioral and pharmacological, should be combined to enhance the probability of achieving dream lucidity in a controlled environment. In addition, the eye-signal technique to flag LD and neuroimaging techniques could be used in conjunction.

LD is a complex state full of opportunities for the dream self. Understanding neural systems and brain areas enables the induction of this state effectively, allowing for the study of neural correlates of self-consciousness. Finally, it is important to note that to study the pharmacological induction of LD, a close interaction of biomedical research with anthropology and the history of ancient people who used oneiric plants is necessary. This collaboration of neuropsychology and social sciences can be fruitful for studies that aim to understand human consciousness.

Supplementary Materials: The following supporting information can be downloaded at: https://www.mdpi.com/article/10.3390/brainsci14050426/s1, File S1: we describe the circadian and homeostatic processes that regulate sleep, and the main neurotransmitters that control REM sleep" [97–114].

Author Contributions: Writing—original draft preparation, A.A.O.; writing—review and editing, all authors. All authors have read and agreed to the published version of the manuscript.

Funding: This research was funded by CAPES, CNPq, and UFABC.

Acknowledgments: We thank AMORA in the name of Lucemere Mota for general support.

Conflicts of Interest: The authors declare no conflicts of interest. The funders had no role in writing this manuscript.

References

1. La Berge, S.P. Lucid Dreaming as a Learnable Skill: A Case Study. *Percept. Mot. Ski.* **1980**, *51*, 1039–1042. [CrossRef]
2. La Berge, S.P.; Nagel, L.E.; Dement, W.C.; Zarcone, V.P. Lucid Dreaming Verified by Volitional Communication during Rem Sleep. *Percept. Mot. Ski.* **1981**, *52*, 727–732. [CrossRef] [PubMed]
3. Mota-Rolim, S.A. On Moving the Eyes to Flag Lucid Dreaming. *Front. Neurosci.* **2020**, *14*, 361. [CrossRef] [PubMed]
4. Stumbrys, T.; Erlacher, D. Lucid dreaming during NREM sleep: Two case reports. *Int. J. Dream Res.* **2012**, *5*, 151–155. [CrossRef]
5. Mota Rolim, S.A.; Brandão, D.S.; Andrade, K.C.; de Queiroz, C.M.T.; Araujo, J.F.; de Araujo, D.B.; Ribeiro, S. Neurophysiological Features of Lucid Dreaming during N1 and N2 Sleep Stages: Two Case Reports. *Sleep Sci.* **2015**, *8*, 215. [CrossRef]

6. Gackenbach, J.; Moorecroft, W.; Alexander, C.; LaBerge, S. "Consciousness" During Sleep in a TM Practitioner: Heart Rate, Respiration, and Eye Movement. *Lucidity Lett.* **1987**, *6*.
7. Mason, L.I.; Orme-Johnson, D. Transcendental consciousness wakes up in dreaming and deep sleep. *Int. J. Dream Res.* **2010**, *3*, 28–32. [CrossRef]
8. Parker, S.; Bharati, S.V.; Fernandez, M. Defining yoga-nidra: Traditional accounts, physiological research, and future directions. *Int. J. Yoga Ther.* **2013**, *23*, 11–16. [CrossRef]
9. Mota-Rolim, S.A.; Bulkeley, K.; Campanelli, S.; Lobão-Soares, B.; de Araujo, D.B.; Ribeiro, S. The Dream of God: How Do Religion and Science See Lucid Dreaming and Other Conscious States During Sleep? *Front. Psychol.* **2020**, *11*, 555731. [CrossRef] [PubMed]
10. Dresler, M.; Wehrle, R.; Spoormaker, V.I.; Koch, S.P.; Holsboer, F.; Steiger, A.; Obrig, H.; Sämann, P.G.; Czisch, M. Neural correlates of dream lucidity obtained from contrasting lucid versus non-lucid REM sleep: A combined EEG/fMRI case study. *Sleep* **2012**, *35*, 1017–1020. [CrossRef] [PubMed]
11. LaBerge, S.; Rheingold, H. *Exploring the World of Lucid Dreaming*, 4/27/94 ed.; Ballantine Books: New York, NY, USA, 1991.
12. Hobson, J.A.; Pace-Schott, E.F. The cognitive neuroscience of sleep: Neuronal systems, consciousness and learning. *Nat. Rev. Neurosci.* **2002**, *3*, 679–693. [CrossRef] [PubMed]
13. Filevich, E.; Dresler, M.; Brick, T.R.; Kühn, S. Behavioral/Cognitive Metacognitive Mechanisms Underlying Lucid Dreaming. *J. Neurosci.* **2015**, *35*, 1082–1088. [CrossRef] [PubMed]
14. Baird, B.; Castelnovo, A.; Gosseries, O.; Tononi, G. Frequent lucid dreaming associated with increased functional connectivity between frontopolar cortex and temporoparietal association areas. *Sci. Rep.* **2018**, *8*, 17798. [CrossRef] [PubMed]
15. Baird, B.; Mota-Rolim, S.A.; Dresler, M. The cognitive neuroscience of lucid dreaming. *Neurosci. Biobehav. Rev.* **2019**, *100*, 305–323. [CrossRef] [PubMed]
16. Stumbrys, T.; Erlacher, D.; Schädlich, M.; Schredl, M. Induction of lucid dreams: A systematic review of evidence. *Conscious. Cogn.* **2012**, *21*, 1456–1475. [CrossRef] [PubMed]
17. Houghton, P.J.; Ren, Y.; Howes, M.-J. Acetylcholinesterase inhibitors from plants and fungi. *Nat. Prod. Rep.* **2006**, *23*, 181–199. [CrossRef] [PubMed]
18. Mukherjee, P.K.; Kumar, V.; Mal, M.; Houghton, P.J. Acetylcholinesterase inhibitors from plants. *Phytomed. Int. J. Phytother. Phytopharm.* **2007**, *14*, 289–300. [CrossRef]
19. Khan, H.; Marya; Amin, S.; Kamal, M.A.; Patel, S. Flavonoids as acetylcholinesterase inhibitors: Current therapeutic standing and future prospects. *Biomed. Pharmacother.* **2018**, *101*, 860–870. [CrossRef] [PubMed]
20. Lima, J.A.; Hamerski, L. Chapter 8—Alkaloids as Potential Multi-Target Drugs to Treat Alzheimer's Disease. In *Studies in Natural Products Chemistry*; Atta-ur-Rahman, Ed.; Elsevier: Amsterdam, The Netherlands, 2019; Volume 61, pp. 301–334. [CrossRef]
21. Zemek, F.; Drtinova, L.; Nepovimova, E.; Sepsova, V.; Korabecny, J.; Klimes, J.; Kuca, K. Outcomes of Alzheimer's disease therapy with acetylcholinesterase inhibitors and memantine. *Expert Opin. Drug Saf.* **2014**, *13*, 759–774. [CrossRef] [PubMed]
22. Overview | Donepezil, Galantamine, Rivastigmine and Memantine for the Treatment of Alzheimer's Disease | Guidance | NICE. Available online: https://www.nice.org.uk/guidance/ta217 (accessed on 25 November 2023).
23. Rösler, M.; Anand, R.; Cicin-Sain, A.; Gauthier, S.; Agid, Y.; Dal-Bianco, P.; Stähelin, H.B.; Hartman, R.; Gharabawi, M.; Bayer, T. Efficacy and safety of rivastigmine in patients with Alzheimer's disease: International randomised controlled trial. *BMJ* **1999**, *318*, 633–638. [CrossRef] [PubMed]
24. Lazarova, M.I.; Tsekova, D.S.; Tancheva, L.P.; Kirilov, K.T.; Uzunova, D.N.; Vezenkov, L.T.; Tsvetanova, E.R.; Alexandrova, A.V.; Georgieva, A.P.; Gavrilova, P.T.; et al. New Galantamine Derivatives with Inhibitory Effect on Acetylcholinesterase Activity. *J. Alzheimer's Dis.* **2021**, *83*, 1211–1220. [CrossRef] [PubMed]
25. Maelicke, A.; Albuquerque, E.X. Allosteric modulation of nicotinic acetylcholine receptors as a treatment strategy for Alzheimer's disease. *Eur. J. Pharmacol.* **2000**, *393*, 165–170. [CrossRef] [PubMed]
26. File:Galanthus Nivalis.jpg—Simple English Wikipedia, the Free Encyclopedia. Available online: https://commons.wikimedia.org/wiki/File:Galanthus_nivalis.jpg (accessed on 25 November 2023).
27. Gott, J.A.; Stücker, S.; Kanske, P.; Haaker, J.; Dresler, M. Acetylcholine and metacognition during sleep. *Conscious. Cogn.* **2024**, *117*, 103608. [CrossRef]
28. LaBerge, S. Substances That Enhance Recall and Lucidity during Dreaming. U.S. Patent 20040266659A1, 30 December 2004. Available online: https://patents.google.com/patent/US20040266659A1/en (accessed on 25 November 2023).
29. LaBerge, S.; LaMarca, K.; Baird, B. Pre-sleep treatment with galantamine stimulates lucid dreaming: A double-blind, placebo-controlled, crossover study. *PLoS ONE* **2018**, *13*, e0201246. [CrossRef] [PubMed]
30. Sparrow, G.; Hurd, R.; Carlson, R.; Molina, A. Exploring the effects of galantamine paired with meditation and dream reliving on recalled dreams: Toward an integrated protocol for lucid dream induction and nightmare resolution. *Conscious. Cogn.* **2018**, *63*, 74–88. [CrossRef] [PubMed]
31. Kern, S.; Appel, K.; Schredl, M.; Pipa, G. No effect of α-GPC on lucid dream induction or dream content. *Somnologie* **2017**, *21*, 180–186. [CrossRef]
32. Schilström, B.; Ivanov, V.B.; Wiker, C.; Svensson, T.H. Galantamine enhances dopaminergic neurotransmission in vivo via allosteric potentiation of nicotinic acetylcholine receptors. *Neuropsychopharmacology* **2007**, *32*, 43–53. [CrossRef] [PubMed]
33. Kjaer, T.W.; Bertelsen, C.; Piccini, P.; Brooks, D.; Alving, J.; Lou, H.C. Increased dopamine tone during meditation-induced change of consciousness. *Cogn. Brain Res.* **2002**, *13*, 255–259. [CrossRef] [PubMed]

34. Saraswati, S.S. *Yoga Nidra*, 6th ed.; Yoga Publications Trust: Munger, India, 2002.
35. Pandi-Perumal, S.R.; Spence, D.W.; Srivastava, N.; Kanchibhotla, D.; Kumar, K.; Sharma, G.S.; Gupta, R.; Batmanabane, G. The Origin and Clinical Relevance of Yoga Nidra. *Sleep Vigil.* **2022**, *6*, 61–84. [CrossRef] [PubMed]
36. Datta, K.; Mallick, H.N.; Tripathi, M.; Ahuja, N.; Deepak, K.K. Electrophysiological Evidence of Local Sleep During Yoga Nidra Practice. *Front. Neurol.* **2022**, *13*, 910794. [CrossRef] [PubMed]
37. Frohlich, J.; Mediano, P.A.M.; Bavato, F.; Gharabaghi, A. Paradoxical pharmacological dissociations result from drugs that enhance delta oscillations but preserve consciousness. *Commun. Biol.* **2023**, *6*, 654. [CrossRef] [PubMed]
38. Lou, H.C.; Rømer Thomsen, K.; Changeux, J.P. The Molecular Organization of Self-awareness: Paralimbic Dopamine-GABA Interaction. *Front. Syst. Neurosci.* **2020**, *14*, 3. [CrossRef]
39. Sodré, M.E.; Wießner, I.; Irfan, M.; Schenck, C.H.; Mota-Rolim, S.A. Awake or Sleeping? Maybe Both. . . A Review of Sleep-Related Dissociative States. *J. Clin. Med.* **2023**, *12*, 3876. [CrossRef] [PubMed]
40. Joensson, M.; Thomsen, K.R.; Andersen, L.M.; Gross, J.; Mouridsen, K.; Sandberg, K.; Østergaard, L.; Lou, H.C. Making sense: Dopamine activates conscious self-monitoring through medial prefrontal cortex. *Hum. Brain Mapp.* **2015**, *36*, 1866–1877 [CrossRef] [PubMed]
41. Murphy, D.L. Mental effects of L-dopa. *Annu. Rev. Med.* **1973**, *24*, 209–216. [CrossRef]
42. Trenado, C.; Boschheidgen, M.; Rübenach, J.; N'Diaye, K.; Schnitzler, A.; Mallet, L.; Wojtecki, L. Assessment of Metacognition and Reversal Learning in Parkinson's Disease: Preliminary Results. *Front. Hum. Neurosci.* **2018**, *12*, 343. [CrossRef] [PubMed]
43. Cipolli, C.; Bolzani, R.; Massetani, R.; Murri, L.; Muratorio, A. Dream structure in Parkinson's patients. *J. Nerv. Ment. Dis.* **1992**, *180*, 516–523. [CrossRef] [PubMed]
44. Partinen, M. Sleep disorder related to Parkinson's disease. *J. Neurol.* **1997**, *244*, S3–S6. [CrossRef] [PubMed]
45. Sharf, B.; Moskovitz, C.; Lupton, M.D.; Klawans, H.L. Dream phenomena induced by chronic levodopa therapy. *J. Neural Transm.* **1978**, *43*, 143–151. [CrossRef] [PubMed]
46. Moskovitz, C.; Moses, H.; Klawans, H.L. Levodopa-induced psychosis: A kindling phenomenon. *Am. J. Psychiatry* **1978**, *135*, 669–675. [CrossRef]
47. Borek, L.L.; Kohn, R.; Friedman, J.H. Phenomenology of dreams in Parkinson's disease. *Mov. Disord.* **2007**, *22*, 198–202. [CrossRef] [PubMed]
48. De Gennaro, L.; Lanteri, O.; Piras, F.; Scarpelli, S.; Assogna, F.; Ferrara, M.; Caltagirone, C.; Spalletta, G. Dopaminergic system and dream recall: An MRI study in Parkinson's disease patients. *Hum. Brain Mapp.* **2016**, *37*, 1136–1147. [CrossRef]
49. Pinter, M.M.; Pogarell, O.; Oertel, W.H. Efficacy, safety, and tolerance of the non-ergoline dopamine agonist pramipexole in the treatment of advanced Parkinson's disease: A double blind, placebo controlled, randomised, multicentre study. *J. Neurol. Neurosurg. Psychiatry* **1999**, *66*, 436–441. [CrossRef] [PubMed]
50. Haas, M.F.; Latchman, J.; Guastella, A.M.; Craig, D.S.; Chang, Y.D. Lucid Dreams Associated with Pregabalin: Implications for Clinical Practice. *J. Pain Palliat. Care Pharmacother.* **2022**, *36*, 194–199. [CrossRef] [PubMed]
51. Mousailidis, G.; Papanna, B.; Salmon, A.; Sein, A.; Al-Hillawi, Q. Pregabalin induced visual hallucinations—A rare adverse reaction. *BMC Pharmacol. Toxicol.* **2020**, *21*, 16. [CrossRef] [PubMed]
52. Biehl, J. Foods and substances influencing (lucid) dreams. *Int. J. Dream Res.* **2022**, *15*, 224–234. [CrossRef]
53. Sergio, W. Use of DMAE (2-dimethylaminoethanol) in the induction of lucid dreams. *Med. Hypotheses* **1988**, *26*, 255–257. [CrossRef]
54. Malík, M.; Tlustoš, P. Nootropics as Cognitive Enhancers: Types, Dosage and Side Effects of Smart Drugs. *Nutrients* **2022**, *14*, 3367. [CrossRef]
55. Richter, J. Dietary Supplement and a Method to Enhance Sleep and Lucid Dreaming. U.S. Patent US8092840B2, 10 January 2012.
56. Zhdanova, I.V.; Wurtman, R.J. Efficacy of melatonin as a sleep-promoting agent. *J. Biol. Rhythm.* **1997**, *12*, 644–650. [CrossRef]
57. Stone, B.M.; Turner, C.; Mills, S.L.; Nicholson, A.N. Hypnotic activity of melatonin. *Sleep* **2000**, *23*, 663–669. [CrossRef]
58. Kahan, T.; Hays, J.; Hirashima, B.; Johnston, K. Effects of melatonin on dream bizarreness among male and female college students. *Psychology* **2000**, *2*, 74–83.
59. Ebben, M.; Lequerica, A.; Spielman, A. Effects of pyridoxine on dreaming: A preliminary study. *Percept. Mot. Ski.* **2002**, *94*, 135–140. [CrossRef] [PubMed]
60. Aspy, D.J.; Madden, N.A.; Delfabbro, P. Effects of Vitamin B6 (Pyridoxine) and a B Complex Preparation on Dreaming and Sleep. *Percept. Mot. Ski.* **2018**, *125*, 451–462. [CrossRef] [PubMed]
61. Laios, K.; Lytsikas-Sarlis, P.; Manes, K.; Kontaxaki, M.-I.; Karamanou, M.; Androutsos, G. Drugs for mental illnesses in ancient greek medicine. *Psychiatriki* **2019**, *30*, 58–65. [CrossRef] [PubMed]
62. Mathianaki, K.; Tzatzarakis, M.; Karamanou, M. Poppies as a sleep aid for infants: The "Hypnos" remedy of Cretan folk medicine. *Toxicol. Rep.* **2021**, *8*, 1729–1733. [CrossRef] [PubMed]
63. Presley, C.C.; Lindsley, C.W. DARK Classics in Chemical Neuroscience: Opium, a Historical Perspective. *ACS Chem. Neurosci.* **2018**, *9*, 2503–2518. [CrossRef] [PubMed]
64. Scarborough, J. The opium poppy in Hellenistic and Roman medicine. In *Drugs and Narcotics in History*; Teich, M., Porter, R., Eds.; Cambridge University Press: Cambridge, UK, 1995; pp. 4–23. [CrossRef]
65. Tekiner, H.; Kosar, M. The opium poppy as a symbol of sleep in Bertel Thorvaldsen's relief of 1815. *Sleep Med.* **2016**, *19*, 123–125. [CrossRef] [PubMed]

66. Laux-Biehlmann, A.; Mouheiche, J.; Vérièpe, J.; Goumon, Y. Endogenous morphine and its metabolites in mammals: History, synthesis, localization and perspectives. *Neuroscience* **2013**, *233*, 95–117. [CrossRef] [PubMed]

67. Mayagoitia, L.; Díaz, J.L.; Contreras, C.M. Psychopharmacologic analysis of an alleged oneirogenic plant: Calea zacatechichi. *J. Ethnopharmacol.* **1986**, *18*, 229–243. [CrossRef] [PubMed]

68. Díaz, J.L. Ethnopharmacology and taxonomy of Mexican psychodysleptic plants. *J. Psychedelic Drugs* **1979**, *11*, 71–101. [CrossRef] [PubMed]

69. Mata, R.; Contreras-Rosales, A.J.; Gutiérrez-González, J.A.; Villaseñor, J.L.; Pérez-Vásquez, A. Calea ternifolia Kunth, the Mexican "dream herb", a concise review. *Botany* **2022**, *100*, 261–274. [CrossRef]

70. Arya, A.; Chahal, R.; Rao, R.; Rahman, M.H.; Kaushik, D.; Akhtar, M.F.; Saleem, A.; Khalifa, S.M.A.; El-Seedi, H.R.; Kamel, M.; et al. Acetylcholinesterase Inhibitory Potential of Various Sesquiterpene Analogues for Alzheimer's Disease Therapy. *Biomolecules* **2021**, *11*, 350. [CrossRef] [PubMed]

71. Fu, Y.; Zhao, W. Polyesterified Sesquiterpenoids from the Seeds of Celastrus paniculatus as Lifespan-Extending Agents for the Nematode Caenorhabditis elegans. *J. Nat. Prod.* **2020**, *83*, 505–515. [CrossRef] [PubMed]

72. Bhagya, V.; Jaideep, S. Neuropharmacological and Cognitive Effects of Celastrus paniculatus—A Comprehensive Review. *Int. J. Pharm. Sci. Rev. Res.* **2020**, *65*, 92–97. [CrossRef]

73. Nagpal, K.; Garg, M.; Arora, D.; Dubey, A.; Grewal, A.S. An extensive review on phytochemistry and pharmacological activities of Indian medicinal plant Celastrus paniculatus Willd. *Phytother. Res.* **2022**, *36*, 1930–1951. [CrossRef] [PubMed]

74. Hirst, M. Dreams and Medicines: The Perspective of Xhosa Diviners and Novices in the Eastern Cape. *Indo-Pac. J. Phenomenol.* **2005**, *5*, 1–22. [CrossRef]

75. Stepnik, K.; Kukula-Koch, W. In Silico Studies on Triterpenoid Saponins Permeation through the Blood–Brain Barrier Combined with Postmortem Research on the Brain Tissues of Mice Affected by Astragaloside IV Administration. *Int. J. Mol. Sci.* **2020**, *21*, 2534. [CrossRef] [PubMed]

76. Larhsini, M.; Marston, A.; Hostettmann, K. Triterpenoid saponins from the roots of Silene cucubalus. *Fitoterapia* **2003**, *74*, 237–241. [CrossRef] [PubMed]

77. Bechkri, S.; Alabdul Magid, A.; Sayagh, C.; Berrehal, D.; Harakat, D.; Voutquenne-Nazabadioko, L.; Kabouche, Z.; Kabouche, A. Triterpene saponins from Silene gallica collected in North-Eastern Algeria. *Phytochemistry* **2020**, *172*, 112274. [CrossRef]

78. Lacaille-Dubois, M.A.; Hanquet, B.; Cui, Z.H.; Lou, Z.C.; Wagner, H. Acylated triterpene saponins from Silene jenisseensis. *Phytochemistry* **1995**, *40*, 509–514. [CrossRef]

79. Sobiecki, J.F. Psychoactive Ubulawu Spiritual Medicines and Healing Dynamics in the Initiation Process of Southern Bantu Diviners. *J. Psychoact. Drugs* **2012**, *44*, 216–223. [CrossRef] [PubMed]

80. Ekiert, H.; Pajor, J.; Klin, P.; Rzepiela, A.; Slesak, H.; Szopa, A. Significance of *Artemisia Vulgaris* L. (Common Mugwort) in the History of Medicine and Its Possible Contemporary Applications Substantiated by Phytochemical and Pharmacological Studies. *Molecules* **2020**, *25*, 4415. [CrossRef] [PubMed]

81. Bashir, A.; Nabi, M.; Tabassum, N.; Afzal, S.; Ayoub, M. An updated review on phytochemistry and molecular targets of Withania somnifera (L.) Dunal (Ashwagandha). *Front. Pharmacol.* **2023**, *14*, 1049334. [CrossRef]

82. Singh, M.; Ramassamy, C. In vitro screening of neuroprotective activity of Indian medicinal plant Withania somnifera. *J. Nutr. Sci.* **2017**, *6*, e54. [CrossRef] [PubMed]

83. Prakash, J.; Chouhan, S.; Yadav, S.K.; Westfall, S.; Rai, S.N.; Singh, S.P. Withania somnifera Alleviates Parkinsonian Phenotypes by Inhibiting Apoptotic Pathways in Dopaminergic Neurons. *Neurochem. Res.* **2014**, *39*, 2527–2536. [CrossRef] [PubMed]

84. Cheah, K.L.; Norhayati, M.N.; Yaacob, L.H.; Rahman, R.A. Effect of Ashwagandha (Withania somnifera) extract on sleep: A systematic review and meta-analysis. *PLoS ONE* **2021**, *16*, e0257843. [CrossRef] [PubMed]

85. Gopukumar, K.; Thanawala, S.; Somepalli, V.; Rao, T.S.S.; Thamatam, V.B.; Chauhan, S. Efficacy and Safety of Ashwagandha Root Extract on Cognitive Functions in Healthy, Stressed Adults: A Randomized, Double-Blind, Placebo-Controlled Study. *Evid.-Based Complement. Altern. Med.* **2021**, *2021*, 8254344. [CrossRef]

86. Choudhary, D.; Bhattacharyya, S.; Bose, S. Efficacy and Safety of Ashwagandha (Withania somnifera (L.) Dunal) Root Extract in Improving Memory and Cognitive Functions. *J. Diet. Suppl.* **2017**, *14*, 599–612. [CrossRef] [PubMed]

87. Yancheva, S.; Ihl, R.; Nikolova, G.; Panayotov, P.; Schlaefke, S.; Hoerr, R. Ginkgo biloba extract EGb 761®, donepezil or both combined in the treatment of Alzheimer's disease with neuropsychiatric features: A randomised, double-blind, exploratory trial. *Aging Ment. Health* **2009**, *13*, 183–190. [CrossRef] [PubMed]

88. Mehla, J.; Gupta, P.; Pahuja, M.; Diwan, D.; Diksha, D. Indian Medicinal Herbs and Formulations for Alzheimer's Disease, from Traditional Knowledge to Scientific Assessment. *Brain Sci.* **2020**, *10*, 964. [CrossRef] [PubMed]

89. de Macêdo, T.C.F.; Ferreira, G.H.; de Almondes, K.M.; Kirov, R.; Mota-Rolim, S.A. My Dream, My Rules: Can Lucid Dreaming Treat Nightmares? *Front. Psychol.* **2019**, *10*, 2618. [CrossRef] [PubMed]

90. Kraehenmann, R. Dreams and Psychedelics: Neurophenomenological Comparison and Therapeutic Implications. *Curr. Neuropharmacol.* **2017**, *15*, 1032–1042. [CrossRef] [PubMed]

91. Mota, N.B.; Resende, A.; Mota-Rolim, S.A.; Copelli, M.; Ribeiro, S. Psychosis and the control of lucid dreaming. *Front. Psychol.* **2016**, *7*, 294. [CrossRef] [PubMed]

92. Dresler, M.; Wehrle, R.; Spoormaker, V.I.; Steiger, A.; Holsboer, F.; Czisch, M.; Hobson, J.A. Neural correlates of insight in dreaming and psychosis. *Sleep Med. Rev.* **2015**, *20*, 92–99. [CrossRef] [PubMed]

93. Tan, S.; Fan, J. A systematic review of new empirical data on lucid dream induction techniques. *J. Sleep Res.* **2023**, *32*, e13786. [CrossRef] [PubMed]

94. Stumbrys, T.; Erlacher, D. Mindfulness and Lucid Dream Frequency Predicts the Ability to Control Lucid Dreams. *Imagin. Cogn. Pers.* **2016**, *36*, 229–239. [CrossRef]

95. Stumbrys, T.; Erlacher, D.; Malinowski, P. Meta-Awareness During Day and Night: The Relationship Between Mindfulness and Lucid Dreaming. *Imagin. Cogn. Pers.* **2015**, *34*, 415–433. [CrossRef]

96. Baird, B.; Riedner, B.A.; Boly, M.; Davidson, R.J.; Tononi, G. Increased lucid dream frequency in long-term meditators but not following mindfulness-based stress reduction training. *Psychol. Conscious. Theory Res. Pract.* **2019**, *6*, 40–54. [CrossRef] [PubMed]

97. Reichert, C.F.; Deboer, T.; Landolt, H.P. Adenosine, caffeine, and sleep–wake regulation: State of the science and perspectives. *J. Sleep Res.* **2022**, *31*, e13597. [CrossRef] [PubMed]

98. Scammell, T.E.; Arrigoni, E.; Lipton, J.O. Neural Circuitry of Wakefulness and Sleep. *Neuron* **2017**, *93*, 747–765. [CrossRef] [PubMed]

99. Iber, C.; Ancoli-Israel, S.; Chesson, A.; Quan, S.F. *The AASM Manual for the Scoring of Sleep and Associated Events: Rules, Terminology, and Technical Specification*; American Academy of Sleep Medicine: Westchester, NY, USA, 2007.

100. Adamantidis, A.R.; Gutierrez Herrera, C.; Gent, T.C. Oscillating circuitries in the sleeping brain. *Nat. Rev. Neurosci.* **2019**, *20*, 746–762. [CrossRef] [PubMed]

101. Siclari, F.; Valli, K.; Arnulf, I. Dreams and nightmares in healthy adults and in patients with sleep and neurological disorders. *Lancet Neurol.* **2020**, *19*, 849–859. [CrossRef]

102. Patel, A.K.; Reddy, V.; Shumway, K.R.; Araujo, J.F. *Physiology, Sleep Stages*; StatPearls: Treasure Island, FL, USA, 2022.

103. Ratna, D.; Mondal, A.C.; Mallick, B.N. Modulation of dopamine from ventral tegmental area neurons by the LC-REM-OFF and PPT-REM-ON neurons in REMS regulation in freely moving rats. *Neuropharmacology* **2023**, *237*, 109621. [CrossRef]

104. Hobson, J.A. REM sleep and dreaming: Towards a theory of protoconsciousness. *Nat. Rev. Neurosci.* **2009**, *10*, 803–814. [CrossRef] [PubMed]

105. Pace-Schott, E.F.; Hobson, J.A. The Neurobiology of Sleep: Genetics, cellular physiology and subcortical networks. *Nat. Rev. Neurosci.* **2002**, *3*, 591–605. [CrossRef] [PubMed]

106. Vazquez, J.; Baghdoyan, H.A. Basal forebrain acetylcholine release during REM sleep is significantly greater than during waking. *Am. J. Physiol. Regul. Integr. Comp. Physiol.* **2001**, *280*, R598–R601. [CrossRef] [PubMed]

107. Vanini, G.; Torterolo, P. Sleep-Wake Neurobiology. In *Advances in Experimental Medicine and Biology*; Springer: Cham, Switzerland, 2021; Volume 1297, pp. 65–82. [CrossRef]

108. Gottesmann, C. The involvement of noradrenaline in rapid eye movement sleep mentation. *Front. Neurol.* **2011**, *2*, 81. [CrossRef] [PubMed]

109. Lu, J.; Bjorkum, A.A.; Xu, M.; Gaus, S.E.; Shiromani, P.J.; Saper, C.B. Selective activation of the extended ventrolateral preoptic nucleus during rapid eye movement sleep. *J. Neurosci. Off. J. Soc. Neurosci.* **2002**, *22*, 4568–4576. [CrossRef] [PubMed]

110. Carboni, E.; Silvagni, A. Dopamine reuptake by norepinephrine neurons: Exception or rule? *Crit. Rev. Neurobiol.* **2004**, *16*, 121–128. [CrossRef] [PubMed]

111. Solms, M. Dreaming and REM sleep are controlled by different brain mechanisms. *Behav. Brain Sci.* **2000**, *23*, 843–850. [CrossRef] [PubMed]

112. Mota-Rolim, S.; Ribeiro, S. Bases Biológicas da Atividade Onírica. 2013, pp. 201–227. Available online: https://repositorio.ufrn.br/bitstream/1/11723/1/SidartaRibeiro_ICE_Bases_biologicas_atividade_onirica.pdf (accessed on 22 March 2024).

113. Perogamvros, L.; Baird, B.; Seibold, M.; Riedner, B.; Boly, M.; Tononi, G. The Phenomenal Contents and Neural Correlates of Spontaneous Thoughts across Wakefulness, NREM Sleep, and REM Sleep. *J. Cogn. Neurosci.* **2017**, *29*, 1766–1777. [CrossRef] [PubMed]

114. Csiernik, R.; Pirie, M. A primer on sleeping, dreaming, and psychoactive agents. *J. Soc. Work Pract. Addict.* **2023**, *23*, 3–23. [CrossRef]

Review

The Epistemic Limits of Impactful Dreams: Metacognition, Metaphoricity, and Sublime Feeling

Don Kuiken

Department of Psychology, University of Alberta, P217 Biological Sciences Building, Edmonton, AB T6G 2E9, Canada; dkuiken@ualberta.ca

Abstract: Taxonomic studies of dreams that continue to influence the dreamer's thoughts and feelings after awakening have distinguished three types of impactful dreams: nightmares, existential dreams, and transcendent dreams. Of these, existential dreams and transcendent dreams are characterized by recurrent metacognitive appraisal of the epistemic tension between complementary (a) metaphoric (A "is" B) assertions and (b) literal (A "is not" B) assertions. Metacognitive appraisal of such complementary metaphoric and literal assertions is detectable as the felt sense of inexpressible realizations. The poesy of such inexpressible realizations depends upon the juxtaposition of a metaphoric topic and vehicle that are both "semantically dense" but at an abstract level "distant" from each other. The result is "emergence" of attributes of the metaphoric vehicle that are sufficiently abstract to be attributes also of the metaphoric topic. The cumulative effect of successive metaphoric/literal categorical transformations produces a higher-level form of metacognition that is consistent with a neo-Kantian account of sublime feeling. Sublime feeling occurs as either sublime disquietude (existential dreams) or as sublime enthrallment (transcendent dreams). The aftereffects of these two dream types are thematically iterative "living metaphors" that have abstract (but not "totalizing") ontological import.

Keywords: metacognition; metaphor; sublime feeling; impactful dreams

Citation: Kuiken, D. The Epistemic Limits of Impactful Dreams: Metacognition, Metaphoricity, and Sublime Feeling. *Brain Sci.* **2024**, *14*, 528. https://doi.org/10.3390/brainsci14060528

Academic Editor: Maurizio Gorgoni

Received: 5 April 2024
Revised: 16 May 2024
Accepted: 20 May 2024
Published: 22 May 2024

1. Introduction

Following spontaneously recalled dreams, dreamers sometimes sense immediately what the dream "means", and the dream's "meaning" continues to influence the dreamer's thoughts and feelings for some time afterwards. In seminal studies of these exceptional oneiric events, Jung [1] (p. 117) referred to "big" dreams; Hartmann [2] (p. 54) focused on "intense" dreams; and States [3] (p. 238) described dreams with "magnitude". The lingering effects of these exceptional dreams may depend upon the kind of semantic complexity that literary scholars attribute to metaphoric (poetic) expression [4]. By implication, oneiric access to semantic complexity may depend upon the same mode of metacognition that initiates waking metaphoric (poetic) expression. While dreaming in general (regardless of sleep stage) lacks the metacognitive nuances of waking cognition [5], impactful dreams may nonetheless involve the same modes of metacognition that support waking aesthetic experience.

In some instances, dream impact seems to derive from the dream per se: the manifest dream presents compelling meanings that defy literal paraphrase. However, the effects of impactful dreams often exceed their manifest content and involve (a) their state-transitional endings (immediate carryover effects) and (b) their lingering aftereffects (response to dream reminders). The interdependent components of this temporal sequence may comprise a *gestalt* [6] that arises from epistemic tension between the metaphoric "is" and literal "is not" of a sequence of semantically resonant dream images [7,8]. Through a process that extends from the manifest dream to the dream ending and waking reminders, the effects of this epistemic tension plausibly include representation of compelling abstract-ontological ("totalizing") categories.

1.1. Impactful Dream Types: Manifest Content

It is difficult to determine whether these exceptional (hereafter "impactful") dreams are of different types ("species" [9]) or whether they occupy extreme points on a selected continuum (e.g., "dream-like fantasy" [10]; "immersion", [11]). Although there are taxometric procedures for systematically differentiating continuous from typological constructs [12], researchers studying impactful dreams have usually not availed themselves of that psychometric technology. However, using an established approach to typological clarity [13], a series of taxonomic studies has revealed three types of impactful dreams, each distinguished by features involving feelings and emotions, narrative themes, movement characteristics, sensory anomalies, spontaneous transformations, effectuality/ineffectuality, and concluding affective tone (see Table 1).

Table 1. The contrasting features of nightmares, existential dreams, and transcendent dreams *.

	Nightmares	Existential Dreams	Transcendent Dreams
Feelings and Emotions	Fear/anxiety	Sadness/despair	Ecstasy/awe
Narrative Themes	Harm avoidance	Separation/loss	(Magical) goal attainment
Movement Characteristics	Energetic (evasive) movement	Movement inhibition (fatigue)	Graceful movement (floating)
Sensory Anomalies	---	Unusual light/dark contrasts	Extraordinary sources of light
Spontaneous Transformations	Physical metamorphoses	Spontaneous feeling shifts	Spontaneous perspective shifts
Concluding Affective Tone	Intense dream endings	Intense dream endings	Intense dream endings
Metacognitive Stance	---	Dual Perspectives	Dual Perspectives

* Table entries capture the gist of (a) two phenomenological studies involving a combination of first-person rating scales and open-ended dream description [14,15] and (b) three studies involving rating scales for the features identified in the original phenomenological studies [16–18].

1.1.1. Primary Emotions

The results of these studies suggest a tripartite classification based on three primary emotions, including affectively congruent narrative themes (fear/anxiety with harm avoidance, sadness/despair with response to loss, ecstasy/awe with magical accomplishment). This pattern makes it tempting to distinguish impactful dreams simply according to their predominant emotional tone and affectively congruent narrative themes. However, an attempt to differentiate impactful dreams in this way would elide other distinctive features of these dream types. Potentially elided features include the spontaneous transformations that characterize all three impactful dream types and the capacity for dual perspectives that is specific to existential and transcendent dreams.

1.1.2. Spontaneous Transformations

Unlike mundane dreams, impactful dreams include transformations of dream content (physical metamorphoses in nightmares, spontaneous feeling shifts in existential dreams, and spontaneous perspective shifts in transcendent dreams) It is conceivable that these transformations are merely part of the imaginal "drift" that follows REM sleep decoupling of the dorsal attentional network from the anterior portions of the default network [19]. However, while the default network model addresses the decline in attention to the external world that separates dreaming from waking cognition and explains the imaginal simulation of a seemingly "real" world, it does not address the role of attentional adjustments in the transformations that characterize REM sleep. Contemporary neurophysiological models have tried to address these attentional adjustments.

The activation synthesis model [20] originally suggested a single oneiric attentional adjustment. Ponto-geniculo-occipital (PGO) activation during REM sleep purportedly generates diffuse activation of the dorsal attentional network without regulation by forebrain executive networks. The result of this disjunction is chaotic ("bizarre") dream narratives. The AIM extension of this model [21] retained focus on a single attentional adjustment but did allow that PGO waves during REM sleep do not simply produce radically chaotic dream narratives. Instead, PGO activation triggers discontinuities, and mnemonic elements following a discontinuity were transformations of a prior mnemonic element (e.g., a car becomes a bike). Rittenhouse et al. [22]; see also [23] began to document these transformations, concluding that, "despite the threat to coherence posed by bizarre discontinuities, significant coherence is maintained by associational constraints" (p. 100).

In his contemporaneous neurophysiological model, Morrison [24] also portrayed PGO activation during REM sleep as the phasic marker of a single attentional system: "the startle network". However, he subsequently refined this account to address the "critical behavioral response of orienting, with startle being a less organized expression of a response to an unexpected stimulus" [25] (p. 171). Although these two forms of "alertness" (startle and orienting) are amply documented in the attention literature [26,27], the oneiric transformations they generate during REM sleep have not been clearly differentiated [28].

Differentiating these types of transformation may require consideration of contrasting patterns of activity within the autonomic nervous system (ANS). Recent evidence confirms that, in response to loud tones, post-traumatic and idiopathic nightmares are associated with susceptibility to startle and increased skin conductance, while idiopathic nightmares are associated specifically with increased skin conductance [29]. This pattern suggests that increased sympathetic activation—with decreased parasympathetic activation—is the ANS substrate of susceptibility to attentional startle and full-bodied arousal during nightmares [30]. In contrast, increased parasympathetic activation—with decreased sympathetic activation—may be the ANS substrate of the heart rate deceleration and postural-kinaesthetic inhibition that precedes attentional reorienting. Such neuromuscular inhibition is the distinctive "pause" that precedes the orienting response—and that distinctively marks existential and transcendent dreams.

Kuiken and colleagues [17] articulated such a dual process model, although empirical support remains limited to studies that rely on self-report rather than neurophysiological evidence of ANS-modulated attentional adjustments. First, the balance of sympathetic and parasympathetic ANS activation plausibly influences the sensory modalities that predominate in the transformations within each dream type. The physical metamorphoses of nightmares involve transformations in the distal (primarily visuospatial) sensory modalities that are salient during the hyperarousal of post-traumatic and perhaps idiopathic nightmares [31]. In contrast, the spontaneous transformations of existential and transcendent dreams involve proximal sensory modalities (affective and kinaesthetic modalities in existential dreams; postural and vestibular modalities in transcendent dreams).

Second, the balance of sympathetic and parasympathetic ANS activation plausibly influences the kind of transformation that is characteristic of each dream type. Sympathetic ANS activation may support the physical metamorphoses of nightmares. These transformations are ampliative, i.e., progressively startling, alarming, or even catastrophic. For example, in one prototypic nightmare [14], a dreamer who was attacked by a metal dog reported panic when the dog came to life and "that's when I woke up very scared". In contrast, parasympathetic ANS activation supports the quasi-metaphoric transformations of existential and transcendent dreams. These transformations involve categorial reorienting, as though the dream is implicitly addressing the question, "What is this?" And, the implicit answer is, "Not X, but a categorial transformation X'". If the orienting response is intense, a relatively remote conceptual neighbor may be activated. For example, thoughts of "aphids on the leaves of my tomatoes" may be transformed into a semantically resonant image of "spiders on my hands" [32].

1.1.3. Dual Perspectives

A distinctive form of metacognition may give the spontaneous transformations of existential and transcendent dreams their epistemic import. There is replicable evidence [14,15,17,18] that existential dreams and transcendent dreams involve a form of reflective awareness in which dreamers sense themselves from two points of view (e.g., "I became split into two parts; I was able to experience the dream world from either perspective"). Such dual attunement not only departs from the single-mindedness of typical dreams [33] but also from the concurrent awareness of dreaming that is the allure of lucid dreaming [34]. The proposal offered here is that the dual attunement evident in existential and transcendent dreams enables metacognitive attunement to the metaphoricity of existential and transcendent dreams.

Metacognition during REM and N2 sleep occurs much less frequently than during wakefulness [5]. Nonetheless, oneiric metacognition is strikingly evident when dreamers become aware of dreaming while dreaming [35] and routinely evident when dreamers reflect on their ongoing thoughts, feelings, and behavior [18,35]. In contrast to these self-reflective moments, the distinctive form of metacognition in existential and transcendent dreams involves an epistemic tension that Searle [8] argued is a fundamental aspect of metaphor comprehension (see also [36]). He proposed that metaphor comprehension involves dual attunement to (a) a *metaphoric* (A "is" B) assertion and (b) a complementary *literal* (A "is not" B) assertion. For example, to say that my surgeon is a butcher brings attention to the metaphoric truth value of a statement (e.g., my surgeon [metaphorically] "is" a butcher) and simultaneously to its literal falsehood (e.g., my surgeon [literally] "is not" a butcher).

The conflict between these assertions induces a reflectively accessible ("felt") tension [7] that has epistemic import. The present proposal is that metacognitive attunement to (a) a metaphoric (A "is" B) assertion and (b) a complementary literal (A "is not" B) assertion contributes to the impact of existential and transcendent dreams.

1.1.4. Manifest Content: Implications

The manifest content of taxonomically delimited impactful dreams affirms the complexity required to explain their influence on waking thoughts and feelings. Emphasis on spontaneous transformations, rather than on the dream's dominant emotional tone, motivates examination of how a metacognitive tension during existential and transcendent dreams supports their distinctive metaphoricity. This proposal contrasts with formulations that acknowledge continuity between cognition during wakefulness and during REM dreaming [19]—and restrict dream metaphoricity to retrospective reflection [37,38]. The distinctive form of metacognition in existential and transcendent dreams underscores the lively metaphoricity of dream cognition per se.

2. Intrinsic Oneiric Metaphoricity

2.1. Spreading Activation and the Metaphoricity of Dream Cognition

To clarify the metaphoricity of impactful dreams, it is useful to consider first the "creativity" of typical REM dreams. Theories of creativity often emphasize problem solutions that result from the generation of remote associations between existing concepts. That basic account has generated discussions of oneiric creativity that emphasize the increased accessibility of associatively remote lexical items during REM sleep [23] and temporally remote memories during late night sleep [39]. Available research indicates that (a) normatively weak lexical associations, ranging from hierarchical semantic relations to mere contiguity, are strengthened following REM sleep [23]; (b) normatively weak associations to emotion words are strengthened when they are primed by presentation of those emotion words prior to REM sleep [40]; and (c) an "association" that captures the (schematic) gist of an array of semantically related words, ranging from part-whole to contextual relations, is "remembered" following REM sleep [41,42]. These results have been presented as evidence that memory traces activate relatively distant, uncommon, and novel semantic associations during REM sleep.

It has been tempting (cf. [43]) to generalize this spreading activation model to explain higher order "creative" phenomena. However, such generalization is problematic partly because, in the studies just mentioned, the weak associations strengthened during REM sleep range from those based on mere contiguity (word co-occurrence) to those involving basic semantic relations (e.g., noun-noun relations, adjective-noun relations). Generalization of the spreading activation model would be more compelling if the weak associations under investigation differentially and systematically included (a) categorial taxonomic relations (e.g., hypernyms, hyponyms, class co-membership); (b) part-whole relations (e.g., characteristic features, traits); (c) thematic relations (e.g., characteristic possessions, locations); and (d) figurative modifier-modified relations [44]. Generalization of the spreading activation model would be even more compelling if the weak associations under investigation included not only the figurative modifier-modified relations of mundane non-literary metaphors (e.g., "genes are blueprints") but also the figurative modifier-modified relations of unconventional poetic metaphors (e.g., "death is a fat fly").

2.2. Carryover Effects and Mundane Dream Metaphoricity

To clarify the poetic metaphoricity of impactful dreams, it is useful to begin by considering the mundane metaphoricity of typical REM dreams. It is well established that awakenings from REM and NREM sleep are followed by a brief period (20–40 min) of changes in affect, perception, cognition, and memory ("carryover effects"). The carryover effects of typical REM dreaming include readiness to identify rudimentary figurative relations. Cai et al. [45] provided evidence that REM sleep reactivation of analogical relations (e.g., "chips" are related to "salty" as "candy" is related to "sweet") influenced post-awakening performance on a task requiring identification of word combinations that included a root term from the analogy (e.g., sweetheart, sweet sixteen, sweet cookies). This task involves identification of modifier-modified compounds that can be paraphrased as conventional nominal metaphors (e.g., in your heart you are sweet, being sixteen is sweet).

However, the relevance of this version of the remote associates task (RAT) for assessing metaphoricity is obscured by the additional challenge of identifying three compounds involving the same root term. This additional interpretive complexity may make the RAT a useful measure of a certain kind of creative problem solving [46–48], especially the inference-driven interpretation that is characteristic of the evaluative component of creative cognition (e.g., recognizing relations among response alternatives, reaching single solutions to complex problems). However, the Cai et al. [45] results do not reflect the carryover of metaphoricity per se.

2.3. Carryover Effects and Impactful Dream Metaphoricity

There is accumulating evidence [49,50] that, rather than separately or additively, the interactive combination of generative and evaluative processes predicts creativity. For example, generating as many different associations as possible to target words (associative fluency) predicts creativity [51,52]; providing as many associations as possible that are unrelated to target words (associative restraint) also predicts creativity [52]. However, the carryover effects of existential and possibly transcendent dreams (but not nightmares) include neither associative fluency nor associative restraint separately but rather their interactive combination [53]. Moreover, in the latter studies, the interactive combination of associative fluency and associative restraint was unrelated to performance on the RAT that Cai et al. [45] used to assess the metaphoricity of typical REM dreaming. By implication, the metaphoricity of existential and transcendent dreams does not involve recognizing relations among response alternatives and reaching single solutions to complex problems as measured by the RAT. Instead, the metaphoricity of impactful dreams may involve a more nuanced interplay of associative activation and associative restraint, perhaps sufficiently nuanced to explain the "involuntary poetry" of impactful dreams [3].

2.4. Metacognitive (Noetic) Feelings

Metaphor generation [54,55], lyrical improvisation [56], and musical improvisation [57,58] involve rapidly oscillating—or even simultaneous—activation of generative and evaluative neural networks [59]. Moreover, tightly woven interplay between generative and evaluative neural networks may prompt "automatic-affective appraisals" [50]. Reference to automatic-affective appraisals invites closer consideration of the tension between the metaphoric (generative) "is" and literal (evaluative) "is not" of oneiric metaphoricity. Metacognitive appraisal of this metaphoric/literal tension is one type of noetic (epistemic) feeling; others include (a) feelings of knowing [60]; (b) tip-of-the-tongue feelings [61]; (c) feelings of familiarity [62]; and (d) feelings of 'déjà vu' [63]. Metacognitive appraisal of metaphoric/literal tension can, with some care and caution, be distinguished from feelings toward independently perceived or imagined intentional objects, such as (a) external states of affairs (e.g., "I feel fond of X") or (b) bodily states and dispositions (e.g., "I feel tired"). However, metacognitive appraisal of the tension between a metaphoric "is" and literal "is not" is also experienced differently—and functions differently—during dreaming than during waking.

2.4.1. A Metacognitive "Felt Sense" of Metaphoric/Literal Tension

Feelings and emotions in existential and transcendent dreams involve proximal sensory modalities that "soften" the way they are "given" in experience. The affective-kinaesthetic modalities of existential dreams and the postural-vestibular modalities of transcendent dreams give dream feelings sensuous palpability (e.g., sadness that involves sensed weakness or inability to move, joy that involves sensed floating or flying). Metacognitive metaphoric/literal tension during dreaming may similarly occur as a softly felt affective moment, a "felt sense" [64] of tension between a metaphoric "is" and literal "is not".

2.4.2. Attentional Reorienting, Category Transformation, and Metaphoric/Literal Tension

Although attention network theory [27] distinguishes three independent systems (vigilant anticipation of selected stimuli; reorienting adjustment to unanticipated stimuli; controlled selection from available stimuli), the attentional network that distinctively shapes existential and transcendent dreams is the orienting response system. On one hand, recent evidence indicates that, during creative response to constrained abstract problems (e.g., multiple uses tasks, interpreting analogies), sustained selective attention is complemented by an executive function (updating working memory) that facilitates access to "different problem solutions" [65]. On the other hand, during a creative response to extended imaginative problems (e.g., metaphor comprehension), sustained selective attention is complemented by an executive function (shifting between mental sets) that facilitates perspective shifts to "different dimensions" [65] and "levels of analysis" [66].

Perspectival shifts to different dimensions and levels of analysis of the same (or a similar) categorial object have been observed within the oddball paradigm that is commonly used to examine the orienting response during waking [65,66]. During dreaming, the primary function of oneiric metacognitive appraisal may be to monitor relations between the successive categorical representations that spontaneously occur during PGO generation and ANS modulation of the perspectival shifts triggered by the orienting response. These shifts involve categorial reorienting, experienced as the transition from one representation of the category (e.g., aphids on my tomatoes) to semantically resonant representation of the same or a similar category (e.g., spiders on my hands).

2.4.3. Openness to Experience and Metaphoric/Literal Tension

The contrast between (a) attentional openness to alternative solutions for constrained abstract problems and (b) attentional openness to different levels or dimensions of sensuous-affective categories is congruent with the distinction between creative instrumental and creative experiential sets [67]. For some creative individuals, shifts in attention occur as part of an instrumental exploration of different problem solutions. For others, shifts in attention occur as part of an experiential transition in which attention is reoriented toward different dimensions or levels of sensuous-affective categorial representations. Different facets of trait openness to experience (Intellect, Openness) reflect this contrast [68,69]; moreover, trait absorption (Tellegen Absorption Scale; TAS) [70] is most closely related to the Openness facet [71]. Understandably, then, the TAS predicts the inexpressible realizations that emerge from existential dreams [72]—although that relationship is not evident for transcendent dreams.

2.4.4. Performative Improvisation and Metacognitive Appraisal

Metacognitive appraisal of metaphoric sameness and literal difference may involve attunement to transitions from one sensible aesthetic representation to semantically resonant aesthetic representations during improvisatory performance [73–76]. The performative aspect of metacognitive attunement to extended metaphor is arguably the generative "life" of extended metaphors that Ricoeur [7] calls "living metaphor". Such metacognitive attunement may also be the generative core of literary authorship. Poets—and some dreamers—often reshape dream imagery to "carry forward" intimations of its affective tone and categorial nuance long after awakening from an impactful dream [3,77]. Rather than veridical and detailed description of a remembered dream, imagery from the manifest dream subtly foreshadows expressive departures from mundane reflection, routine remembering, and everyday reasoning. The dream prefigures and the poet extends the dream's meaning by elaborating its figurative configurations beyond what can, at best, be partly captured by a carefully framed literal paraphrase.

3. Temporally Extended Metaphoric Interplay

Continued metacognitive appraisal of the tension between metaphoric and complementary literal assertions supports the integration of successive metaphoric/literal representations. While it is possible to identify metaphoric categorial representations analytically, it is especially important to articulate their progressive integration. To clarify this objective, what follows is a linguistic and then an oneiric example of such temporally extended metaphoricity.

3.1. A Generic Linguistic Example

The loci of such progressions and interplay are usually identifiable, as in the following excerpt from Borges' [78] essay, "A New Refutation of Time". This excerpt is useful here because it is neither a literary text nor a dream report. In these lines, an understanding of time is metaphorically (and progressively) represented during reflective consideration of its moving, devouring, and immolating insistence:

> Time is a river that carries me along,
> and I am the river;
> It is a tiger that devours me,
> and I am the tiger;
> It is a fire that consumes me,
> and I am the fire.

In this excerpt, several local figurative features stand out. First, each line begins with a simple nominal metaphor (e.g., "Time is a river"). Second, each nominal metaphor is followed by elaborative (i.e., explicative, perhaps even ampliative) modulation of an initial metaphoric vehicle (e.g., ". . . it [the river] carries me along"). Third, each nominal metaphor with its elaborative modulation is followed by compound metaphoric modulation involving a nominal first-person metaphor (e.g., "I am the river") that has both the initial metaphoric vehicle (e.g., "a river") and, indirectly, the initial metaphoric topic (e.g., "time") as its metaphoric topics (e.g., "I am" river-like time).

And yet, beyond these local figurative features, a higher order figurative structure is also evident. Although Borges originally wrote this passage as prose, its three-part "verse" structure (accentuated by the line breaks added here) supports (a) three separate first-person compound metaphoric modulations ("I am" [river-like] time; "I am" [tiger-like] time; "I am" [fire-like] time) and (b) an integrative first-person compound metaphoric modulation ("I am" [river-like, tiger-like, fire-like] time). Are the separate (iterative) and inclusive (integrative) effects of these metaphoric modulations independently evident to actual readers of these lines? Some empirical evidence indicates that they are. In an empirical study of young adult readers, Kuiken and Douglas [79] (Study 1) found that the separate (iterative) first-person compound forms ("I am [river-like] time"; "I am [tiger-like] time"; and "I am [fire-like] time") prompted reports of inexpressible realizations (e.g., "I sensed something that I could not find a way to express"). They also found that the inclusive (integrative) compound form ("I am" [river-like, tiger-like, fire-like] time) distinctively precipitated inexpressible realizations with a concomitant sense of release. The latter result is consistent with informal observations [64] that such an ineffable realization is a "felt shift" accompanied by a momentary "sense of release" as though there is a subtle "letting go", a subdued and softened "Oh, that's there". This metacognitive appraisal marks an apex moment of sublime feeling.

3.2. An Oneiric Example

Comparably compounded metaphoric modulations often occur in impactful dreams. In the following prototypic existential dream (from [80], p. 136), a three-episode structure (analogous to the Borges three-verse structure) progressively—and metaphorically—discloses the dreamer's felt sense of the coercive character of inauthentically intimate relationships.

> ***The Hotel Dream.*** [I was in] a hotel in southern Alberta or someplace. I was traveling by myself, I think, and I remember worrying about rapists in the hall and this sort of thing. I remember thinking, "Well, I'll have to be brave because I'm by myself". So I took this room in a secluded area of the hotel . . . and anyway it seemed to work out. // And then this hotel seemed to be in France. My family was with me, and we were all in a room together. We were packing to leave. I was very organized; I had all of my stuff ready to go . . . My family was very disorganized and I was having to help them. I didn't want to. I thought, "Well, they can do it themselves; I'm not responsible for their packing". But it was almost impossible not to help them because I needed another bag or two and I had things stored in a particular drawer and they had dumped all their stuff in there, too. So in order for me to get this packing done, I had to help them anyway // [Then I think I had gone off on my own for a while] and I came back [to the hotel again.] I got a phone call, an overseas phone call from my Dad . . . He had gotten a doctor's report, and the doctor said that he [had an ailment that] would never heal. And I had plans about my whole family moving to France . . . but he just told me how sick he was and that he would never heal. And there was some stupid person on the phone . . . some practical sounding person, who was sticking her nose in there. I kept telling her to shut up . . . and I was really upset and crying very hard. My Dad said that he wanted to talk to my Mom. So my Mom came to the phone [and she thought that it wasn't practical to live in France]. She seemed to think that it was better to stay in Canada. I was surprised by her ability to say what was best for me . . . and I remember trying to talk her into it. I was overwhelmed by the fact that my father would never heal. I couldn't be with him and also stay in France. I woke up crying. I was just really sad. I felt this sadness just coming out of the bottom of my soul, from way down deep some place.

In this dream, three successive episodes (separated here by //) present semantically resonant portrayals of concern about coercive intimacy (worry about rapists in an Alberta hotel hallway; unwelcome assistance to filial fellow travelers in a hotel in France; aversion to a series of familiar voices who seem to know what is "best" for the dreamer). Although less simply than in the Borges lines that present compound metaphoric representations of "time", this three-episode structure presents compound metaphoric representations of the coercive character of inauthentically intimate relationships.

Given persistent self-representation in dream narratives, this dream presents a series of three separate first-person compound metaphoric modulations ("I am here" [metaphorically] living with the possibility of rapists in the hallway; "I am here" [metaphorically] living in opposition to de facto filial collaboration; "I am here" [metaphorically] living with dispersed (parental) presumptions about what is best for me). As in the Borges lines, the result is a repetitive structure (a recurrent theme) that involves three separate first-person compound metaphoric modulations. Beyond that, an integrative higher order first-person compound metaphoric modulation is established by the series of first-person metaphoric modulations moving from (a) "I am here" [metaphorically] living with the possibility of rapists in the hallway to (b) "I am here" [metaphorically] living in opposition to de facto filial collaboration and then to (c) "I am here" [metaphorically] living with dispersed (parental) presumptions about what is best for me. This sequence is more clearly ampliative than the sequence of Borges lines, moving the dreamer toward a dramatic and mournful sense of release: She concludes, "I felt this sadness just coming out of the bottom of my soul, from way down deep some place". This second level of metacognitive metaphoric/literal appraisal marks an apex moment of sublime feeling.

More complete articulation of the unfolding metaphoric structure of this dream would require the scrutiny derived from additional (open-ended) questions about nuances in the dream narrative [81]. Perhaps, however, the preceding analysis is sufficient to clarify how the entirety of this dream is analogous to the entirety of the highly structured Borges lines in the way it moves toward first-person metaphoric articulation of a theme and toward a poignant moment of sublime feeling—all within the dream and prior to waking reflection.

3.3. A Hierarchy of Figurative Relations

The preceding account relies upon a hierarchy of figurative relations: (a) detectable image resonance; (b) detectable unidirectional metaphoricity; (c) detectable bidirectional metaphoricity; and (d) metaphors of personal identification. The first level of metacognition moves toward detectable bidirectional metaphoricity and emergent meanings; the second level of metacognition moves toward metaphors of personal identification and sublime feeling.

3.3.1. Detectable Semantic Resonance

In dream reports [82,83], narrative discontinuities often juxtapose neighboring instances of the same categorial concept (e.g., a car and a bike as means of transportation [22]; a person and a statue as human forms [84]). During impactful dreams, the dreamer may "sense" semantic resonance between successive neighboring representations—even without the syntactic structure that might, in another context, link them as the topic and vehicle of nominal metaphors (as occurs with modifying-modifier compounds such as "statue people", "flute birds" [85]). The possibility of such accentuated—and detectable—resonance is reinforced by evidence that awakening from existential dreams, transcendent dreams, and nightmares is followed by increased readiness to include metaphoric vehicles and topics in the same conceptual category (e.g., "genes" and "blueprints" taken from the nominal metaphoric assertion "genes are blueprints"). Post-dream detection of such semantic resonance is specific to vehicles and topics taken from *conventional* literary and *conventional* non-literary metaphors [53] (Study 2). Individuals participating in dream research may sense this kind of semantic resonance as indication that their dream was "impactful", i.e., that it affected—and continues to affect—their thoughts and feelings after awakening.

3.3.2. Detectable Unidirectional Metaphoricity

In empirical studies of dreaming, conceptual metaphor theory has often been used to explain how a metaphoric vehicle is unidirectionally "mapped" onto a metaphoric topic [86]. According to a summary provided by Malinowski and Horton [38] (p. 10), (a) dreams can concretely picture something abstract from waking life; (b) dreams can be about emotional aspects of the dreamer's life; and (c) it may be necessary to talk with the dreamer about current life events to understand the metaphor. In general, the function of unidirectional metaphoric reference to waking life is to generate novel self-understanding during wakefulness [37].

The nature of unidirectional mapping must be considered carefully; two aspects of this process entail interaction between the vehicle and topic prior to such mapping. First, the determination of which vehicle attributes are mapped onto the topic depends upon the selection of salient attributes of the vehicle that were not salient attributes of the topic prior to unidirectional mapping [87]. Second, vehicle attributes may be transformed prior to unidirectional mapping onto the topic. There is no consensus about how such attribute transformations are derived, although likely sources include polysemy, analogy, and extended metaphor [88]. These forms of interplay between vehicle and topic determine the nature of unidirectional mapping—but disclose nothing novel about the vehicle.

3.3.3. Detectable Bidirectional Metaphoricity

In contrast, the present framework focuses on the disclosive potential of bidirectional metaphoric structures in existential and transcendent dreams. A primary source of this potential is the distinctive bidirectionality (A "is" B and B "is" A) of poetic metaphors. While Lakoff and Turner [89] dismissed bidirectional mapping, a tradition extending from Richards [90] and Black [91] to Fauconnier and Turner [92] and Goodblatt and Glicksohn [6] provides articulation of the creative potential of bidirectional metaphor comprehension. In contrast to unidirectional mappings (A "is" B; "death is a fat fly"), the bidirectional epistemic import of the vehicle and topic (A "is" B and B "is" A; "death is a fat fly" and "a fat fly is death") leads to the disclosure of emergent meanings [93]. Emergent meanings are attributes of the topic and vehicle that are not salient for either category considered separately but that become salient for both categories during metaphor comprehension (e.g., apropos death and fat flies, "lurking contamination"). There is now secure empirical evidence that emergent meanings derive from poetic metaphors [94–97], although it remains unclear how the integration of (A "is" B) and (B "is" A) generates emergent meanings.

To address the unfolding generation of oneiric emergent meanings, it is useful to examine the domain interaction model of metaphor comprehension [98,99]. The domain interaction model describes the juxtaposition of a topic and vehicle that are both "semantically dense" (concrete, embodied, self-relevant [100]) but that, at an abstract (hypernemic, taxonomic) level, are "distant" from each other. Using this model, it is possible to identify attributes of the metaphoric vehicle that are sufficiently abstract to also be attributes of the metaphoric topic and, in that way, support bidirectional metaphor comprehension. This model of metaphor comprehension suggests that the inclusiveness of an abstract (and ad hoc) ontological category may subsume two or more semantically dense regional ontological categories. The result is an abstract-ontological concept that is "inclusive" but not "totalizing". Both existential and transcendent dreams generate that kind of metaphoric outcome.

3.3.4. Metaphors of Personal Identification

A fourth important figurative relation is the self-relevant metaphoricity that provides a context for inexpressible realizations with a felt release. The iterative modulations that characterize the extended metaphors of existential and transcendent dreams (e.g., in the Hotel Dream) generate an unfolding sense of a meaning-finding self that Cohen [101] calls metaphors of personal identification. Consistent with Cohen's account, there is fledgling evidence that engagement with nominal metaphors facilitates recognition of subsequent personifying metaphors [102] and that the cumulative effect of recurrent personifying metaphors is a sense of self as the author of unfolding metaphor extension and explication [103]. For example, in response to "all prayer is grief flying", extended metaphoric modulation contributes to tacit endorsement of the following locutions: "I am [the ascent of] prayer"; "I am [the anguish of] grief". The progressive integration of compound metaphors of personal identification—and their poignantly sublime conclusion during, for example, the Hotel Dream—contrasts with the unidirectional metaphoric representations that characterize retrospective attempts to identify waking concerns [37,38]. Moreover, the sublime aesthetic-epistemic effects of this progressive thematic integration during the dream differ from the individual personal insights derived from retrospective dream reflection [104].

4. Metacognition at the Abstract Ontological Limits of Sublime Feeling

Articulating a theory of oneiric movement toward metaphors of personal identification motivates comparison with Kant's discussion of the radical limit experiences that constitute sublime feeling. Those radical limit experiences occur within a mode of metacognition that integrates *successive* metaphoric/literal metaphoric structures. Given that the tenets of Kant's account have been altered by scholars working in the continental tradition [105], the present framework is best considered neo-Kantian.

Kant [106] proposed that interplay between a succession of sensible "aesthetic ideas" and insensible "rational concepts" moves toward "symbolic" (metaphoric) approximation of an abstract-ontological concept. Husserl [107] similarly described how interplay between a succession of intuitable (sensible) past, present, and anticipated "moments" move toward approximation of an abstract-ontological type. However, Kant does and Husserl does not address the metaphoricity of evolving abstract-ontological types. Consequently, Kant does, but Husserl does not, address the radical limits of "totalizing" abstract-ontological types.

4.1. Affective Awakenings

It is useful to begin by considering Husserl's relatively accessible account of abstract-ontological types. In his (late) genetic phenomenology, Husserl developed the notion of a preconceptual type, which is comparable to the Kantian notion of a schema [108]. Whereas Kant argued that a schematic a priori is independent of nearly all sensible intuition, Husserl, already in *Logical Investigations II* [109] (§27), argued for the articulation of an "ideational abstraction" (*ideierende Abstraktion*) through a fully intuitable "eidetic variation" (*eidetische Variation*). A preconceptual type is initially generated through a temporal succession of adumbrations that is synthesized to provide a singular, intuitively given categorial object (e.g., "this" is "a doll"). In turn, articulation of an initial intuitively given categorial object enables an active explication of another intuitively given categorial object. One possible outcome is that the adumbrations originally gathered into a sensible intuition of the preconceptual type are "run through" again during confirming or disconfirming explication of the original type ("this" is [or is not] "another doll"). A second possibility is that the adumbrations may "give" another categorical object that differs from a previous determination, motivating alteration of the original preconceptual type ("this" is "not a doll" but rather "a mannequin"). A third possibility is that the adumbrations may not just alter but expand the original preconceptual type ("this" is a preconceptual type that subsumes both "mannequins and dolls").

Husserl [109] (§27) gave special consideration to affectively "awakened" associations in which preconceptual types are in conflict. During a momentary "awakening", he argued, two perceptual types "permeate" each other. Their permeation is apprehended as conflict, although both objects share a tenor of appearance (*Erscheinungsgehalt*). For Husserl, the tension in this permeation is resolved in a literal judgment (this is not a doll but rather a mannequin). He does not address the possibility that this permeation persists as a modifier-modified compound that provides adumbration of a relatively abstract ontological type that includes both doll-like mannequins and mannequin-like dolls.

4.2. Symbolic Hypotyposis

Both Kant and Husserl were alert to the ontology-expanding import of successions of associated categorial concepts (Kant as symbolic approximations of insensible *a priori* concepts; Husserl as the extended explication of permeating preconceptual types). In Kant's account, orienting attunement toward a succession of aesthetic ideas initiates the "spontaneous" movement of "aesthetic ideas" toward "totalizing" concepts [110]. Kant argues that it does so by initiating a form of metaphoricity through which aesthetic ideas "approximate" rational conceptions of the "world as a whole". He argues that the movement (*Schwung*) of the productive imagination toward totalizing concepts entails "symbolic" presentations that are "in accordance with an analogy" [106] (CPJ 5: 352). Such "symbolic hypotyposis", he says, is pivotal when analogy offers the only way to exhibit (or present) rational concepts. Because Kant's theory of symbolic hypotyposis resembles Aristotle's theory of metaphor, it is possible to construe his account as a general theory of metaphor. However, Pillow [111] more specifically proposed that Kant's theory of symbols resembles the "interactionist" theories of poetic metaphor [6,90–92]. Clarification of this theoretical possibility follows a path leading from Kant's discussion of symbolic hypotyposis to Ricoeur's discussion of "living metaphor". Toward the end of this theoretical path, Ricoeur [7] presented living metaphor

> "... as a mode of discourse that functions at the intersection of two domains, metaphorical and speculative. It is therefore a composite discourse, and as such cannot but feel the opposite pull of two rival demands. On one side, interpretation seeks the clarity of the [abstract ontological] concept; on the other, it hopes to preserve the dynamism of meaning that the [abstract ontological] concept holds and pins down. This is the situation Kant considers in the celebrated paragraph 49 of the Critique of the Faculty of Judgment. ... [The interplay] in which imagination and understanding engage assumes a task assigned by the Ideas of reason, to which no [abstract ontological] concept is equal. But where the understanding fails, imagination still has the power of 'presenting' the Idea (*Darstellung*). It is this 'presentation' of the Idea by the imagination that forces conceptual thought to think more. Creative imagination is nothing other than this demand put to [abstract ontological] conceptual thought." (p. 303)

Ricoeur goes on to describe (a) how poetic metaphoricity entails "discourse [that is] turned back upon itself and spurred on by [another] metaphorical utterance"; (b) how such discourse is iteratively extended (perhaps indefinitely) to constitute a poetic theme; and (c) how poetic thematization approximates (without attaining) abstract ontological determination. Such iteratively extended and unfolding thematization does not simply reduce poesy to the explicative identification of abstract ontological concepts. Rather, such thematization moves toward further metaphoric modulation that persistently—but always inadequately—is sensed as an approximation of abstract ontological concepts.

4.3. Implications

Although it is difficult to investigate the chained images that give epistemic import to the themes emerging from extended metaphors, doing so is pivotal in studies of the unfolding metaphoricity of a literary text [112,113], the unfolding metaphoricity of the Borges lines, or the unfolding metaphoricity of the Hotel Dream. In the Borges excerpt, constitution of a relatively abstract (ad hoc) ontological type occurs as the permeations through which "time" is successively apprehended as a moving river, a devouring tiger, and an immolating fire (and then the integrative sense of this succession). These permeations are not literally "the same" but, considered metaphorically, they are "the same". Moreover, this succession of permeations may provide adumbrations of an abstract ontological type that subsumes a network of subordinate ontological types (a network that subsumes the regional ontology of linear time, the regional ontology of moving rivers, the regional ontology of devouring tigers, and the regional ontology of immolating fires).

Similarly, constitution of a relatively abstract (ad hoc) ontological type may occur as the permeations through which "coercive intimacy" is successively apprehended as "worry about rapists"; as "unwelcome assistance to filial fellow travelers"; and as "aversion to a series of demanding voices" (and then as the integrative sense of this succession). Again, these permeations are not literally "the same" but, considered metaphorically, they are "the same". Also, this succession of permeations potentially provides adumbrations of an abstract ontological type that subsumes a network of subordinate ontological types (a network that subsumes the regional ontology of coercive intimacy; the regional ontology of lurking rapists; the regional ontology of oblivious siblings; and the regional ontology of demanding voices).

5. Sublime Feeling: Disquietude and Enthrallment

The preceding framework begins to clarify the difference between (a) metacognition of the metaphoric/literal tension within local modifier-modified syntactic structures and (b) metacognition of the extending, expanding, and perhaps "totalizing" tensions that are the substrate of sublime feeling.

On one hand, within a local modifier-modified syntactic structure, a resolvable tension emerges between what metaphorically "is" and literally "is not". While metaphor comprehension resists assimilative subsumption within familiar literal categories, it also invites accommodative subsumption within unfamiliar—and often ad hoc—semantically resonant categories. While an accommodation gradient extends from quasi-metaphoric resonance detection to unidirectional vehicle-topic mapping and the emergent meanings of bidirectional vehicle-topic mapping, an accommodating category expansion is usually possible.

On the other hand, successively extending, expanding, and perhaps "totalizing" metaphoric/literal syntactic structures may generate an irresolvable tension between abstract ontological thematization of what metaphorically "is" and literally "is not". A radical epistemic limit may result from a succession of thematically coherent but conceptually inadequate quasi-metaphoric resonances, unidirectional vehicle-topic mappings, bidirectional metaphoric (emergent) meanings, and (eventually) "totalizing" abstract-ontological concepts. Within such "totalizing" metaphoric extension and expansion, accommodating category generation is often impossible [114], except as a further metaphoric extension of an abstract-ontological theme [7].

During existential and transcendent dreams, the dreamer may experience such extended and expanded but conceptually inadequate meanings as a cumulative (even "totalizing") thematic tension between what metaphorically "is" and literally "is not". However, whether such thematization moves toward further metaphoric modulation in "living metaphors" and whether further ("living") metaphoric modulation is disquieting or enthralling depends upon dream type.

5.1. Sublime Disquietude

Recurrent and inevitably inadequate "symbolic" (metaphoric) approximations of "existential" (autobiographical) abstract-ontological concepts characterize oneiric sublime disquietude. That hypothesis is supported by studies [16,18] indicating that the carryover effects of existential dreams include disquieting doubt [115] (e.g., "... doubt became an important part of what it means to be honest about my existential concerns"). Also, the self-perceptual depth that consistently is reported following existential dreams [14,16,18,73] relies upon items that are not only self-relevant (e.g., "... I became sensitive to aspects of my life that I usually ignore") but also relevant for humans in general (e.g., "... my sense of life seemed less superficial"). Consequently, the index of sublime disquietude used in ongoing research [116] incorporates the interactive combination of rated disquietude, self-perceptual depth, and inexpressible realizations. The carryover effects of existential dreams consistently involve sublime disquietude in this sense.

5.2. Sublime Enthrallment

In contrast, recurrent and inevitably inadequate "symbolic" (metaphoric) approximations of "transcendent" abstract-ontological concepts characterize sublime enthrallment.

Sublime enthrallment arises through recurrent metaphoric/literal intimations of the possibilities and limits of ontological intersubjectivity. Intimations of ontological intersubjectivity (sensed as "awe", "respect") hover between (a) non-utilitarian (respectful) attunement to mutual intentionality within a dichotomous relationship [117] (*Ich und Du*) and (b) non-utilitarian (respectful) attunement to the "inner subjectivity" of beings in general (i.e., inclusively "totalizing" intersubjectivity).

The ontological intersubjectivity that characterizes sublime enthrallment is suggested by studies [16,18] indicating that the carryover effects of transcendent dreams include the apprehension of distributed liveliness (e.g., "... I had the sense that everything around me was somehow alive"). Also, the self-perceptual depth that is reported consistently after transcendent dreams [16,18] reflects not only poignant self-referential "depth" (e.g., "... I became sensitive to aspects of my life that I usually ignore") but also the disclosure of "depth" in the experience of humans more generally (e.g., "... my sense of life seemed less superficial"). Consequently, the index of sublime enthrallment used in ongoing research [116] incorporates the interactive combination of rated reverence, self-perceptual depth, and inexpressible realizations. The carryover effects of transcendent dreams consistently involve sublime enthrallment in this sense.

In his portrayal of dialogical I-Thou relations, Buber emphasized the pivotal character of the "fresh running springs" of poetic language (cited in [118] p. 17), a locution that echoes Ricoeur's account of "living metaphor" Although Stace [119] worried that the language of poetry was too "conceptual" to capture the ecstatic experience of the "inner subjectivity" of beings in general, psychometric studies have isolated a component of extravertive mystical experience ("distributed liveliness") that is also evident in spiritual (secular) poetry [120] and in dreams [121,122]. Even so, these studies—and their authors—do not address the complexities of poetic metaphor, especially the extended and ontologically expanded themes that are disclosed in an unfolding succession of poetic metaphors.

Oneiric Sublime Enthrallment: An Example

At this point, it may be helpful to examine the metaphoric form of a prototypic transcendent dream [80] (p. 129–130):

> ***The Snake Dream.*** I dream that I am climbing a stairs, coming to a landing. There are windows on both sides [and] high walls. It's very bright. Four people are there, and one person in particular is ... interacting with me. He hands me a box that is gift-wrapped and ... about shoe size. I start opening the box and then I [see it has] a cover [with] snakes [on it] ... Obviously there is a feeling of friendship between us or a feeling of comfortable joking ... and I ask, "Oh, you wouldn't dare, would you?" And, he said, "Well, you'll have to open it and find out". // So I open the box and, when I open the box, this multitude, this large number of snakes come out of the box. They're very thin, and they're long like ribbons. They're black, shiny, with yellow heads. They all wrap around my lower right leg like a cuff ... I'm trying to push them away, to kick them off, feeling panicky, trying to ... pry them off. And the more I try, the more I'm getting nowhere, the tighter they are wrapping themselves -- // until I suddenly calm down and start looking at them and see how they fit together and how they are not dangerous snakes. I start looking at them and I notice the color of their backs and how they seem to be so quiet and peaceful. So, I start touching them gently and, as I'm touching them, I'm talking to them. And, eventually I say, "OK, now you have to leave", and they do leave... // When the snakes did leave, I felt like I was light, I was lighter in weight, but there was also a sense of release...

In this dream, successive episodes (separated here by //) provide semantically resonant representations through which the dreamer becomes more respectfully aware of the dangerous beauty of the snakes' participation in distributed ontological intersubjectivity. In the first episode, the snakes' presence is suggested by a pictorial image on the cover of a box that contains them; metaphorically their sentience is a teasingly hidden gift. In the second episode, the colorfully present snakes are disclosed in a threatening form that motivates the dreamer's anxious resistance; metaphorically their sentience is antagonistic. In the third episode, the same colorfully presented snakes are quiet and peaceful, motivating the dreamer's touch and dialogue; metaphorically their sentience is that of mutual intentionality. The snakes share the subjectivity of the dreamer; metaphorically their awareness—and the dreamer's—is mutual and respectful. The culminating moment of "release" is not appreciation of a gift or relief from antagonistic threat—but rather non-utilitarian regard. The snakes retain their beauty (as in the decorated box); they retain their danger (they still "must leave"); but they also have the capacity for dialogue. Like the Hotel Dream, each successive metaphor deepens and enlivens without erasing its predecessor.

6. Conclusions

The results of the preceding analysis indicate that existential dreams transition into sublime disquietude and transcendent dreams transition into sublime enthrallment (while nightmares transition into post-awakening vigilance). This pattern prompts a reconsideration of when, if at all, nightmares generate poetic metaphors—and their aesthetic-epistemic aftereffects (contra Barcaro [4] and Hartmann & Kunzendorf [123]). Instead, this pattern is consistent with the notion that the transition into sublime feeling is precipitated by a distinctive form of metacognition specifically within existential and transcendent dreams. The two levels of metacognition within these two dream types may enable attunement to tension between the unfolding metaphoric and literal truth value of resonant categorial dream objects. The felt sense of that tension in these two dream types may disclose the unfolding epistemic—and aesthetic—significance of successive resonant images, culminating in the depth of sublime feeling.

REM Sleep Neurophysiology. The present account identifies oneiric sublime feeling as a distinctive REM sleep carryover effect. Examining the neurophysiology of that effect is a challenge because the infrequent and aperiodic character of impactful dreams effectively precludes their observation within a sleep laboratory. A comparable methodological limitation prevails in studies of sleep paralysis, although a few laboratory case studies indicate that isolated sleep paralysis episodes occupy an intermediate position between the cortical and neuromuscular activation of wakefulness and the cortical activation and muscle atonia of REM sleep. The result is that the dreamer is "partially aware" of the sleep environment [124], p. 723. Similarly, we suggest, during existential and transcendent dreams, the dreamer is partially aware of dream images that seem metaphorically real (present) but literally not real (not present).

There may be another reason to compare impactful dreams and sleep paralysis episodes. Cheyne and colleagues [125,126] reported three types of sleep paralysis episodes: (a) intruder episodes (multi-sensory hallucinations of a threatening figure); (b) incubus episodes (multi-sensory hallucinations of chest pressure, suffocation, and pain); and vestibular-motor episodes (illusory movement and out-of-body experiences). This tripartite typology is analogous to the tripartite typology of impactful dreams. Perhaps, then, existential dreams, transcendent dreams, and nightmares are muted or modified forms of the same tripartite psychobiological network that Cheyne et al. argue is the substrate of intruder, incubus, and vestibular-motor sleep paralysis episodes. If so, it would be useful to compare not only the underlying psychobiological networks of sleep paralysis episodes and impactful dreams but also the nuances of metacognitive epistemic attunements that accompany these exceptional oneiric moments.

Such comparisons—and any further examination of the hypotheses examined here—calls for psychometric ingenuity. One option is to use recently developed quantitative procedures to assess the semantic density (concreteness, embodiment, self-relevance) of basic level dream categories [100]. The complementary requirement is the development of systematic methods for assessing the semantic distance between dream categories and their relative level of abstraction [99,100]. Using enhanced computational procedures with category pairs identified as resonant by the dreamer may enable the prediction of metaphoric bidirectionality and emergent meanings [93].

Comparisons and hypothesis testing in this domain also call for a balance between psychometric rigor and connoisseurship. It will be important to attempt further articulation of the content categories that differentiate the metacognitive appraisals, inexpressible realizations, and moments of sublime feeling that are grounded in "living" aesthetic-epistemic metaphors. This may not be the place to expect operational definitions alone but rather the kind of connoisseurship that provides reliable articulation of the metaphoric structure of the Borges lines, existential dreams (e.g., the Hotel Dream), and transcendent dreams (e.g., the Snake Dream).

Funding: This research received no external funding.

Conflicts of Interest: The author declares no conflicts of interest.

References

1. Jung, C.G. *The Development of Personality*; Read, H.E., Ed.; Princeton University Press: Princeton, NJ, USA, 1964; Volume 17, ISBN 978-0-691-09763-3.
2. Hartmann, E. The Central Image Makes "Big" Dreams Big: The Central Image as the Emotional Heart of the Dream. *Dreaming* **2008**, *18*, 44–57. [CrossRef]
3. States, B.O. *Seeing in the Dark: Reflections on Dreams and Dreaming*; Yale University Press: New Haven, CT, USA, 1997; ISBN 0-300-06910-3.
4. Barcaro, U. A Reflection on Ernest Hartmann's Equation between the Central Image in Dreams and the Objective Correlative in Poetry. *Dreaming* **2022**, *33*, 93–103. [CrossRef]
5. Perogamvros, L.; Baird, B.; Seibold, M.; Riedner, B.; Boly, M.; Tononi, G. The Phenomenal Contents and Neural Correlates of Spontaneous Thoughts across Wakefulness, NREM Sleep, and REM Sleep. *J. Cogn. Neurosci.* **2017**, *29*, 1766–1777. [CrossRef] [PubMed]
6. Goodblatt, C.; Glicksohn, J. *The Gestalts of Mind and Text*, 1st ed.; Routledge: London, UK, 2022; ISBN 978-0-429-32953-1.
7. Ricoeur, P. *The Rule of Metaphor: Multi-Disciplinary Studies of the Creation of Meaning in Language*; Routledge: London, UK; Taylor & Francis Group: New York, NY, USA, 2004.
8. Searle, J. Metaphor. In *Metaphor and Thought*; Ortony, A., Ed.; Cambridge University Press: Cambridge, UK, 1979; pp. 92–123.
9. Hunt, H.T. *The Multiplicity of Dreams: Memory, Imagination, and Consciousness*; Yale University Press: New Haven, CT, USA, 1989; ISBN 0-300-04330-9.
10. Foulkes, D.; Spear, P.S.; Symonds, J.D. Individual Differences in Mental Activity at Sleep Onset. *J. Abnorm. Psychol.* **1966**, *71*, 280–286. [CrossRef] [PubMed]
11. Windt, J.M. How Deep Is the Rift between Conscious States in Sleep and Wakefulness? Spontaneous Experience over the Sleep–Wake Cycle. *Phil. Trans. R. Soc. B* **2021**, *376*, 20190696. [CrossRef]
12. Borsboom, D.; Rhemtulla, M.; Cramer, A.O.J.; van der Maas, H.L.J.; Scheffer, M.; Dolan, C.V. Kinds versus Continua: A Review of Psychometric Approaches to Uncover the Structure of Psychiatric Constructs. *Psychol. Med.* **2016**, *46*, 1567–1579. [CrossRef] [PubMed]
13. Bailey, K.D. *Typologies and Taxonomies: An Introduction to Classification Techniques*; Quantitative Applications in the Social Sciences; Sage Publications: Thousand Oaks, CA, USA, 1994; ISBN 978-0-8039-5259-1.
14. Busink, R.; Kuiken, D. Identifying Types of Impactful Dreams: A Replication. *Dreaming* **1996**, *6*, 97–119. [CrossRef]
15. Kuiken, D.; Sikora, S. The Impact of Dreams on Waking Thoughts and Feelings. In *The Functions of Dreaming*; Moffitt, A., Kramer, M., Hoffmann, R., Eds.; SUNY Series in Dream Studies; State University of New York Press: Albany, NY, USA, 1993; pp. 419–476. ISBN 978-0-7914-1297-8.
16. Kuiken, D.; Lee, M.-N.; Eng, T.; Singh, T. The Influence of Impactful Dreams on Self-Perceptual Depth and Spiritual Transformation. *Dreaming* **2006**, *16*, 258–279. [CrossRef]
17. Kuiken, D.; Lee, M.-N.; Northcott, P. Predicting Impactful Dreams: The Contributions of Absence-Related Melancholy and Absence-Related Depression. *Dreaming* **2023**, *33*, 164–186. [CrossRef]
18. Lee, M.-N.; Kuiken, D. Continuity of Reflective Awareness across Waking and Dreaming States. *Dreaming* **2015**, *25*, 141–159. [CrossRef]
19. Domhoff, G.W. *The Neurocognitive Theory of Dreaming: The Where, How, When, What, and Why of Dreams*; The MIT Press: Cambridge, MA, USA, 2022; ISBN 978-0-262-37088-2.
20. Hobson, J.A.; McCarley, R.W. The Brain as a Dream State Generator: An Activation-Synthesis Hypothesis of the Dream Process. *Am. J. Psychiatry* **1977**, *134*, 1335–1348. [CrossRef] [PubMed]
21. Gott, J.A.; Liley, D.T.J.; Hobson, J.A. Towards a Functional Understanding of PGO Waves. *Front. Hum. Neurosci.* **2017**, *11*, 89. [CrossRef] [PubMed]

22. Rittenhouse, C.D.; Stickgold, R.; Hobson, J.A. Constraint on the Transformation of Characters, Objects, and Settings in Dream Reports. *Conscious. Cogn.* **1994**, *3*, 100–113. [CrossRef]
23. Stickgold, R.; Rittenhouse, C.D.; Hobson, J.A. Dream Splicing: A New Technique for Assessing Thematic Coherence in Subjective Reports of Mental Activity. *Conscious. Cogn.* **1994**, *3*, 114–128. [CrossRef]
24. Bowker, R.M.; Morrison, A.R. The Startle Reflex and PGO Spikes. *Brain Res.* **1976**, *102*, 185–190. [CrossRef] [PubMed]
25. Morrison, A.R. You Have Interpreted the PGO Waves of REM Sleep as Activation of the Startle Network of the Brain. What Is Your Theory of the Function of off-Line Startle and What Impact, If Any, Does This Activation Have upon Dreaming? In *Dream Consciousness*; Tranquillo, N., Ed.; Vienna Circle Institute Library; Springer International Publishing: New York, NY, USA, 2014; Volume 3, pp. 171–173. ISBN 978-3-319-07295-1.
26. Graham, F.K. Distinguishing among Orienting, Defense, and Startle Reflexes. In *The Orienting Reflex in Humans: An International Conference Sponsored by the Scientific Affairs Division of the North Atlantic Treaty Organization Leeuwenhorst Congress Center*; Kimmel, H.D., Van Olst, E.H., Orlebeke, J.F., Eds.; Routledge: London, UK, 1979; pp. 137–167. ISBN 978-1-00-317140-9.
27. Petersen, S.E.; Posner, M.I. The Attention System of the Human Brain: 20 Years After. *Annu. Rev. Neurosci.* **2012**, *35*, 73–89. [CrossRef] [PubMed]
28. Kuiken, D. Does Morrisson's PGO Wave/Startle Hypothesis Help Us Explain Such Robust Dream Features as Surprise and Scene Shift? In *Dream Consciousness*; Tranquillo, N., Ed.; Vienna Circle Institute Library; Springer International Publishing: New York, NY, USA, 2014; Volume 3, pp. 153–155. ISBN 978-3-319-07295-1.
29. Mäder, T.; Oliver, K.I.; Daffre, C.; Kim, S.; Orr, S.P.; Lasko, N.B.; Seo, J.; Kleim, B.; Pace-Schott, E.F. Autonomic Activity, Posttraumatic and Nontraumatic Nightmares, and PTSD after Trauma Exposure. *Psychol. Med.* **2023**, *53*, 731–740. [CrossRef] [PubMed]
30. Tanev, K.S.; Orr, S.P.; Pace-Schott, E.F.; Griffin, M.; Pitman, R.K.; Resick, P.A. Positive Association Between Nightmares and Heart Rate Response to Loud Tones: Relationship to Parasympathetic Dysfunction in PTSD Nightmares. *J. Nerv. Ment. Dis.* **2017**, *205*, 308–312. [CrossRef] [PubMed]
31. Mathew, A.S.; Lotfi, S.; Bennett, K.P.; Larsen, S.E.; Dean, C.; Larson, C.L.; Lee, H.-J. Association between Spatial Working Memory and Re-Experiencing Symptoms in PTSD. *J. Behav. Ther. Exp. Psychiatry* **2022**, *75*, 101714. [CrossRef] [PubMed]
32. Kuiken, D.; Bears, M.; Miall, D.S.; Smith, L. Eye Movement Desensitization Reprocessing Facilitates Attentional Orienting. *Imagin. Cogn. Personal.* **2001**, *21*, 3–20. [CrossRef]
33. Rechtschaffen, A. The Single-Mindedness and Isolation of Dreams. *Sleep* **1978**, *1*, 97–109. [CrossRef]
34. Windt, J.M.; Voss, U. Spontaneous Thought, Insight, and Control in Lucid Dreams. In *The Oxford Handbook of Spontaneous Thought: Mind-Wandering, Creativity, and Dreaming*; Christoff, K., Fox, K.C.R., Eds.; Oxford University Press: Oxford, UK, 2018; Volume 1, p. 47.
35. Kahan, T.L.; Sullivan, K.T. Assessing Metacognitive Skills in Waking and Sleep: A Psychometric Analysis of the Metacognitive, Affective, Cognitive Experience (MACE) Questionnaire. *Conscious. Cogn.* **2012**, *21*, 340–352. [CrossRef] [PubMed]
36. Giora, R.; Givoni, S.; Heruti, V.; Fein, O. The Role of Defaultness in Affecting Pleasure: The Optimal Innovation Hypothesis Revisited. *Metaphor. Symb.* **2017**, *32*, 1–18. [CrossRef]
37. Blagrove, M.; Edwards, C.; van Rijn, E.; Reid, A.; Malinowski, J.; Bennett, P.; Carr, M.; Eichenlaub, J.-B.; McGee, S.; Evans, K.; et al. Insight from the Consideration of REM Dreams, Non-REM Dreams, and Daydreams. *Psychol. Conscious. Theory Res. Pract.* **2019**, *6*, 138–162. [CrossRef]
38. Malinowski, J.E.; Horton, C.L. Metaphor and Hyperassociativity: The Imagination Mechanisms behind Emotion Assimilation in Sleep and Dreaming. *Front. Psychol.* **2015**, *6*, 1132. [CrossRef] [PubMed]
39. Picard-Deland, C.; Konkoly, K.; Raider, R.; Paller, K.A.; Nielsen, T.; Pigeon, W.R.; Carr, M. The Memory Sources of Dreams: Serial Awakenings across Sleep Stages and Time of Night. *Sleep* **2023**, *46*, zsac292. [CrossRef] [PubMed]
40. Carr, M.; Nielsen, T. Morning REM Sleep Naps Facilitate Broad Access to Emotional Semantic Networks. *Sleep* **2015**, *38*, 433–443. [CrossRef] [PubMed]
41. Lewis, P.A.; Durrant, S.J. Overlapping Memory Replay during Sleep Builds Cognitive Schemata. *Trends Cogn. Sci.* **2011**, *15*, 343–351. [CrossRef] [PubMed]
42. Payne, J.D.; Schacter, D.L.; Propper, R.E.; Huang, L.-W.; Wamsley, E.J.; Tucker, M.A.; Walker, M.P.; Stickgold, R. The Role of Sleep in False Memory Formation. *Neurobiol. Learn. Mem.* **2009**, *92*, 327–334. [CrossRef] [PubMed]
43. Zadra, A.; Stickgold, R. *When Brains Dream: Understanding the Science and Mystery of Our Dreaming Minds*; W.W. Norton & Company, Inc.: New York, NY, USA, 2022; ISBN 978-1-324-02029-5.
44. Fogel, S.M.; Ray, L.B.; Sergeeva, V.; De Koninck, J.; Owen, A.M. A Novel Approach to Dream Content Analysis Reveals Links between Learning-Related Dream Incorporation and Cognitive Abilities. *Front. Psychol.* **2018**, *9*, 1398. [CrossRef] [PubMed]
45. Cai, D.J.; Mednick, S.A.; Harrison, E.M.; Kanady, J.C.; Mednick, S.C. REM, Not Incubation, Improves Creativity by Priming Associative Networks. *Proc. Natl. Acad. Sci. USA* **2009**, *106*, 10130–10134. [CrossRef] [PubMed]
46. Mednick, S.A. The Remote Associates Test. *J. Creat. Behav.* **1968**, *2*, 213–214. [CrossRef]
47. Lee, C.S.; Huggins, A.C.; Therriault, D.J. A Measure of Creativity or Intelligence? Examining Internal and External Structure Validity Evidence of the Remote Associates Test. *Psychol. Aesthet. Creat. Arts* **2014**, *8*, 446–460. [CrossRef]
48. Mussel, P.; McKay, A.S.; Ziegler, M.; Hewig, J.; Kaufman, J.C. Predicting Creativity Based on the Facets of the Theoretical Intellect Framework: Creativity and the Theoretical Intellect Framework. *Eur. J. Personal.* **2015**, *29*, 459–467. [CrossRef]

49. Beaty, R.E.; Jung, R.E. Interacting Brain Networks Underlying Creative Cognition and Artistic Performance. In *The Oxford Handbook of Spontaneous Thought: Mind-Wandering, Creativity, and Dreaming*; Christoff, K., Fox, K.C.R., Eds.; Oxford University Press: Oxford, UK, 2018; Volume 1, p. 17.
50. Zamani, A.; Mills, C.; Girn, M.; Christoff, K. A Closer Look at Transitions between the Generative and Evaluative Phases of Creative Thought. In *The Routledge International Handbook of Creative Cognition*; Routledge: London, UK, 2023; pp. 453–474. ISBN 978-1-00-300935-1.
51. Mednick, S. The Associative Basis of the Creative Process. *Psychol. Rev.* **1962**, *69*, 220–232. [CrossRef]
52. Benedek, M.; Neubauer, A.C. Revisiting Mednick's Model on Creativity-Related Differences in Associative Hierarchies. Evidence for a Common Path to Uncommon Thought. *J. Creat. Behav.* **2013**, *47*, 273–289. [CrossRef] [PubMed]
53. Kuiken, D.; Porthukaran, A.; Albrecht, K.-A.; Douglas, S.; Cook, M. Metaphoric and Associative Aftereffects of Impactful Dreams. *Dreaming* **2018**, *28*, 59–83. [CrossRef]
54. Beaty, R.E.; Silvia, P.J.; Benedek, M. Brain Networks Underlying Novel Metaphor Production. *Brain Cogn.* **2017**, *111*, 163–170. [CrossRef] [PubMed]
55. Benedek, M.; Beaty, R.; Jauk, E.; Koschutnig, K.; Fink, A.; Silvia, P.J.; Dunst, B.; Neubauer, A.C. Creating Metaphors: The Neural Basis of Figurative Language Production. *NeuroImage* **2014**, *90*, 99–106. [CrossRef]
56. Liu, S.; Chow, H.M.; Xu, Y.; Erkkinen, M.G.; Swett, K.E.; Eagle, M.W.; Rizik-Baer, D.A.; Braun, A.R. Neural Correlates of Lyrical Improvisation: An fMRI Study of Freestyle Rap. *Sci. Rep.* **2012**, *2*, 834. [CrossRef] [PubMed]
57. Pinho, A.L.; Ullén, F.; Castelo-Branco, M.; Fransson, P.; de Manzano, Ö. Addressing a Paradox: Dual Strategies for Creative Performance in Introspective and Extrospective Networks. *Cereb. Cortex* **2015**, *26*, 3052–3063. [CrossRef] [PubMed]
58. Limb, C.J.; Braun, A.R. Neural Substrates of Spontaneous Musical Performance: An fMRI Study of Jazz Improvisation. *PLoS ONE* **2008**, *3*, e1679. [CrossRef] [PubMed]
59. Girn, M.; Mills, C.; Roseman, L.; Carhart-Harris, R.L.; Christoff, K. Updating the Dynamic Framework of Thought: Creativity and Psychedelics. *NeuroImage* **2020**, *213*, 116726. [CrossRef]
60. Koriat, A. Metacognition and Consciousness (Chapter 11). In *The Cambridge Handbook of Consciousness*; Zelazo, P.D., Moscovitch, M., Thompson, E., Eds.; Cambridge University Press: Cambridge, UK, 2007; pp. 289–325.
61. Brown, S.R. Tip-of-the-Tongue Phenomena: An Introductory Phenomenological Analysis. *Conscious. Cogn.* **2000**, *9*, 516–537. [CrossRef] [PubMed]
62. Whittlesea, B.W.; Williams, L.D. The Discrepancy-Attribution Hypothesis: I. The Heuristic Basis of Feelings of Familiarity. *J. Exp. Psychol. Learn. Mem. Cogn.* **2001**, *27*, 3–13. [CrossRef] [PubMed]
63. Cleary, A.; Brown, A.S. *The Déjà vu Experience*, 2nd ed.; Routledge: New York, NY, USA, 2021; ISBN 978-1-00-040170-7.
64. Gendlin, E.T. *Experiencing and the Creation of Meaning: A Philosophical and Psychological Approach to the Subjective*; Northwestern University Studies in Phenomenology and Existential Philosophy; Northwestern University Press: Evanston, IL, USA, 1997; ISBN 0-8101-1427-5.
65. Zabelina, D.L.; Friedman, N.P.; Andrews-Hanna, J. Unity and Diversity of Executive Functions in Creativity. *Conscious. Cogn.* **2019**, *68*, 47–56. [CrossRef] [PubMed]
66. Zabelina, D.; Saporta, A.; Beeman, M. Flexible or Leaky Attention in Creative People? Distinct Patterns of Attention for Different Types of Creative Thinking. *Mem. Cogn.* **2016**, *44*, 488–498. [CrossRef] [PubMed]
67. Tellegen, A. Practicing the Two Disciplines for Relaxation and Enlightenment: Comment on "Role of the Feedback Signal in Electromyograph Biofeedback: The Relevance of Attention" by Qualls and Sheehan. *J. Exp. Psychol. Gen.* **1981**, *110*, 217–226. [CrossRef]
68. DeYoung, C.G. Openness/Intellect: A Dimension of Personality Reflecting Cognitive Exploration. In *APA Handbook of Personality and Social Psychology: Personality Processes and Individual Differences*; Mikulincer, P.R., Shaver, P.R., Cooper, M.L., Larsen, R.J., Eds.; American Psychological Association: Washington, DC, USA, 2015; Volume 4, pp. 369–399.
69. Oleynick, V.C.; DeYoung, C.G.; Hyde, E.; Kaufman, S.B.; Beaty, R.E.; Silvia, P.J. Openness/Intellect: The Core of the Creative Personality. In *The Cambridge Handbook of Creativity and Personality Research*; Feist, G.J., Reiter-Palmon, R., Kaufman, J.C., Eds.; Cambridge University Press: Cambridge, UK, 2017; pp. 9–27. ISBN 978-1-316-22803-6.
70. Tellegen, A.; Atkinson, G. Openness to Absorbing and Self-Altering Experiences ("Absorption"), a Trait Related to Hypnotic Susceptibility. *J. Abnorm. Psychol.* **1974**, *83*, 268–277. [CrossRef] [PubMed]
71. DeYoung, C.G.; Grazioplene, R.G.; Peterson, J.B. From Madness to Genius: The Openness/Intellect Trait Domain as a Paradoxical Simplex. *J. Res. Personal.* **2012**, *46*, 63–78. [CrossRef]
72. Kuiken, D.; Nielsen, T. Individual Differences in Orienting Activity Mediate Feeling Realization in Dreams: I. Evidence from Retrospective Reports of Movement Inhibition. *Dreaming* **1996**, *6*, 201–217. [CrossRef]
73. Bashwiner, D.; Bacon, D. Musical Creativity and the Motor System. *Curr. Opin. Behav. Sci.* **2019**, *27*, 146–153. [CrossRef]
74. Colombetti, G. *The Feeling Body: Affective Science Meets the Enactive Mind*; The MIT Press: Cambridge, MA, USA, 2014; ISBN 978-0-262-01995-8.
75. Gallagher, S.; Gallagher, J. Acting Oneself as Another: An Actor's Empathy for Her Character. *Topoi* **2020**, *39*, 779–790. [CrossRef]
76. Sheets-Johnstone, M. From Movement to Dance. *Phenom. Cogn. Sci.* **2012**, *11*, 39–57. [CrossRef]
77. Knudson, R.M.; Minier, S. The On-Going Significance of Significant Dreams: The Case of the Bodiless Head. *Dreaming* **1999**, *9*, 235–245. [CrossRef]

78. Borges, J.L. *Labyrinths: Selected Stories and Other Writing*; Yates, D.A., Ed.; Penguin Books: Harmonsworth, UK, 2000; ISBN 978-0-14-118484-5.

79. Kuiken, D.; Douglas, S. Living Metaphor as the Site of Bidirectional Literary Engagement. *Sci. Study Lit.* **2018**, *8*, 47–76. [CrossRef]

80. Kuiken, D. Dreams and Feeling Realization. *Dreaming* **1995**, *5*, 129–157. [CrossRef]

81. Petitmengin, C.; Remillieux, A.; Valenzuela-Moguillansky, C. Discovering the Structures of Lived Experience: Towards a Micro-Phenomenological Analysis Method. *Phenom. Cogn. Sci.* **2019**, *18*, 691–730. [CrossRef]

82. Revonsuo, A.; Salmivalli, C. A Content Analysis of Bizarre Elements in Dreams. *Dreaming* **1995**, *5*, 169–187. [CrossRef]

83. Revonsuo, A.; Tarkko, K. Binding in Dreams—The Bizarreness of Dream Images and the Unity of Consciousness. *J. Conscious. Stud.* **2002**, *9*, 3–24.

84. Schweickert, R.; Xi, Z. Metamorphosed Characters in Dreams: Constraints of Conceptual Structure and Amount of Theory of Mind. *Cogn. Sci.* **2010**, *34*, 665–684. [CrossRef] [PubMed]

85. Al-Azary, H.; Gagné, C.L.; Spalding, T.L. Flute Birds and Creamy Skies: The Metaphor Interference Effect in Modifier–Noun Phrases. *Can. J. Exp. Psychol. Rev. Can. Psychol. Expérimentale* **2021**, *75*, 175–181. [CrossRef] [PubMed]

86. Lakoff, G. How Metaphor Structures Dreams: The Theory of Conceptual Metaphor Applied to Dream Analysis. *Dreaming* **1993**, *3*, 77–98. [CrossRef]

87. Ortony, A. Beyond Literal Similarity. *Psychol. Rev.* **1979**, *86*, 161–180. [CrossRef]

88. Vega-Moreno, R. Metaphor Interpretation and Emergence. *UCL Work. Pap. Linguist.* **2004**, *16*, 297–322.

89. Lakoff, G.; Turner, M. *More than Cool Reason: A Field Guide to Poetic Metaphor*; University of Chicago Press: Chicago, IL, USA, 1989.

90. Richards, I.A. *The Philosophy of Rhetoric*; The Mary Flexner Lectures on the Humanities; Oxford University Press: London, UK, 1936; ISBN 978-0-19-500715-2.

91. Black, M. *Models and Metaphors: Studies in Language and Philosophy*; Cornell University Press: Ithaca, NY, USA, 1962.

92. Fauconnier, G.; Turner, M. Conceptual Blending, Form and Meaning. *Rech. Commun.* **2003**, *19*, 57–86. [CrossRef]

93. Reid, J.N.; Katz, A. The RK Processor: A Program for Analysing Metaphor and Word Feature-Listing Data. *Behav. Res. Methods* **2022**, *54*, 174–195. [CrossRef] [PubMed]

94. Becker, A.H. Emergent and Common Features Influence Metaphor Interpretation. *Metaphor. Symb.* **1997**, *12*, 243–259. [CrossRef]

95. Terai, A.; Goldstone, R.L. An Experimental Examination of Emergent Features in Metaphor Interpretation Using Semantic Priming Effects. In Proceedings of the Thirty-Fourth Annual Conference of the Cognitive Science Society, Sapporo, Japan, 1–4 August 2012; Cognitive Science Society: Sapporo, Japan, 2012; pp. 2399–2404.

96. Tourangeau, R.; Rips, L. Interpreting and Evaluating Metaphors. *J. Mem. Lang.* **1991**, *30*, 452–472. [CrossRef]

97. Utsumi, A. The Role of Feature Emergence in Metaphor Appreciation. *Metaphor. Symb.* **2005**, *20*, 151–172. [CrossRef]

98. Katz, A.N.; Al-Azary, H. Principles That Promote Bidirectionality in Verbal Metaphor. *Poet. Today* **2017**, *38*, 35–59. [CrossRef]

99. Tourangeau, R.; Sternberg, R.J. Understanding and Appreciating Metaphors. *Cognition* **1982**, *11*, 203–244. [CrossRef] [PubMed]

100. Westbury, C.; Wurm, L.H. Is It You You're Looking for?: Personal Relevance as a Principal Component of Semantics. *Ment. Lex.* **2022**, *17*, 1–33. [CrossRef]

101. Cohen, T. Identifying with Metaphor: Metaphors of Personal Identification. *J. Aesthet. Art Crit.* **1999**, *57*, 399–409. [CrossRef]

102. Dorst, A.G.; Mulder, G.; Steen, G.J. Recognition of Personifications in Fiction by Non-Expert Readers. *Metaphor. Soc. World* **2011**, *1*, 174–200. [CrossRef]

103. Bruhn, M.J. Harmonious Madness: The Poetics of Analogy at the Limits of Blending Theory. *Poet. Today* **2011**, *32*, 619–662. [CrossRef]

104. Blagrove, M.; Lockheart, J. *The Science and Art of Dreaming*, 1st ed.; Routledge: London, UK, 2023; ISBN 978-1-00-303755-2.

105. Cazeaux, C. *Metaphor and Continental Philosophy: From Kant to Derrida*; Routledge Studies in Twentieth-Century Philosophy; Routledge: New York, NY, USA; London, UK, 2007; ISBN 978-0-415-32400-7.

106. Kant, I. *Critique of the Power of Judgment*, 1st ed.; The Cambridge Edition of the Works of Immanuel Kant; Guyer, P., Wood, A.W., Eds.; Cambridge University Press: Cambridge, UK, 2009; ISBN 978-0-521-34892-8.

107. Husserl, E. *Experience and Judgment: Investigations in a Genealogy of Logic*; Northwestern University Press: Evanston, IL, USA, 1973; ISBN 978-0-8101-0396-2.

108. Lohmar, D. Husserl's Type and Kant's Schemata. In *New Husserl: A Critical Reader*; Welton, D., Ed.; Indiana University Press: Bloomington, IN, USA, 2003; pp. 92–124. ISBN 978-0-253-21601-4.

109. Husserl, E. *Logical Investigations*; International Library of Philosophy; Routledge: London, UK, 1970; Volume 2, ISBN 978-0-415-24189-2.

110. Tinguely, J.J. *Kant and the Reorientation of Aesthetics: Finding the World*, 1st ed.; Series: Routledge Studies in Eighteenth-Century Philosophy; Routledge: New York, NY, USA, 2018; Volume 14, ISBN 978-1-315-11269-5.

111. Pillow, K. Jupiter's Eagle and the Despot's Hand Mill: Two Views on Metaphor in Kant. *J. Aesth. Art Crit.* **2001**, *59*, 193–209. [CrossRef]

112. Natanson, M.A. *The Erotic Bird: Phenomenology in Literature*; Princeton University Press: Princeton, NJ, USA, 1998; ISBN 0-691-01219-9.

113. Crowell, S. Phenomenology Is the Poetic Essence of Philosophy: Maurice Natanson on the Rule of Metaphor. *Res. Phenomenol.* **2005**, *35*, 270–289. [CrossRef]

114. Arcangeli, M.; Dokic, J. At the Limits: What Drives Experiences of the Sublime. *Br. J. Aesthet.* **2021**, *61*, 145–161. [CrossRef]

115. Batson, C.D.; Schoenrade, P.A. Measuring Religion as Quest: 1) Validity Concerns. *J. Sci. Study Relig.* **1991**, *30*, 416. [CrossRef]
116. Kuiken, D. The Contrasting Effects of Nightmares, Existential Dreams, and Transcendent Dreams. In *Dream Research: Contributions to Clinical Practice*; Kramer, M., Glucksberg, M.L., Eds.; Routledge: New York, NY, USA; Taylor & Francis Group: Abingdon, UK, 2015; pp. 174–187.
117. Buber, M. *I and Thou*; A Touchstone Book, 46. Print; Simon & Schuster Touchstone: New York, NY, USA, 1971; ISBN 978-0-684-71725-8.
118. Perlina, N. Bakhtin and Buber: Problems of Dialogic Imagination. *Stud. 20th 21st Century Lit.* **1984**, *9*, 3. [CrossRef]
119. Stace, W.T. *Mysticism and Philosophy*, 1st ed.; Macmillan: Houndmills, Basingstoke, UK, 1960; ISBN 978-0-333-08274-4.
120. Ghorbani, N.; Watson, P.J.; Ebrahimi, F.; Chen, Z.J. Poets and Transliminality: Relationships with Mystical Experience and Religious Commitment in Iran. *Psychol. Relig. Spiritual.* **2019**, *11*, 141–147. [CrossRef]
121. Sears, R.E. The Construction, Preliminary Validation, and Correlates of a Dream-specific Scale for Mystical Experience. *Sci. Study Relig.* **2015**, *54*, 134–155. [CrossRef]
122. Sears, R.E.; Hood, R.W. Dreaming Mystical Experience among Christians and Hindus: The Impact of Culture, Language, and Religious Participation on Responses to the Dreaming Mysticism Scale. *Ment. Health Relig. Cult.* **2016**, *19*, 833–845. [CrossRef]
123. Hartmann, E.; Kunzendorf, R.G. Thymophor in Dreams, Poetry, Art, and Memory: Emotion Translated into Imagery as a Basic Element of Human Creativity. *Imagin. Cogn. Personal.* **2013**, *33*, 165–191. [CrossRef]
124. Mainieri, G.; Maranci, J.-B.; Champetier, P.; Leu-Semenescu, S.; Gales, A.; Dodet, P.; Arnulf, I. Are Sleep Paralysis and False Awakenings Different from REM Sleep and from Lucid REM Sleep? A Spectral EEG Analysis. *J. Clin. Sleep Med.* **2021**, *17*, 719–727. [CrossRef] [PubMed]
125. Cheyne, J.A.; Rueffer, S.D.; Newby-Clark, I.R. Hypnagogic and Hypnopompic Hallucinations during Sleep Paralysis: Neurological and Cultural Construction of the Night-Mare. *Conscious. Cogn.* **1999**, *8*, 319–337. [CrossRef]
126. Cheyne, J.A.; Girard, T.A. Paranoid Delusions and Threatening Hallucinations: A Prospective Study of Sleep Paralysis Experiences. *Conscious. Cogn.* **2007**, *16*, 959–974. [CrossRef]

Review

Impact of Pre-Sleep Visual Media Exposure on Dreams: A Scoping Review

Ajar Diushekeeva [1], Santiago Hidalgo [2] and Antonio Zadra [1,3,*] 

[1] Department of Psychology, Université de Montréal, C.P. 6128, Succursale Centre-ville, Montréal, QC H3C 3J7, Canada; ajar.diushekeeva@umontreal.ca

[2] Department of Art History and Cinematographic Studies, Université de Montréal, C.P. 6128, Succursale Centre-ville, Montréal, QC H3C 3J7, Canada; santiago.hidalgo@umontreal.ca

[3] Center for Advanced Research in Sleep Medicine, CIUSSS-NÎM—Hôpital du Sacré-Cœur de Montréal, 5400 Gouin Blvd Ouest, Montréal, QC H4J 1C5, Canada

* Correspondence: antonio.zadra@umontreal.ca; Tel.: +1-514-343-6626

Abstract: A body of experimental research has aimed to investigate processes underlying dream formation by examining the effects of a range of pre-sleep stimuli and events on subsequent dream content. Given its ever-growing presence and salience in people's everyday lives, pre-sleep media consumption stands out as a key variable that could influence people's dreams. We conducted a scoping review to evaluate the experimental evidence of the effects of pre-sleep exposure to visual media on dream content. A systematic search on PubMed, PsycInfo, and Web of Science using terms related to moving visual media and dreams yielded 29 studies meeting the inclusion criteria. Overall, we found modest yet varied effects of pre-sleep exposure to visual media on dream content, with rates of stimulus-related incorporation ranging from 3% to 43% for REM dream reports, 4% to 30% for NREM sleep mentation reports, and between 11% and 35% for home dream reports. Our review highlights the large methodological heterogeneity and gaps across studies, the general difficulty in influencing dream content using pre-sleep exposure to visual media, and suggests promising venues for future research to advance our understanding of how and why digital media may impact people's dreams.

Keywords: dreams; dreaming; media consumption; REM sleep; NREM sleep; dream incorporation; dream content

Citation: Diushekeeva, A.; Hidalgo, S.; Zadra, A. Impact of Pre-Sleep Visual Media Exposure on Dreams: A Scoping Review. *Brain Sci.* **2024**, *14*, 662. https://doi.org/10.3390/brainsci14070662

Academic Editors: Luigi De Gennaro, Sergio A. Mota-Rolim, Brigitte Holzinger and Michael Nadorff

Received: 30 May 2024
Revised: 24 June 2024
Accepted: 24 June 2024
Published: 29 June 2024

1. Introduction

Dream formation draws upon a rich tapestry of sources encompassing semantic, episodic, and autobiographical memories, from current concerns to recent and remote experiences, psychological history, and sociocultural background [1–3]. Much scholarly attention in the field of dream research has been devoted to understanding how certain experiences, such as events from the preceding day, impact dreams [4–6]. To better understand how various events impact people's dreams and uncover the underlying processes and potential functions of dreaming, researchers have attempted to influence dream contents by systematically manipulating pre-sleep and during-sleep experiences.

Within this line of research, different modalities of sensory stimulations have been administered during sleep (see [4,7,8] for reviews). These include somatosensory stimuli such as water sprayed on the skin [9], pressure stimuli [10], thermal stimuli [11], and electrical stimuli [12]; olfactory stimuli like pleasant and unpleasant odors [13]; visual stimuli like flashes of light [9]; auditory stimuli such as tones [14]; verbal stimuli such as tape recordings of persons' names [15], and neutral versus meaningful words [16]; and even transcranial direct current stimulation [17].

Similarly, a range of stimuli and experiences have been experimentally presented prior to people's sleep (see [4,18] for reviews), including group therapy sessions [19], periods of free association [20], hypnotic suggestions [21], studying [22], mental task performance [23], physical activity [22], fluid deprivation [24], food deprivation [11], social isolation [25,26], visual inverting prisms [27], and film viewing [28].

In addition to during-sleep sensory stimulation and pre-sleep priming, recent approaches to "dream engineering" (i.e., using techniques and technologies to manipulate dreams) have been employed, including dream incubation (i.e., pre-sleep rehearsal of content aiming to incubate desired dream features) and targeted memory reactivation (i.e., pairing pre-sleep and within-sleep stimulation to elicit reactivation of specific content; see [1] for a review). A substantial body of research has thus examined whether external stimuli can consistently alter sleep mentation, while deploying dream-incorporated stimuli as an experimental paradigm for investigating the mechanisms of dream production [4,8].

Media use is a particularly intriguing type of pre-sleep event; all the more so given the now-pervasive presence of digital media in our daily lives, including during our bedtime routines. For instance, one representative survey found that more than half of Canadians report that checking their smartphones is the last thing they do before going to sleep [29]. There is also a growing body of evidence suggesting that virtual experiences during our waking life (e.g., video game play; [30]) infiltrate our dreams, aligning with the continuity hypothesis of dreaming, which posits that our dreams embody and reflect our waking concerns and experiences [31–33].

Several correlational studies have noted associations between various measures of dream content and preceding media use, including overall media consumption, level of daytime exposure, and exposure prior to sleep onset. For example, in one study that sampled 3167 children aged 6–18 years, 74% of children reported that their dreams reflected what they had viewed on television or in films [34], while in another study, 53% of participants of a wide age range reported experiencing television-related dreams [35]. In pediatric populations, watching television and playing computer games has been linked with the frequency of TV-related dreams [36] as well as with unpleasant and pleasant dreams featuring TV content [37,38]. Another study found an association between the consumption of violent and sexual media before bedtime and the occurrence of violent and sexual dreams [39]. Moreover, playing video games has been shown to be associated with the incorporation of video game content into dreams [40], self-rated violence in dreams [41], and lucid dreaming [42]. Overall, social media use, especially engagement, has been linked to the prevalence of social media dreams [35,43]. Interestingly, children may be more susceptible to the effects of media use, as many report that watching TV has a more pronounced effect on their dreams than other daytime experiences [34]. In contrast, adolescents report that both watching TV and engaging in other daily activities impact their dreams in equal measure, and adults declare that daily activities have a greater influence on their dreams compared to watching TV [44]. While this correlational research provides valuable observations on the relationship between visual media use and dream content, it does not allow for causal inferences. Moreover, self-reported retrospective estimates of such effects may capture participants' beliefs about how media manifests in their dream life, rather than measuring the actual impact of media exposure on dreaming.

Turning to experimental research, a number of studies have investigated how filmic stimuli, mainly stressful, arousing, and neutral films, influence various measures of dream content. These studies have yielded varying degrees of consistency in terms of how and to what extent these stimuli are incorporated into people's dreams. Many of these experiments were conducted between the 1960s and 1990s, and as a result, the kinds of media stimuli employed in these studies may not accurately reflect the current media landscape. Some newer research has utilized interactive visual stimuli, including the computer game *Tetris* and an alpine ski visuomotor simulator game, demonstrating a significant degree of direct incorporation of game-related imagery into sleep-onset mentation while providing insights into the role of sleep in memory consolidation and learning [45–47]. Since then, a few other studies have examined the effects of immersive and interactive visual stimuli, such as video games and virtual reality, on dreams during REM sleep, the sleep stage most consistently and robustly associated with vivid, emotionally salient, and narratively driven dreams [48].

As digital media continues to shape and transform every aspect of our waking lives, it would be surprising if these stimuli did not exert an influence on our dream lives. Moving visual media, including activities such as streaming movies or shows, watching short-form videos, and gaming, represents one of the most pervasive and immersive forms of modern media consumption. Given its ubiquity facilitated by the use of smartphones, moving visual media emerges as a compelling factor with the potential to shape our dream experiences. To help unravel the intricate relationships between various kinds of pre-sleep media uses and subsequent dream content, we undertook a comprehensive examination of the existing literature. More specifically, we conducted a scoping review to systematically chart, describe, and synthesize the experimental research on the effects of exposure to moving visual media prior to sleep on subsequent dream content.

2. Materials and Methods

We adopted Arksey and O'Malley's [49] methodological framework for conducting scoping reviews, as well as consulting the updated methodological recommendations outlined by the Joanna Briggs Institute [50]. The review process encompassed five stages, which are described below.

2.1. Stage 1: Identifying the Research Question(s)

This scoping review aimed to summarize and examine the experimental evidence regarding the impact of visual media stimulus exposure prior to sleep on ensuing dream content and features. How have visual media stimuli (e.g., in terms of methodologies) been utilized to influence dream content? What were the overarching objectives behind such endeavors, and which theoretical perspectives have informed this topic? What outcomes (e.g., incorporation rates of stimuli into dreams) have been reported? What approaches seem to be most efficacious or promising?

2.2. Stage 2: Identifying Relevant Studies

To identify eligible studies, we performed a systematic search of three databases: PubMed, APA PsycNet (PsycInfo and PsycArticles), and Web of Science Core Collection. The initial search was conducted in June 2021, followed by a subsequent search update across all databases in July 2023. The initial search strategy was developed within PubMed, using Medical Subject Headings (MeSH) terms and keywords such as "film", "movie", "video", "television", or "virtual reality". This set of search terms was then combined using the Boolean operator "AND" to a second set of search terms focusing on "dreams" or "sleep mentation". The search strategy formulated in PubMed was then adapted to suit the syntax of other databases. The search strategy encompassed peer-reviewed articles from any time period but limited itself to those published in English. A detailed search strategy is provided in Appendix A. Additionally, we manually checked the reference lists of selected publications (i.e., backward citation tracking) as well as their citing references via Google Scholar (i.e., forward citation tracking) to identify any potentially relevant articles that might not have been captured by the database searches.

2.3. Stage 3: Study Selection: Inclusion and Exclusion Criteria

All the retrieved records were imported into a reference management software, *Zotero* 5.0, where duplicates in the library were detected and eliminated. We proceeded with screening the titles and abstracts of these records and excluded those extraneous to our topic. A total of 152 articles were retained for full-text review. The inclusion criteria covered experimental studies that investigated the effects of pre-sleep exposure to moving visual stimuli, such as films, videos, video games, and virtual reality, on subsequent dreaming, regardless of whether this constituted the primary focus of the study. Studies sampling adult, adolescent, and pediatric populations were considered eligible. The following exclusion criteria were applied: studies that were not experimental, lacked controlled exposure to moving visual media during wakefulness prior to sleep, or involved exposure to *static* visual stimuli (e.g., photographs), as our review focused on exposure to *moving* visual media since this form of media is most intimately tied to the question of how modern pre-sleep media consumption (e.g., TV shows, streaming services, etc.) may influence people's dreams. We also excluded studies that did not assess dream content as an outcome or failed to report sufficient dream-related results. Only published articles were included; conference proceedings, abstracts, and dissertations were excluded, though they are briefly mentioned in the discussion if relevant. Studies employing duplicate participant samples were also excluded, while only the most relevant study for each distinct sample was retained to avoid redundancy and ensure the inclusion of unique data. A total of 29 articles satisfying the criteria were selected for data extraction. Figure 1 provides a representation of the screening process, in accordance with the Preferred Reporting Items for Systematic Review and Meta-Analysis Extension for Scoping Reviews (PRISMA-ScR) framework [51].

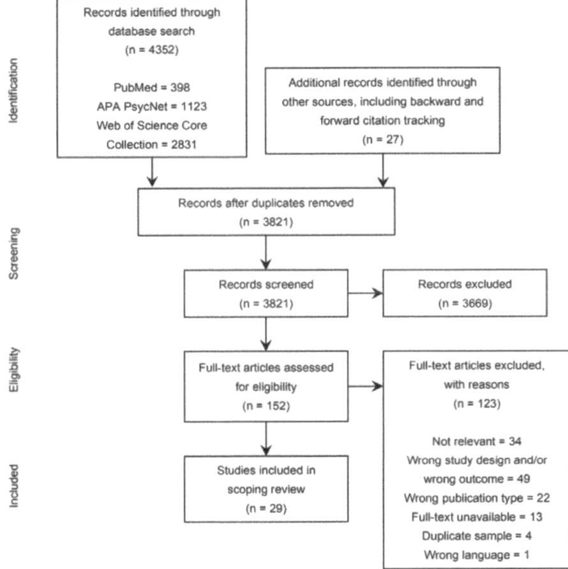

Figure 1. Preferred Reporting Items for Systematic Reviews and Meta-Analyses extension for Scoping Reviews (PRISMA-ScR) flow chart.

2.4. Stage 4: Charting the Data

The data charting process involved extracting the following information from the selected studies: bibliometric characteristics (author(s), year of publication, title, location); methodological data (sample size and characteristics, study design, pre-sleep stimulus exposure, measurement(s) of additional variables, study aims and hypotheses); outcomes (findings pertaining to dream content and proportion of stimulus-incorporation into dreams).

2.5. Stage 5: Collating, Summarizing, and Reporting Results

Some of the extracted data, including study characteristics and stimulus incorporation rates, were organized and summarized in tabular form. Table 1 presents all the included studies, which are organized into three groups: those conducted in a laboratory, those carried out at home, and those conducted in both laboratory and home settings. Further synthesis is provided in narrative form under the following themes: study and sample characteristics, theoretical frameworks, methodological characteristics, and relevant outcomes.

Table 1. Study characteristics and stimulus incorporation rates of included studies (N = 29).

Study	Population	Stimulus	Sleep Stage			Incorporation Rate (% Content-Filled Dream Reports) [a]	Incorporation Rate (% Participants) [b]
			REM	NREM	N/A		
Laboratory-Based Studies							
Foulkes and Rechtschaffen (1964) [28]	24 students (13 m, 11 f)	Violent Western film Comedic Western film Duration: 30 min.	✓	✓		5% of dream reports 5.5% of REM reports 3.8% of NREM reports	N/S
Witkin and Lewis (1965) [52]	3 male nights workers	Birth film Subincision film Neutral travelogue film Suggestion session	✓			N/S	N/S
Foulkes et al. (1967) [53]	32 boys (6–12 y.o.)	Violent Western film Baseball documentary Duration: 10 min.	✓			8% of dream reports	N/S
Cartwright et al. (1969) [54]	10 male students	Two erotic films Duration: 10 min. each	✓			N/S	N/S
Foulkes et al. (1971) [55]	40 boys (10–12 y.o.)	Violent Western film Nonviolent Western film	✓			N/S	N/S
Goodenough et al. (1975) [56]	28 male night workers	Birth film Subincision film Neutral travelogue film Another neutral travelogue film Duration: 11 min. each	✓			N/S	N/S
De Koninck and Koulack (1975) [57]	24 male students	*It Didn't Have to Happen*, film depicting workshop accidents Duration: 13 min.	✓			N/S	N/S
Lauer et al. (1987) [58]	11 men	Massacre film Prison film Neutral animal documentary Duration: 90 min.	✓			37.5% of dream reports 33.3% of initial REM reports 42.9% of final REM reports	54.5% of participants

Table 1. *Cont.*

Study	Population	Stimulus	Sleep Stage			Incorporation Rate (% Content-Filled Dream Reports) [a]	Incorporation Rate (% Participants) [b]
			REM	NREM	N/A		
Kuiken et al. (1990) [59]	12 students (6 m, 6 f)	*Where is Dead?*, film about death and grief *Dreamspeaker*, film about a boy who is committed to a mental institution	✓			N/S	N/S
Stickgold et al. (2000) [45]	27 volunteers (*Tetris* novices, *Tetris* experts, and amnesics)	*Tetris* Duration: 7 h in total		N1		7.4% of mentation reports by amnesics 7.2% of mentation reports by normals	63% of participants
Wamsley et al. (2010a) [60]	99 students (44 m, 55 f)	Virtual maze task Duration: 45 min.		N1 N2		N/S	8% of participants Among those, 6% during N1, 2% during N2 55% of participants in the questionnaire protocol group
Kussé et al. (2012) [47]	43 healthy volunteers (19 m, 24 f)	*Tetris* Duration: 6 h in total		N1 N2		10% of mentation reports Among those, 3.3% contained direct incorporations, 6.8% indirect incorporations 11.2% of N1 reports 6.5% of N2 reports	81% of participants
Stamm et al. (2014) [61]	65 healthy volunteers (37 m, 28 f)	Virtual maze task Duration: 35 min.	✓	N1 N2		8.5% of dream reports Among those, 3.7% contained direct incorporations, 5% indirect incorporations	36.9% of participants

Table 1. *Cont.*

Study	Population	Stimulus	Sleep Stage			Incorporation Rate (% Content-Filled Dream Reports) [a]	Incorporation Rate (% Participants) [b]
			REM	NREM	N/A		
Nefjodov et al. (2016) [62]	13 students (9 m, 4 f)	*Wii Fit* video game with a Wii Balance Board Duration: Two hours	✓			5.6% of dream reports, according to external ratings 19.4% of dream reports, according to subjective ratings	15.4% of participants, according to external ratings 54% of participants, according to subjective ratings
Wamsley et al. (2016) [63]	100 students (40 m, 60 f)	Virtual maze task Duration: 35 min.	✓	N1 N2		1.9% of dream reports	11.8% of participants
Fogel et al. (2018) [64]	24 healthy volunteers (4 m, 20 f)	Virtual maze task (resembling the video game *Team Fortress*) *Grand Slam Tennis* video game for Wii Duration: 30 min.		N1		N/S	N/S
Solomonova et al. (2018) [65]	40 volunteers (meditators and nonmeditators; 20 m, 20 f)	*Wii Fit* video game (*Balance Bubble*) with a Wii Balance Board Duration: 7 min.	✓	N1 N2		19.1% of dream reports 25% of REM reports 17.5% of N1 reports 18.8% of N2 reports	15% of participants
Wamsley & Stickgold (2019) [66]	39 students (13 m, 26 f)	Virtual maze task Duration: 35 min.	✓	N1 N2		6.4% of dream reports 12.5% of morning REM reports 7.1% of N1 reports 2.1% of N2 reports 12.5% of morning N2 reports	47.1% of participants
Klepel & Schredl (2019) [67]	22 students (5 m, 17 f)	Sequence from the film *Four Rooms*, without sound Duration: 5 min.	✓	✓		N/S	N/S

Table 1. *Cont.*

Study	Population	Stimulus	Sleep Stage			Incorporation Rate (% Content-Filled Dream Reports) [a]	Incorporation Rate (% Participants) [b]
			REM	NREM	N/A		
Home-Based Studies							
Powell et al. (1995) [68]	20 volunteers, mostly students (10 m, 10 f)	Film depicting the ceremonial slaughter of a buffalo Duration: 30 min.			✓	N/S	47.4% of participants
Davidson et al. (2005) [69]	24 volunteers, mostly students (8 m, 16 f)	9/11 media coverage video Introductory psychology lecture video Duration: 20 min. each			✓	N/S	N/S
Wamsley et al. (2010b) [46]	43 students (16 m, 27 f)	*Alpine Race II*, a visuomotor skiing video game Duration: 45 min.		N1 N2		29.5% of mentation reports Among those, 23.6% contained imagery, 6% contained thoughts; 23.3% contained direct incorporations; 6.2% contained indirect incorporations 1.28% of morning dream reports 47% of mentation reports on the first post-stimulus night	65% of participants (on the first post-stimulus night)
Gackenbach et al. (2011) [70]	40 gamer students (34 m, 6 f)	*Mirror's Edge* video game Duration: 26 min.			✓	N/S	N/S
Davidson & Lynch (2012) [71]	75 students (10 m, 65 f)	9/11 media coverage video Introductory psychology lecture video			✓	N/S	N/S
Solomonova et al. (2015) [72]	26 healthy volunteers (10 m, 16 f)	VR maze task Duration: 23 min.			✓	11% of dream reports (on the first post-stimulus day)	61.5% of participants

Table 1. *Cont.*

Study	Population	Stimulus	Sleep Stage			Incorporation Rate (% Content-Filled Dream Reports) [a]	Incorporation Rate (% Participants) [b]
			REM	NREM	N/A		
Flockhart & Gackenbach (2017) [73]	76 male students (high-end gamers and low-end gamers)	Fearful film sequence from *Misery* *Far Cry 3*, a combat video game *Minecraft*, a creative video game Scholarly search task on the computer Duration: 10 min.			✓	N/S	N/S
Gott et al. (2021) [74]	39 students (10 m, 29 f)	VR-assisted training of lucid dreaming, with dream-like video games Duration: 9 h in total over 12 sessions			✓	N/S	N/S
Ribeiro et al. (2021) [75]	57 students (13 m, 44 f)	VR spatial memory task			✓	35.3% of dream reports	22.2% of participants
Laboratory and Home-Based Studies							
Picard-Deland et al. (2020) [76]	137 volunteers, mostly students (52 m, 84 f)	VR flying task Duration: 15 min.	✓	N2 N3		7.1% of dream reports (flying dreams) collected in the laboratory 3.1% of REM reports 20% of NREM reports 10.6% of dream reports (flying dreams) on the first post-laboratory night 4.1% of dream reports (flying dreams) across all post-laboratory nights	4.4% of participants (at the laboratory) 22.1% of participants (across all post-laboratory nights)

Note. Laboratory-based studies involved within-laboratory dream collection via scheduled awakenings. Home-based studies involved dream collection at home. Laboratory and home-based studies involved both within-laboratory and at-home dream collection. N/S = Not specified. [a] The incorporation rate (% of dream reports) was determined by dividing the number of dream reports containing stimulus-related content by the total number of content-filled dream reports. Content-filled dream reports were defined as those arising from awakenings that resulted in dream recall and contained at least some content. The incorporation rate for specific sleep stages was computed by dividing the number of stimulus-related dream reports from a specific sleep stage by the total number of content-filled reports obtained from that same stage. [b] The incorporation rate (% participants) was calculated by dividing the number of participants who incorporated the stimulus in at least one dream report by the total number of participants who underwent the appropriate experimental treatment.

3. Results

Database searching yielded a total of 4352 records. After the removal of duplicates, 3794 records were screened based on their titles and abstracts. Among these, 125 records were retained, while an additional 27 records were identified through manual searching (i.e., backward and forward citation tracking), leading to a total of 152 records that underwent full-text screening. The reasons for exclusion are provided in Figure 1. The final selection consisted of 29 studies fulfilling the inclusion criteria.

3.1. Study and Sample Characteristics

The studies were published between 1964 and 2021. The majority of the studies were conducted in the United States (n = 12) [28,45,46,52–56,60,61,63,66] and Canada (n = 9) [57,59,64,65,68,70,72,73,76], while the rest were conducted in Germany (n – 3) [58,62,67], Australia (n = 2) [69,71], Belgium (n = 1) [47], The Netherlands (n = 1) [74], and France (n = 1) [75]. The sample sizes ranged between 3 and 137 participants, totaling 1193 participants across all the studies (51.2% male, 48.7% female). Gender information was missing in one study [45]. The age range spanned from 6 to 62 years, with a weighted average age of 21 years computed from the mean age data provided by seventeen studies. Four studies did not disclose any age-related information, while eight studies only reported age ranges. A significant portion of studies recruited university students (18/29; 62%), while only two studies investigated pediatric populations [53,55]. Nearly all the studies focused on healthy populations, with only one notable exception that recruited patients with medial temporal lobe damage in addition to healthy participants [45]. A few studies included distinctive populations, including night shift workers, *Tetris* experts, frequent video game players, and meditators [45,52,56,65,70,73]. Several studies lacked information regarding screening criteria. Table 1 outlines all the included publications along with their sample characteristics.

3.2. Theoretical Frameworks

The reviewed studies adopted a broad range of theoretical frameworks. Certain studies primarily sought to assess the effects of stimuli on the content of dreams, whereas others made use of this experimental paradigm to address other research questions while tangentially observing such effects. The first wave of studies addressing the effects of various stimuli on dream content, conducted in the 1960s and 1970s, broadly aimed to explore how experimentally controlled pre-sleep stimuli could be utilized to influence the dream formation process [28,53]. Some also aimed to investigate how emotionally charged and stressful experiences during wakefulness were integrated and transformed within subsequent dreams [52,54,56], while one study specifically endeavored to examine the effects of violent media on children's dreams [55].

Several ensuing studies focused on the potential adaptive functions of dreaming. For instance, De Koninck and Koulack [57], as well as Lauer and colleagues [58], examined two contrasting hypotheses regarding the functions of dreams: the mastery hypothesis, which suggests that dreaming offers a means of gaining mastery over stressors, and the compensation hypothesis, which argues that dreaming plays a compensatory role by introducing content that is lacking from our waking life. In a similar way, Davidson and colleagues [69,71] set out to test Hartmann's connectionist theory [77], which postulates that dreaming helps us process emotional concerns by forming new connections. One study in particular drew on Revonsuo's threat simulation theory of dreaming [78] to examine how video game play may protect against nightmares [73]. Many contemporary studies (12/29; 41%) aimed to test the idea that dreaming contributes to sleep-dependent memory consolidation, usually by examining whether dreaming about a pre-sleep learning task is associated with improved post-sleep memory performance for that task (see [79] for a meta-analysis).

Finally, other studies examined specific features of dream formation without necessarily investigating the functions of dreaming. For instance, two studies focused on the temporal patterns of incorporation of memory elements into dreams [68,72], one evaluated how expressing feelings about pre-sleep events influenced their incorporation into dreams [59], another examined how video game immersion affected incorporation [70], while two others used virtual reality tasks to induce lucid dreaming and flying dreams [74,76].

3.3. Methodological Characteristics

3.3.1. Types of Pre-Sleep Stimuli

Just under half of the studies (13/29; 45%) solely employed filmic stimuli, often comparing stressful films with neutral ones. One study among these described using a soundless film clip but did not provide a rationale for omitting the sound [67]. The use of film sequences was most common in studies from the 1960s to the 1990s, with only one recent study exclusively employing a film sequence [67].

Another group of studies utilized either video games, virtual tasks, or virtual reality (VR) tasks as stimuli (16/29; 55%). Two studies employed the classic puzzle video game *Tetris* [45,47], and four utilized virtual maze tasks in which the participants navigated a three-dimensional environment to solve a maze [60,61,63,66]. Another three studies employed visuomotor video games requiring the participants to engage in whole-body movements, with two of those choosing the *Wii Fit* video game [62,65] and another opting for an alpine skiing video game [46]. One study selected a first-person action–adventure video game on a *PlayStation 3* console, implementing four conditions varying in fidelity (i.e., high immersion versus low immersion) and interactivity (playing versus watching only) [70]. Four studies used VR tasks, including a VR maze task [72], a VR spatial memory task [75], a VR flying task [76], and VR video games developed for lucid dreaming training [74]. Solomonova and colleagues' VR maze task resembled the virtual maze tasks mentioned above, but it incorporated VR goggles, thereby making it even more immersive [72]. Lastly, two studies employed more than one type of task. In one of these studies, a virtual maze task was compared with a visuomotor tennis video game on the Wii console [64], while the other compared three conditions: a first-person shooter combat video game, a non-combat creative video game, and a non-video academic searching task on the computer, with all conditions including exposure to a fearful movie sequence [73]. Table 1 offers a summary of the types of stimuli employed in each study included in the present review.

3.3.2. Stimulus Exposure

In twenty-seven of the twenty-nine studies (93%), the participants were exposed to stimuli in a laboratory setting. In two studies, the participants engaged with the stimuli at home, where they viewed the videos on a DVD [69,71]. Usually, the participants were exposed to stimuli for durations ranging from 5 to 90 min over the course of one to three sessions. However, there were three exceptions where the participants experienced significantly longer periods of exposure. In two studies involving *Tetris*, the participants played the game for 6 to 7 h over the course of three consecutive days [45,47]. In a third study, the participants underwent VR lucid dreaming training for a total of 9 h, spread across twelve sessions over four weeks [74]. Three studies did not specify the duration of stimulus exposure [52,55,75].

3.3.3. Sleep Context

In two-thirds of the studies (19/29; 66%), the participants slept in a laboratory environment. Of these nineteen studies, thirteen involved overnight sleep periods, four involved daytime naps [47,60,64,65], and two involved full daytime sleep sessions in male night workers [52,56]. Conversely, the participants slept in the comfort of their own homes in nine of the twenty-nine studies (31%). Notably, Picard-Deland and colleagues' study featured both laboratory daytime naps and home sleep [76]. Table 1 categorizes studies according to three sleep settings, namely laboratory-based studies, home-based studies, and concomitant lab- and home-based studies.

3.3.4. Dream Report Collection

Almost all of the laboratory-based studies used standard electroencephalography (EEG) to monitor sleep architecture and collect dream reports following scheduled awakenings. An exception was noted in the study by Stickgold and colleagues, where the Nightcap, a portable sleep monitoring system, was employed in a laboratory setting [45]. The home-based studies required the participants to maintain dream logs at home over a specified period of time, typically ranging from 1 to 14 nights. Exceptionally, one study required the participants to keep a home dream diary over a period of 6 weeks [74]. Only one study conducted in a home setting relied on a sleep monitoring device, Nightcap, to induce awakenings during specific sleep stages [46]. Finally, Picard-Deland and colleagues supplemented laboratory-collected dream reports following scheduled awakenings with home dream logs [76].

3.3.5. Sleep Stages

Out of the twenty-one studies that monitored sleep stages and induced experimental awakenings, nine focused exclusively on REM sleep, five centered solely on NREM sleep (three targeted N1 and N2, and two targeted N1 alone), and seven examined both REM and NREM sleep (see Table 1 for details on the sleep stages focused upon in each of the studies).

3.3.6. Dream Content Analysis

A majority of the studies (23/29; 79%) assessed the incorporation of stimuli into subsequent dreams. However, seven of these twenty-three studies either did not report global incorporation rates or failed to provide sufficient data to determine them [54,57,59,64,67,70,71]. Table 1 presents incorporations rates (expressed in percentages) for the remaining sixteen studies which either explicitly reported this information or included sufficient data to allow for their calculation. Across the studies, incorporation was typically scored by two to three blind judges. However, two of the studies did not specify whether their judges were blind [59,62], while four of the studies did not disclose any information regarding the involvement of judges [45,54,60,64].

Stimulus incorporation was often measured in a categorical fashion by classifying dreams based on the presence or absence of stimuli-related elements within the dream reports. Some studies categorized incorporations into subtypes. For instance, seven studies distinguished between direct or indirect incorporations [46,47,61,63,65,66,72], two studies differentiated between "certain" and "uncertain" incorporations [28,53], one differentiated between "sure" and "probable" incorporations [58], another between "literal", "closely associated", and "distantly associated" incorporations [71], and yet another between "primary" and "secondary" incorporated elements [70]. Three studies further classified incorporations according to the modality of their representation in the dream reports, distinguishing between thoughts, imagery, and visual, kinesthetic, and auditory sensory classes [45–47]. Two studies notably compared external ratings with self-reported ratings of incorporations [62,70].

In contrast, other researchers opted for continuous measures of the degree of incorporation into dream content. For instance, Kuiken and colleagues computed the ratios of the number of incorporations relative to the number of opportunities for incorporation [59], Powell and colleagues used a 10-point incorporation likelihood scale [68], while Gackenbach and colleagues developed their own analysis tool to detect video game elements within dreams [70]. For their part, Fogel and colleagues computed semantic similarity scores between wake task reports and subsequent dream reports [64]. Finally, Klepel and Schredl designed a film–dream similarity scale [67], while Ribeiro and colleagues employed a word matching strategy to derive an incorporation probability score [75].

In addition, a range of scoring scales and systems were used across thirteen of the twenty-nine studies (45%) to analyze diverse aspects of dream content aside from stimulus incorporation. These included scales devised by the research teams themselves (n = 5) [53,55,57,58,65], scales adapted from the Thematic Apperception Test (n = 1) [28], the Hall and Van de Castle [80] coding system (n = 3) [54,57,70], Gottschalk, Gleser, and Springer's [81] anxiety and hostility scales (n = 2) [55,56], Hartmann's [82] central imagery scoring system (n = 2) [69,71], Schredl's [83] scoring scales (n = 1) [62], and Revonsuo and Valli's [84] dream threat scale (n = 1) [73]. The rating process usually involved two to four blind judges, although two studies did not specify the blindness of their raters [62,73], two relied on a single judge [56,70], while one study did not provide information about the use and blinding of the judges [54].

Beyond relying on incorporation ratings by external judges, over a third of the studies (11/29; 38%) asked the participants to provide self-ratings of their dreams [28,55,58–60,62, 70,71,73,74,76]. Finally, one study did not quantify dream content or apply any statistical treatment, instead adopting a qualitative psychoanalytic method to examine the presumed symbolic transformation of filmic stimuli in dream reports [52].

3.4. Narrative Overview of Outcomes

3.4.1. Laboratory-Based Studies

In the 1960s and 1970s, a series of studies investigated the effects of stressful, arousing, and neutral films on dream content. The first study of this kind, by Foulkes and Rechtschaffen, found that when compared to the effects of a comedic Western film, a violent Western film produced REM dream reports that were longer and more imaginative, as rated by external judges, as well as more vivid and emotional, as rated by the participants themselves [28]. The violent film, however, did not result in dreams that were more unpleasant or violent in comparison to the nonviolent film. The authors reported a net incorporation rate of 5%, with clear film elements appearing in 5.5% of the REM reports and 3.8% of the NREM reports. Contrastingly, another study by Foulkes and colleagues with an incorporation rate of 8% revealed that a baseball film elicited more imaginative, hostile, guilty, and well-recalled dream reports in boys than did a violent Western film [53]. While adult participants preferred the comedic film in the former study, the boys in the Foulkes et al. study showed greater interest in the violent film [53]. However, a subsequent study of boys by Foulkes and colleagues found no significant relationship between hostility or other dream variables and the type of film viewed, whether violent or nonviolent [55]. They observed that increased attentional involvement with the films led to greater dream hostility, but they refrained from drawing conclusions regarding the link between interest or involvement levels and dream intensity or hostility.

A study by Witkin and Lewis, which presented qualitative results based on data from only three participants, observed that films about birth and subincision led to dreams "filled with obvious sexual symbolism", while neutral films did not [52]. The researchers also found a higher frequency of "forgotten dreams" (i.e., awakenings with no dream recall) following arousing films compared to neutral ones. In a study by Cartwright and colleagues, erotic films resulted in more one-character dreams and less heterosexuality, with an ensuing rise in two-character dreams on the last night of five consecutive experimental nights [54]. While no direct film incorporation was observed, the authors reported frequent symbolic representations of sexual material. In another study, birth and subincision films provoked increased dream anxiety and decreased social affection compared to neutral films, particularly among the participants who reported more waking anxiety after viewing the films [56]. Finally, De Koninck and Koulack found no significant effects of stressful films depicting workshop accidents on dream content [57].

A subsequent laboratory-based study found that stressful films depicting a cruel massacre and the brutality of the prison system elicited more anxious and aggressive dreams during the first REM period compared to a neutral film [58]. Self-rated dreamer participation was higher following exposure to the stressful films than the neutral film. Overall, 37.5% of the dream reports contained clear incorporations of stressful films, with 33% of the initial REM reports and 43% of the final REM reports incorporating film elements. Close to 55% of the participants integrated film elements into their dreams.

In another study, emotionally engaging films addressing themes of mortality and mental health were shown to participants in two conditions: the "feeling expression" condition, in which the participants were prompted to reflect on the film segment most significant to them, and the "no feeling expression" condition, in which the participants were not provided with the same opportunity to express their feelings [59]. Dreams stemming from the feeling expression condition were found to be more affectively similar to the films but less likely to incorporate film narrative elements, particularly actions, compared to the no feeling expression condition. Dreams from the feeling expression condition also exhibited more self-reference than those from the no feeling expression condition.

More recent studies in the 2000s and 2010s shifted towards the use of video games and virtual tasks as stimuli. Stickgold and colleagues reported that 7% of sleep-onset mentation reports produced by amnesic and normal participants alike incorporated explicit *Tetris* elements that closely mirrored the visual imagery encountered during awake gameplay, with 63% of all the participants experiencing *Tetris*-related hypnagogic imagery [45]. Mentation reports containing *Tetris*-related thoughts were absent in amnesiacs who only reported *Tetris* imagery, while non-amnesic *Tetris* novices reported both. Interestingly, *Tetris* imagery in non-amnesiacs increased across nights, with 90% of imagery reports occurring on the second night, whereas *Tetris* thoughts appeared more frequently on the first night. The incidence of *Tetris* imagery dropped by two-fold over a two-minute period at sleep-onset, while *Tetris* thoughts remained constant during this timeframe. In Kussé and colleagues' study, a remarkable 81% of the participants incorporated *Tetris* elements into sleep mentation [47]. *Tetris* content was incorporated into 10% of the mentation reports, consistently across three days of testing. Among these incorporations, one-third were direct incorporations and two-thirds were indirect. While 11.2% of N1 reports and 6.5% of N2 reports in this study contained *Tetris* elements, this difference did not reach statistical significance. However, there was a significant variation in the occurrence of *Tetris*-related reports across sensory classes, with visual imagery being the most prevalent.

Wamsley and colleagues reported that 8% of the participants produced sleep mentation unambiguously related to a virtual maze task, with 6% of the participants experiencing them during N1 and 2% during N2 [60]. In a subset of participants instructed to report at the end of an uninterrupted sleep opportunity whether they had experienced maze-related sleep mentation, 55% answered affirmatively. In another study, 37% of the participants incorporated content related to a virtual maze task [61]. About 9% of these reports were maze-related, with almost 4% directly associated with the task and 5% solely indirectly associated. The occurrence of maze-related reports was consistent across N1, N2, REM, and morning awakenings. Interestingly, introducing a monetary reward and performance feedback to the task did not affect the incorporation levels in the dream reports. In another study by Wamsley and colleagues, the incorporation rate of a virtual maze task proved to be much lower—1.9% of all reports—with 12% of the participants incorporating maze-related content [63]. A test expectation manipulation (i.e., whether the participants were informed about an upcoming memory test) did not influence incorporation into dream reports. Wamsley and Stickgold observed that 6.4% of mentation reports directly incorporated a virtual maze task across all sleep stages [66]. Specifically, 7.1% of N1 reports, 2.1% of N2 reports, 12.5% of morning N2 reports, and 12.5% of morning REM reports contained maze-related elements. A total of 47% of the participants directly incorporated the maze task.

In a study by Fogel and colleagues that utilized both a virtual maze task and a visuomotor tennis video game for the Wii console, no significant difference was found in the degree of incorporation between the two stimuli [64]. The extent of incorporation into early versus late N1 sleep mentation also did not differ, but incorporation into early N1 mentation was positively associated with reasoning ability, while incorporation into late mentation was positively correlated with verbal ability. In another study involving a visuomotor video game for the Wii console and Wii Balance Board, researchers compared external ratings by judges with participants' self-ratings of incorporation into REM dreams [62]. The external ratings indicated that 5.6% of the dream reports featured balance-related elements, with

15% of the participants judged as having incorporated balancing elements. Additionally, 47.2% of the dream reports were judged to reference the laboratory setting, and 85% of the participants were judged as having incorporated laboratory elements. The participants' self-assessments revealed that almost 20% of the dream reports contained balance-related elements, with 54% of the participants reporting at least one balancing dream.

Another study featuring a visuomotor video game for the Wii console and Wii Balance Board showed that game elements were incorporated into 25% of REM reports, 18.8% of N2 reports, and 17.5% N1 reports, totaling 19.1% of all dream reports [65]. Conversely, laboratory references were found in 75% of REM reports, 37.5% of N2 reports, and 27.5% of N1 reports, accounting for 38.2% of all dream reports.

Finally, a study by Klepel and Schredl exposed participants to a short comedic film sequence and compared their subsequent dream reports with those of control participants who did not view the film [67]. Morning the dream reports obtained from the film-viewing group showed higher film–dream similarity compared to the control group, but no differences were found for the first REM dream reports collected at night. More film incorporations were observed in the morning dream reports than in the first REM reports.

In summary, laboratory-based studies conducted over several decades have yielded a broad range of outcomes, including mixed, and at times contradictory, results. The heterogenous nature of these outcomes is also reflected in the limited number of studies having quantified the proportion of dream reports showing evidence of stimulus-related incorporations, with percentages ranging from a low of 2% to a high of 38%.

3.4.2. Home-Based Studies

In a study by Powell and colleagues, participants were shown a stressful film depicting a ceremonial buffalo slaughter and then asked to keep a dream diary for seven days to investigate the "dream-lag effect", which refers to the delayed incorporation into dreams of waking events often experienced 6–8 days prior [68]. Nearly half (47.4%) of the participants were classified as high incorporators. A temporal U-shaped quadratic trend for film incorporations was observed across seven nights, specifically in high incorporators. Davidson and colleagues compared the effects of viewing 9/11 media coverage and a neutral psychology lecture on subsequent dreams [69]. They found that dreams following the 9/11 video showed higher levels of contextualizing imagery compared to those following the control video. Some dream elements directly reflected aspects of the 9/11 video, while others were thematically related. A measure of subjective stress after exposure to the 9/11 footage, rather than trait empathy, was associated with the presence of contextualizing imagery in dreams. Using the same stimuli, Davidson and Lynch reported that the 9/11 video produced dreams characterized by more intense central imagery and stronger negative emotions compared to the lecture video [71]. The 9/11 video also led to dreams with more literal, closely associated, and distantly associated 9/11 elements, as well as more thematic 9/11 imagery.

In another study, participants played a visuomotor alpine skiing video game, which was featured in 29.5% of mentation reports across all nights [46]. Of these reports, 23.6% were characterized by sensory imagery, while 6% contained game-related thoughts. Visual imagery predominated over other kinds of sensory modalities, although kinesthetic imagery was present in one-third of the reports. Direct and indirect incorporations accounted for 23.3% and 6.2% of game-related reports, respectively. Directly related imagery diminished with increasing time since sleep onset, as monitored by the Nightcap, suggesting that game-related imagery became more abstracted as time into sleep increased. Overall, incorporations were more common during awakenings scheduled earlier in the night, and only a small portion (1.28%) of the morning dream reports were game-related. Across three post-stimulus nights, there was a decline in the incorporations per participant. On the first night, 47% of the mentation reports produced by 65% of the participants were related to the video game. Remarkably, a small group of observers who simply watched others play the game showed similar rates of incorporation as active players. Self-reported task engagement did not seem to influence the incorporation rate.

In a study by Gackenbach and colleagues, participants played an action–adventure video game under one of four conditions varying in fidelity (i.e., wearing immersive goggles versus no goggles) and interactivity (playing the game versus watching a recorded gaming session) levels [70]. The high fidelity–high interactivity condition yielded the highest rate of incorporations according to both the judges' ratings and the participants' self-ratings, particularly on the first, sixth, and seventh nights of the self-reports. Curiously, the low interactivity–low fidelity condition produced the second most self-reported incorporations on these nights. The high fidelity–low interactivity condition showed the lowest primary incorporations and the highest laboratory incorporations, as scored by the judges. Self-rated emotional engagement during gameplay was not associated with incorporation into dreams. In another study by Flockhart and Gackenbach, high-end gamers and low-end gamers were all exposed to a fearful movie sequence and engaged in one of three activities: a combat video game, a creative video game, and a scholarly search task on the computer [73]. The findings indicated that high-end gamers who played the combat game experienced marginally fewer and less severe threats in their dreams, as evaluated by judges, compared to low-end gamers who played the same game. The high-end gamers also self-reported fewer bad dreams and less fear in their dreams after playing the combat game relative to the other two tasks. Conversely, the low-end gamers self-reported more bad dreams after playing the combat game compared to the other conditions. The participants who played the creative video game experienced higher levels of self-reported dream bizarreness.

In one among a series of contemporary studies utilizing VR tasks, Solomonova and colleagues observed that VR maze incorporations into dreams followed a distinct temporal pattern marked by a peak on days four and five over a ten-day period, while laboratory incorporations revealed a standard U-shaped quadratic pattern characterized by a day-residue effect and a delayed dream-lag effect [72]. A total of 11% of the dream reports featured VR incorporations on the first day post-VR, contrasting with 53% of the dreams referencing the laboratory. Overall, 61.5% of the participants incorporated the VR maze task into their dreams, with 12.5% of dreams collected per participant containing VR elements. Interestingly, VR and laboratory incorporations almost exclusively occurred in separate dreams. Dreams highest in VR incorporations were associated with a relatively more internal rather than external dream locus of control.

In a study by Ribeiro and colleagues, 35.3% of dream reports originating from 22% of participants incorporated items from a VR spatial memory task [75]. The researchers determined that the likelihood of observing the incorporation of these items into dreams of control compared to those who were not exposed to this VR task ranged from 0.18% to 4.23%. In another study, the participants partook in VR-assisted lucid dreaming training involving dream-like video games, amounting to 9 h of VR exposure over the course of four weeks [74]. While VR training led to increases in lucid dreaming compared to no training, it did not fare better than classical lucid dreaming training. There were no differences in dream lucidity or VR incorporation between the nights following VR training and those without preceding VR sessions.

In summary, home-based studies have found stimuli incorporation rates that are comparable to those observed under laboratory conditions, ranging between 11% and 35%. These home-based studies have also facilitated the collection of dream content data over longer periods of time, allowing researchers to investigate the temporal relationship between exposure to stimuli and their subsequent incorporation into dreams.

3.4.3. Laboratory and Home-Based Studies

Finally, Picard-Deland and colleagues had participants complete a VR-flying task followed by a lab-based morning nap while also maintaining home dream journals for five days before and ten days after the VR exposure [76]. The authors found that the VR flying task resulted in an increase in the frequency of flying dreams, from 1.7% of the dream reports at baseline to 7.1% of the reports following VR exposure. Flying was also featured in 3.1% of the REM reports and 20% of the NREM reports collected during the laboratory nap session. A day-residue effect was observed on the first post-VR night, when the incidence of flying dreams peaked at 10.6% of the dream reports. Overall, 4.1% of the dream reports featured flying across the ten nights post-VR. In all, 4.4% of the participants experienced flying dreams in the laboratory, while 22.1% experienced them at home. Flying dreams were more likely to occur in the participants with previous flying and lucid dreaming experiences. The intensity of flying sensations within dreams was positively associated with the degree of immersion-proneness and cybersickness, but not with flying dream frequency.

4. Discussion

The goal of the present scoping review was to examine the experimental evidence pertaining to the effects of moving visual media on dream content. Overall, the results suggest that visual media exposure before sleep has the potential to alter dream content, but the degree and nature of this influence varies significantly. A synthesis of key outcomes and methodologies is presented below, and remaining questions and gaps in the literature are highlighted.

4.1. Effects of Moving Visual Media Stimuli on Dream Content

When considering the body of evidence as a whole, moving visual media stimuli have been shown to exert a moderate influence on dream content. Several studies have noted various changes in dreams following exposure to specific stimuli, including changes in dream recall, their length, emotional content (e.g., hostility, guilt, anxiety, fear), social interactions (e.g., affection, aggression, sexuality), characters, and other attributes (e.g., imaginativeness, vividness, bizarreness, threats, contextualizing imagery). However, these observed effects may not fully capture the extent of the impact of visual media stimuli on dream content, as the identification of such effects depends on each study's specific objectives and hypotheses. In most later studies, researchers were not aiming to identify changes in dream features, so these may have gone unnoticed.

The rates of stimuli incorporation into dream reports appear to be rather modest, ranging from about 2% to 38% across all studies. The proportion of direct versus indirect incorporations also varies across studies, with direct ones typically occurring earlier in the night as well as earlier into the N1 sleep stage [46,47,61]. Most incorporations manifested as visual imagery [46,47]. This variability in incorporation rates could be attributed to several factors, including the heterogeneity in research methodologies, which complicates direct comparisons and prevents the delineation of more robust conclusions. A large number of methodological details differed from one study to the next, such as the duration, timing, and context of stimulus exposure, the protocol for collecting dream reports, and the assessment of incorporation and other dream-related variables. The next section highlights some of the factors that may have contributed to the observed variations in incorporation rates.

4.2. Methodological Differences and Other Factors Contributing to Variance in Outcomes
4.2.1. Types of Stimuli

Hardly any studies have systematically compared different types of stimuli, making it challenging to determine which features of visual media may enhance the likelihood of incorporation into dreams. One notable exception is a study that cleverly manipulated the attributes of a video game by varying its levels of fidelity (i.e., immersive goggles versus no goggles) and interactivity (i.e., playing versus watching) [70]. The study demonstrated that highly immersive and interactive video gaming led to the highest rates of incorporation, as determined both by external judges' evaluations and self-ratings. Similarly, another study, which was not included in the present review, as it has not been published, manipulated a virtual maze task's level of interactivity and visual display [85]. It revealed that high interactivity (i.e., playing as opposed to merely viewing) resulted in more self-reported incorporations, while the type of visual display (VR goggles versus TV screen) did not significantly affect incorporation scores. In contrast, Wamsley and colleagues found that watching others play a visuomotor skiing video game led to similar incorporation rates (19% of observers' dream reports) as actively playing it (24% of players' dream reports), although the small sample size of observers (n = 3) warrants caution [46]. Additionally, a study by Fogel and colleagues compared a virtual maze task with a visuomotor tennis video game and observed comparable incorporation rates for both [64]. This suggests that the involvement of motor movements in gameplay, such as in the tennis game, do not necessarily enhance incorporation. However, because this study did not control or manipulate specific stimulus attributes, this conclusion remains speculative. Briefly put, drawing definitive conclusions about which types of stimuli are most prone to induce incorporations into dreams is difficult due to the lack of systematic comparisons. Levels of interactivity and immersion might play a role, but their relative importance remains unclear, given the inconsistent findings across the studies described above and the lack of a clearcut discrepancy in incorporation rates between studies employing filmic stimuli and those utilizing video games, virtual tasks, and VR.

4.2.2. Stimulus Exposure

The impact of the duration of stimulus exposure, or stimulus "dose", has not been investigated. Without controlled investigations varying exposure durations within the same experiment, it is difficult to hypothesize how duration might influence incorporation. Comparing incorporation rates across studies with extended periods of exposure to those with brief exposure reveals no straightforward relationship between the duration of exposure and the likelihood of incorporation. For example, playing *Tetris* for 6 to 7 h over three days does not appear to facilitate incorporation compared to engaging in virtual maze tasks or other video games for as little as seven minutes [45,47,65]. This aligns with previous research indicating that the amount of time spent on waking-life activities may not be the most reliable predictor of their incorporation into dreams [86].

4.2.3. Perceived Salience and Engagement with Stimuli

The perceived salience of the stimuli and the level of involvement or engagement with it is another factor that might modulate its observed effects on dreams. Research indicates that emotionally charged and personally significant waking-life experiences are more likely to be integrated into dreams [87,88]. However, the studies included in the current review do not provide clear evidence in that regard. In two separate studies using similar filmic stimuli, opposite effects on dream content were observed in terms of the dreams' imaginativeness and emotionality [28,53]. Foulkes and colleagues suggested that this divergence could be attributed to the differing levels of interest in one film stimuli over the other: adult participants in the former study favored a romantic comedy compared to a violent Western, whereas children in the latter study preferred the violent Western over a baseball film [53]. These complementary studies suggest that one's sense of interest or investment in the stimuli may modulate their effects on dream content.

Another study by Foulkes and colleagues manipulated levels of attention allocated to films and observed that greater attention involvement resulted in more dream hostility [55]. Other studies that manipulated variables such as monetary reward, performance feedback, and test expectations in virtual maze tasks did not find that these factors moderated incorporation into dreams [61,63]. Although these variables might be expected to increase task salience and involvement, consequently enhancing incorporation, other factors such as high perceived task difficulty and negative emotional valence might have been at play [61]. Moreover, in some studies, self-rated task engagement and emotional engagement during gameplay were surprisingly not associated with incorporation rates [46,70].

While subjective salience and engagement might play a role in dream incorporation, their impact is not straightforward and may interact with other factors. Novelty is another factor that may impact the extent to which stimuli are incorporated into dreams, as some evidence suggests that novel and highly emotional experiences are preferentially incorporated into dreams compared to recurrent daily activities [87]. Despite this, no study has tried comparing the effects of familiar versus unfamiliar stimuli to explore the impact of novelty.

4.2.4. Experimental Setting

A few studies conducted partially or fully in a laboratory have shown that laboratory-related elements are incorporated into dreams at substantially higher rates than intended stimuli, with incorporation rates of 53% versus 11% [72], 47.2% versus 5.6% [62], and 38.2% versus 19.1% [65]. These findings are consistent with other research indicating that over a third of dream reports collected at a laboratory reference the experimental setting [89,90]. Although the laboratory setting offers the advantage of closely monitoring sleep and collecting reports throughout the night, it may interfere with the incorporation of the intended stimuli [90]. In support of this, one study has shown that laboratory and stimulus incorporations rarely, if ever, appeared within the same dreams [72]. This interference likely arises because the experience of staying in the laboratory while undergoing polysomnography can be more salient, novel, and stressful to the participants than the stimuli they are exposed to, leading to a preferential incorporation of laboratory elements. This phenomenon may be more pronounced during the REM stage, which is most strongly associated with laboratory incorporations [90]. To mitigate this issue, experiments could include an adaptation night, allowing the participants to familiarize themselves with the environment before being exposed to the stimulus.

4.2.5. Sleep Stages

The influence of sleep stages on the likelihood of incorporation into dreams appears to vary across different studies. Two studies have reported consistent rates of incorporation between N1 and N2 sleep, as well as across N1, N2, REM, and morning awakenings [47,61]. Another study has shown significantly greater incorporation in N2 compared to REM sleep [76]. Three additional studies, while monitoring the frequency of incorporations across different sleep stages, did not perform statistical comparisons to confirm whether incorporation rates differed meaningfully between stages. In one study, 6% of participants experienced incorporations during N1 compared to 2% during N2 [60]. In another study, 7.1% of N1 reports, 2.1% of N2 reports, 12.5% of morning N2 reports, and 12.5% of morning REM reports produced incorporations [66]. Yet another study showed incorporations in 25% of REM reports, 17.5% of N1 reports, and 18.8% of N2 reports [65]. These findings suggest that incorporations can occur during all stages of sleep, but it is not possible to determine which stage tends to outperform the others and under what circumstances.

4.2.6. Time-of-Night

While the results from two studies seem to indicate that film incorporations are more common in final REM and morning dream reports compared to those collected during the first REM stage of the night [58,67], findings from another study contradict this trend [46]. The latter found that incorporations occurred more frequently for awakenings scheduled earlier in the night, with morning report incorporations proving to be less common. The rate of incorporation into N1 reports early in the night does not appear to differ from that of N1 reports later in the night, as reported in one study [64]. One study further observed that the occurrence of incorporated imagery seemed to decrease across two minutes into N1, while the incidence of incorporated thoughts remained constant [45]. Another study using the same stimulus found no changes in the incorporation rate over the same two-minute period since N1 onset, although it did not differentiate between imagery and thoughts [47]. Direct incorporations were shown to decrease as time since sleep onset increased in another study, suggesting a process of abstraction during the first minutes of sleep as well as through the night [46].

4.2.7. Time Interval between Stimulus Exposure and Incorporation

A few studies have examined the temporal pattern of incorporation across post-stimulus nights, revealing varying findings. Some studies have identified a marked "day-residue effect", wherein the incidence of incorporations peaked on the first night following exposure to the stimulus [46,68,76]. For instance, Wamsley and colleagues observed incorporations in a remarkable 47% of mentation reports on the first night, followed by a linear decline over the next two nights [46]. This is in line with other research indicating that daytime material tends to appear more frequently in dreams the following night [5].

A study by Powell and colleagues identified both a day-residue effect and a dream-lag effect, where a second peak in incorporations occurred on the sixth and seventh nights, demonstrating a bimodal U-shaped quadratic pattern across seven nights in total [68]. In contrast, another study observed no such dream-lag effect, despite showing a day-residue effect [76]. Curiously, Solomonova and colleagues observed a completely different temporal pattern, with incorporations peaking on days four and five over a ten-day period, speculating that the experience of staying in a laboratory may have competed with the VR experience [72]. Another study noticed a divergent distribution of incorporations for imagery versus thoughts: imagery peaked on the second night, while thoughts peaked on the first night [45]. However, these findings may be confounded by the fact that the participants were exposed to the stimulus across all three experimental nights.

4.2.8. Dream Content Analysis and Scoring of Incorporations

Different scoring methods were used across studies or the assessment of dream-related variables and incorporations. Regarding content analysis in particular, some studies employed established analysis systems like the Hall and Van de Castle [80] scoring scales, while others relied on their own scoring systems. Varying criteria were also used to detect incorporations across studies, and this inconsistency could have impacted the proportion of reported incorporations. The variability in scoring standards, including the levels of criteria stringency, might have contributed to overestimates or underestimates of incorporation. A few studies also did not provide enough detail on what constitutes an "incorporation", making it harder to confidently draw comparisons between studies. In some instances, it was also unclear whether evaluators were effectively blinded to the experimental conditions. Overall, studies could benefit from better standardization of scoring methods.

4.2.9. External Ratings vs. Self-Ratings

Based on the limited number of studies that employed both external judges' ratings and participants' self-ratings of stimuli incorporations in dreams, it appears that self-assessments tend to estimate higher incorporation rates than those provided by judges. For example, Nefjodov and colleagues found that over half of their participants rated 20% of their dream reports as containing incorporations, whereas external ratings identified only a 5.6% incorporation rate [62]. Similarly, in another study, the judges determined that 8% of the participants produced reports with incorporations, while a much larger 55% of the participants self-reported such incorporations [60]. This disparity could be attributed to external raters potentially underestimating the quantity of incorporations due to a more restrictive scoring method. Conversely, participants' ratings might be inflated due to demand characteristics. Each assessment method has its potential biases and limitations, so studies may benefit from using both concurrently [91].

4.2.10. Trait and State Differences

Trait and state correlates that may be associated with incorporation amplitude remain largely unexplored. Nonetheless, a few findings are worth mentioning. Research by Fogel and colleagues indicates that different cognitive abilities may facilitate incorporation into dreams from the N1 stage [64]. Another study suggests that participants' subjective stress response to a stimulus, rather than their empathetic disposition, may play a role in the way stimuli are reflected in dream contents [69]. One study involving high-end and low-end gamers suggests that the level of experience with a stimulus can mediate its impact on dream contents [73]. Another study found that prior dreaming experiences can increase the likelihood of incorporation of specific stimuli; participants with previous experiences of flying and lucid dreaming were more likely to report flying dreams [76]. Immersion-proneness predicted the intensity of flying sensations, showing that susceptibility to immersive media experiences can augment dream experiences [76]. This research highlights that factors such as cognitive abilities, emotional responses, and past experiences can potentially play a role in the incorporation of stimuli into dreams.

4.3. Considerations for Future Research

This section provides recommendations for future research directions in light of key methodological limitations in the field. First, the nature of stimuli, including their characteristics, represents an area of research that warrants clarification and further investigation. While existing studies have used various kinds of stimuli, their salience to participants may have been insufficient, and their impact varied considerably from each participant to the next. Future research should consider the roles of stimuli salience, engagement, and novelty as modulating factors that could potentially influence the occurrence of incorporations into dreams. For instance, no studies to date have allowed participants to select their own media stimuli, especially those personally relevant to them, in order to better investigate the impact of such subjective levels of stimuli salience on dream content and the extent of incorporation.

Further investigations are also needed to clarify the role of media attributes, including degree of immersion and interactivity. While some studies have demonstrated the importance of such attributes, it remains unclear under what circumstances they are most likely to exert their influence.

Future studies should also strive to better reflect current media use habits in their stimuli selection. Many experiments have used stimuli such as virtual maze tasks or disturbing film sequences, which do not represent activities that individuals typically engage in before sleep. To our knowledge, no studies have examined the effects of newer forms of media, such as social media platforms, which are expressly designed to retain user engagement and are used daily by many, particularly before bedtime. Aligning stimuli with up-to-date real-world media usage would ensure greater ecological validity of key experiments, especially those carried out in home settings.

Increasing the standardization of methods for assessing dream content and degree of incorporation is important for facilitating comparisons across studies. Researchers should continue exploring computational linguistics approaches [63] while comparing them with manual scoring methods to determine whether the former can better promote comparability and reproducibility across studies. Employing both external ratings and self-ratings as complementary methods may mitigate inherent biases and limitations in either approach. Moreover, extending the period of dream report collection to at least 14 days post-stimulus would be desirable, as this would allow researchers to capture potential incorporations that may arise later in time (i.e., the dream-lag effect) and to determine temporal patterns of incorporation.

There is also a greater need for laboratory-based studies to better document which sleep stages, and during what sleep cycles or periods of the night, are most likely to show evidence of dream incorporations of pre-sleep media stimuli. Lab studies should also aim to report incorporation rates separately for different sleep stages, as opposed to pooling them together, be it during regular overnight sleep assessments or during daytime naps. Similarly, home-based studies should, if possible, include portable sleep-monitoring devices to better track sleep stage awakenings as well as to carry out experimental awakenings in participants' natural home environments.

Finally, future research should examine how trait-dependent interindividual differences modulate the relationship between visual media stimuli and dream content. Factors such as immersion and absorption proneness in mediated environments may well affect how media experiences are incorporated into dreams, yet our understanding of their contribution remains limited.

5. Conclusions

In conclusion, despite mixed results, the overall evidence compiled throughout this scoping review indicates that engaging with moving visual media (e.g., films, video games, virtual tasks) has moderate effects on dream content, including on the incorporation of media-related elements into dreams. The substantial variability in outcomes across studies highlights the need for further research to clarify the nature and magnitude of these effects and to better identify factors contributing to the observed variations. Notably, there exists a need for more targeted investigations in the field, as most of the reviewed studies were not explicitly designed to examine the impact of pre-sleep media exposure on dream content, but rather reported on these kinds of observations while perusing other research questions. Given the substantial role that various kinds of media now play in people's daily lives, including the growing engagement with virtual environments, it is important to better understand how and why media comes to influence dream content. Moreover, such research efforts could lead to advances in media-assisted engineering of desired dream experiences and even the management of disordered dreaming [1]. Finally, such experimental protocols could provide valuable insights into the mechanisms through which various kinds of waking-life experiences are integrated into people's dreams, thereby contributing to our understanding of the dream formation process and, ultimately, the possible function of dreams.

Author Contributions: A.D.—conceptualization, methodology, data extraction, analysis, drafting, and editing. S.H.—conceptualization, reviewing, and editing. A.Z.—conceptualization, drafting, reviewing, and editing. All authors have read and agreed to the published version of the manuscript.

Funding: Research supported by grant (# 435-2023-0634) to A.Z. and S.H. from the Social Sciences and Humanities Research Council of Canada.

Institutional Review Board Statement: Not applicable.

Informed Consent Statement: Not applicable.

Data Availability Statement: Not applicable.

Conflicts of Interest: The authors have no conflicts of interest to declare.

Appendix A

Search strategy updated on 10 July 2023.

Appendix A.1. PubMed

("Motion Pictures" [MeSH Terms] OR "Television" [MeSH Terms] OR "Virtual Reality" [MeSH Terms] OR "Augmented Reality" [MeSH Terms] OR "film*" [Text Word] OR "movie*" [Text Word] OR "cinema*" [Text Word] OR "video*" [Text Word] OR "motion picture*" [Text Word] OR "moving picture*" [Text Word] OR "documentar*" [Text Word] OR "Television" [Text Word] OR "TV" [Text Word] OR "Virtual Reality" [Text Word] OR "VR" [Text Word] OR "Augmented Reality" [Text Word]) AND ("Dreams" [MeSH Terms] OR "dream*" [Text Word] OR "sleep mentation" [Text Word]).
n = 398 results.

Appendix A.2. APA PsycNet (PsycInfo and PsycArticles)

((IndexTermsFilt: ("Films") OR IndexTermsFilt: ("Television") OR IndexTermsFilt: ("Television Viewing") OR IndexTermsFilt: ("Virtual Reality") OR IndexTermsFilt: ("Augmented Reality")) OR (Any Field: (film*)) OR (Any Field: (movie*)) OR (Any Field: (cinema*)) OR (Any Field: (video*)) OR (Any Field: ("motion picture*")) OR (Any Field: ("moving picture*")) OR (Any Field: (documentar*)) OR (Any Field: (television)) OR

(Any Field: (TV)) OR (Any Field: ("virtual reality")) OR (Any Field: (VR)) OR (Any Field: ("augmented reality"))) AND ((IndexTermsFilt: ("Dreaming") OR IndexTermsFilt: ("Dream Content") OR IndexTermsFilt: ("Dream Recall") OR IndexTermsFilt: ("REM Dreams")) OR (Any Field: (dream*)) OR (Any Field: ("sleep mentation"))).
n = 1123 results (*PsycInfo* = 1064; *PsycArticles* = 57).

Appendix A.3. Web of Science Core Collection

ALL = (film* OR movie* OR cinema* OR video* OR "motion picture*" OR "moving picture*" OR documentar* OR television OR TV OR "virtual reality" OR VR OR "augmented reality") AND TS = (dream* OR "sleep mentation").
n = 2831 results.

References

1. Carr, M.; Haar, A.; Amores, J.; Lopes, P.; Bernal, G.; Vega, T.; Rosello, O.; Jain, A.; Maes, P. Dream Engineering: Simulating Worlds through Sensory Stimulation. *Conscious. Cogn. Int. J.* **2020**, *83*, 102955. [CrossRef] [PubMed]
2. Domhoff, G.W. *The Neurocognitive Theory of Dreaming: The Where, How, When, What, and Why of Dreams*; The MIT Press: Cambridge, MA, USA, 2022; ISBN 978-0-262-54421-4.
3. Picard-Deland, C.; Bernardi, G.; Genzel, L.; Dresler, M.; Schoch, S.F. Memory Reactivations during Sleep: A Neural Basis of Dream Experiences? *Trends Cogn. Sci.* **2023**, *27*, 568–582. [CrossRef] [PubMed]
4. Arkin, A.M.; Antrobus, J.S. The Effects of External Stimuli Applied Prior to and during Sleep on Sleep Experience. In *The Mind in Sleep: Psychology and Psychophysiology*; Ellman, S.J., Antrobus, J.S., Eds.; John Wiley & Sons: Hoboken, NJ, USA, 1991; pp. 265–307.
5. Nielsen, T.A.; Powell, R.A. The Day-Residue and Dream-Lag Effects: A Literature Review and Limited Replication of Two Temporal Effects in Dream Formation. *Dreaming* **1992**, *2*, 67–77. [CrossRef]
6. Vallat, R.; Chatard, B.; Blagrove, M.; Ruby, P. Characteristics of the Memory Sources of Dreams: A New Version of the Content-Matching Paradigm to Take Mundane and Remote Memories into Account. *PLoS ONE* **2017**, *12*, e0185262. [CrossRef] [PubMed]
7. Solomonova, E.; Carr, M. Incorporation of External Stimuli into Dream Content. In *Dreams: Understanding Biology, Psychology, and Culture*; Valli, K., Hoss, R., Eds.; Greenwood Publishing Group: Westport, CT, USA, 2019; pp. 213–218.
8. Salvesen, L.; Capriglia, E.; Dresler, M.; Bernardi, G. Influencing Dreams through Sensory Stimulation: A Systematic Review. *Sleep Med. Rev.* **2024**, *74*, 101908. [CrossRef]
9. Dement, W.; Wolpert, E.A. The Relation of Eye Movements, Body Motility, and External Stimuli to Dream Content. *J. Exp. Psychol.* **1958**, *55*, 543–553. [CrossRef] [PubMed]
10. Nielsen, T.A. Changes in the Kinesthetic Content of Dreams Following Somatosensory Stimulation of Leg Muscles during REM Sleep. *Dreaming* **1993**, *3*, 99–113. [CrossRef]
11. Baldridge, B.J. Physical Concomitants of Dreaming and the Effect of Stimulation on Dreams. *Ohio State Med. J.* **1966**, *62*, 1273–1275. [PubMed]
12. Koulack, D. Effects of Somatosensory Stimulation on Dream Content. *Arch. Gen. Psychiatry* **1969**, *20*, 718. [CrossRef]
13. Schredl, M.; Atanasova, D.; Hörmann, K.; Maurer, J.T.; Hummel, T.; Stuck, B.A. Information Processing during Sleep: The Effect of Olfactory Stimuli on Dream Content and Dream Emotions. *J. Sleep Res.* **2009**, *18*, 285–290. [CrossRef]
14. Burton, S.A.; Harsh, J.R.; Badia, P. Cognitive Activity in Sleep and Responsiveness to External Stimuli. *Sleep* **1988**, *11*, 61–68. [CrossRef] [PubMed]
15. Berger, R.J. Experimental Modification of Dream Content by Meaningful Verbal Stimuli. *Br. J. Psychiatry* **1963**, *109*, 722–740. [CrossRef] [PubMed]
16. Hoelscher, T.J.; Klinger, E.; Barta, S.G. Incorporation of Concern- and Nonconcern-Related Verbal Stimuli into Dream Content. *J. Abnorm. Psychol.* **1981**, *90*, 88–91. [CrossRef]
17. Noreika, V.; Windt, J.M.; Kern, M.; Valli, K.; Salonen, T.; Parkkola, R.; Revonsuo, A.; Karim, A.A.; Ball, T.; Lenggenhager, B. Modulating Dream Experience: Noninvasive Brain Stimulation over the Sensorimotor Cortex Reduces Dream Movement. *Sci. Rep.* **2020**, *10*, 6735. [CrossRef] [PubMed]
18. Baekeland, F. Laboratory Studies of Effects of Presleep Events on Sleep and Dreams. *Int. Psychiatry Clin.* **1970**, *7*, 49–58.
19. Breger, L.; Hunter, I.; Lane, R.W. The Effect of Stress on Dreams. *Psychol. Issues* **1971**, *7*, 213.
20. Baekeland, F. Presleep Mentation and Dream Reports: I. Cognitive Style, Contiguity to Sleep, and Time of Night. *Arch. Gen. Psychiatry* **1968**, *19*, 300. [CrossRef]
21. Tart, C.T. Some Effects of Posthypnotic Suggestion on the Process of Dreaming. *Int. J. Clin. Exp. Hypn.* **1966**, *14*, 30–46. [CrossRef]
22. Hauri, P. Evening Activity, Sleep Mentation, and Subjective Sleep Quality. *J. Abnorm. Psychol.* **1970**, *76*, 270–275. [CrossRef]
23. Koulack, D.; Prevost, F.; De Koninck, J. Sleep, Dreaming, and Adaptation to a Stressful Intellectual Activity. *Sleep* **1985**, *8*, 244–253. [CrossRef]
24. Bokert, E.G. The Effects of Thirst and a Related Verbal Stimulus on Dream Reports. *Diss. Abstr. Int.* **1968**, *28*, 4753.
25. Wood, P.B. *Dreaming and Social Isolation*; The University of North Carolina at Chapel Hill: Chapel Hill, NC, USA, 1962.

26. Tuominen, J.; Olkoniemi, H.; Revonsuo, A.; Valli, K. 'No Man Is an Island': Effects of Social Seclusion on Social Dream Content and REM Sleep. *Br. J. Psychol.* **2022**, *113*, 84–104. [CrossRef] [PubMed]

27. Corsi-Cabrera, M.; Becker, J.; García, L.; Ibarra, R.; Morales, M.; Souza, M. Dream Content after Using Visual Inverting Prisms. *Percept. Mot. Skills* **1986**, *63*, 415–423. [CrossRef] [PubMed]

28. Foulkes, D.; Rechtschaffen, A. Presleep Determination of Dream Content: Effects of Two Films. *Percept. Mot. Skills* **1964**, *19*, 983–1005. [CrossRef] [PubMed]

29. Statistics Canada. Canadian Internet Use Survey. Available online: https://www150.statcan.gc.ca/n1/daily-quotidien/210622/dq210622b-eng.htm (accessed on 1 October 2021).

30. Gackenbach, J.I.; Kuruvilla, B. Video Game Play Effects on Dreams: Self-Evaluation and Content Analysis. *Eludamos J. Comput. Game Cult.* **2008**, *2*, 169–186. [CrossRef] [PubMed]

31. Domhoff, G.W. *The Scientific Study of Dreams: Neural Networks, Cognitive Development, and Content Analysis*; American Psychological Association: Washington, DC, USA, 2003; ISBN 978-1-55798-935-2.

32. Domhoff, G.W. The Invasion of the Concept Snatchers: The Origins, Distortions, and Future of the Continuity Hypothesis. *Dreaming* **2017**, *27*, 14–39. [CrossRef]

33. Schredl, M. Continuity between Waking and Dreaming: A Proposal for a Mathematical Model. *Sleep Hypn.* **2003**, *5*, 38–52.

34. Stephan, J.; Schredl, M.; Henley-Einion, J.; Blagrove, M. TV Viewing and Dreaming in Children: The UK Library Study. *Int. J. Dream Res.* **2012**, *5*, 130–133. [CrossRef]

35. Moverley, M.; Schredl, M.; Göritz, A.S. Media Dreaming and Media Consumption—An Online Study. *Int. J. Dream Res.* **2018**, *11*, 127–134. [CrossRef]

36. Viemerö, V.; Paajanen, S. The Role of Fantasies and Dreams in the TV Viewing-Aggression Relationship. *Aggress. Behav.* **1992**, *18*, 109–116. [CrossRef]

37. Hoedlmoser, K.; Kloesch, G.; Wiater, A.; Schabus, M. Self-Reported Sleep Patterns, Sleep Problems, and Behavioral Problems among School Children Aged 8–11 Years. *Somnology Sleep Res. Sleep Med.* **2010**, *14*, 23–31. [CrossRef] [PubMed]

38. Van den Bulck, J. Media Use and Dreaming: The Relationship Among Television Viewing, Computer Game Play, and Nightmares or Pleasant Dreams. *Dreaming* **2004**, *14*, 43–49. [CrossRef]

39. Van Den Bulck, J.; Çetin, Y.; Terzi, Ö.; Bushman, B.J. Violence, Sex, and Dreams: Violent and Sexual Media Content Infiltrate Our Dreams at Night. *Dreaming* **2016**, *26*, 271–279. [CrossRef]

40. Sestir, M.; Tai, M.; Peszka, J. Relationships between Video Game Play Factors and Frequency of Lucid and Control Dreaming Experiences. *Dreaming* **2019**, *29*, 127–143. [CrossRef]

41. Gackenbach, J.; Kuruvilla, B. The Relationship between Video Game Play and Threat Simulation Dreams. *Dreaming* **2008**, *18*, 236–256. [CrossRef]

42. Gackenbach, J. Electronic Media and Lucid-Control Dreams: Morning after Reports. *Dreaming* **2009**, *19*, 1–6. [CrossRef]

43. Schredl, M.; Göritz, A.S. Social Media, Dreaming, and Personality: An Online Study. *Cyberpsychol. Behav. Soc. Netw.* **2019**, *22*, 657–661. [CrossRef]

44. Lambrecht, S.; Schredl, M.; Henley-Einion, J.; Blagrove, M. Self-Rated Effects of Reading, TV Viewing and Daily Activities on Dreaming in Adolescents and Adults: The UK Library Study. *Int. J. Dream Res.* **2013**, *6*, 41–44.

45. Stickgold, R.; Malia, A.; Maguire, D.; Roddenberry, D.; O'Connor, M. Replaying the Game: Hypnagogic Images in Normals and Amnesics. *Science* **2000**, *290*, 350–353. [CrossRef]

46. Wamsley, E.J.; Perry, K.; Djonlagic, I.; Reaven, L.B.; Stickgold, R. Cognitive Replay of Visuomotor Learning at Sleep Onset: Temporal Dynamics and Relationship to Task Performance. *Sleep J. Sleep Sleep Disord. Res.* **2010**, *33*, 59–68. [CrossRef]

47. Kussé, C.; Bourdiec, A.S.-L.; Schrouff, J.; Matarazzo, L.; Maquet, P. Experience-dependent Induction of Hypnagogic Images during Daytime Naps: A Combined Behavioural and EEG Study. *J. Sleep Res.* **2012**, *21*, 10–20. [CrossRef] [PubMed]

48. Nielsen, T.A. A Review of Mentation in REM and NREM Sleep: "Covert" REM Sleep as a Possible Reconciliation of Two Opposing Models. *Behav. Brain Sci.* **2000**, *23*, 851–866. [CrossRef]

49. Arksey, H.; O'Malley, L. Scoping Studies: Towards a Methodological Framework. *Int. J. Soc. Res. Methodol.* **2005**, *8*, 19–32. [CrossRef]

50. Peters, M.D.; Godfrey, C.; McInerney, P.; Munn, Z.; Tricco, A.C.; Khalil, H. Chapter 11: Scoping Reviews. In *JBI Manual for Evidence Synthesis*; JBI, 2020; ISBN 978-0-648-84880-6. Available online: https://edisciplinas.usp.br/pluginfile.php/7315963/mod_resource/content/1/manual_capitulo_revisao_escopo_JBIMES_2021April.pdf (accessed on 1 October 2021).

51. Tricco, A.C.; Lillie, E.; Zarin, W.; O'Brien, K.K.; Colquhoun, H.; Levac, D.; Moher, D.; Peters, M.D.J.; Horsley, T.; Weeks, L.; et al. PRISMA Extension for Scoping Reviews (PRISMA-ScR): Checklist and Explanation. *Ann. Intern. Med.* **2018**, *169*, 467–473. [CrossRef] [PubMed]

52. Witkin, H.A.; Lewis, H.B. The Relation of Experimentally Induced Presleep Experiences to Dreams: A Report on Method and Preliminary Findings. *J. Am. Psychoanal. Assoc.* **1965**, *13*, 819–849. [CrossRef] [PubMed]

53. Foulkes, D.; Pivik, T.; Steadman, H.S.; Spear, P.S.; Symonds, J.D. Dreams of the Male Child: An EEG Study. *J. Abnorm. Psychol.* **1967**, *72*, 457–467. [CrossRef] [PubMed]

54. Cartwright, R.D.; Bernick, N.; Borowitz, G. Effect of an Erotic Movie on the Sleep and Dreams of Young Men. *Arch. Gen. Psychiatry* **1969**, *20*, 262–271. [CrossRef]

55. Foulkes, D.; Belvedere, E.; Brubaker, T. Televised Violence and Dream Content. *Telev. Soc. Behav.* **1971**, *5*, 59–119.

56. Goodenough, D.R.; Witkin, H.A.; Koulack, D.; Cohen, H. The Effects of Stress Films on Dream Affect and on Respiration and Eye-Movement Activity during Rapid-Eye-Movement Sleep. *Psychophysiology* **1975**, *12*, 313–320. [CrossRef]

57. de Koninck, J.-M.; Koulack, D. Dream Content and Adaptation to a Stressful Situation. *J. Abnorm. Psychol.* **1975**, *84*, 250–260. [CrossRef]

58. Lauer, C.; Riemann, D.; Lund, R.; Berger, M. Shortened REM Latency: A Consequence of Psychological Strain? *Psychophysiology* **1987**, *24*, 263–271. [CrossRef] [PubMed]

59. Kuiken, D.; Rindlisbacher, P.; Nielsen, T. Feeling Expression and the Incorporation of Presleep Events into Dreams. *Imagin. Cogn. Personal.* **1990**, *10*, 157–166. [CrossRef]

60. Wamsley, E.J.; Tucker, M.; Payne, J.D.; Benavides, J.A.; Stickgold, R. Dreaming of a Learning Task Is Associated with Enhanced Sleep-Dependent Memory Consolidation. *Curr. Biol.* **2010**, *20*, 850–855. [CrossRef] [PubMed]

61. Stamm, A.W.; Nguyen, N.D.; Seicol, B.J.; Fagan, A.; Oh, A.; Drumm, M.; Lundt, M.; Stickgold, R.; Wamsley, E.J. Negative Reinforcement Impairs Overnight Memory Consolidation. *Learn. Mem.* **2014**, *21*, 591–596. [CrossRef]

62. Nefjodov, I.; Winkler, A.; Erlacher, D. Balancing in Dreams: Effects of Playing Games on the Wii Balance Board on Dream Content. *Int. J. Dream Res.* **2016**, *9*, 89–92. [CrossRef]

63. Wamsley, E.J.; Hamilton, K.; Graveline, Y.; Manceor, S.; Parr, E. Test Expectation Enhances Memory Consolidation across Both Sleep and Wake. *PLoS ONE* **2016**, *11*, e0165141. [CrossRef]

64. Fogel, S.M.; Ray, L.B.; Sergeeva, V.; De Koninck, J.; Owen, A.M. A Novel Approach to Dream Content Analysis Reveals Links Between Learning-Related Dream Incorporation and Cognitive Abilities. *Front. Psychol.* **2018**, *9*, 1398. [CrossRef]

65. Solomonova, E.; Dubé, S.; Samson-Richer, A.; Blanchette-Carrière, C.; Paquette, T.; Nielsen, T. Dream Content and Procedural Learning in Vipassana Meditators and Controls. *Dreaming* **2018**, *28*, 99–121. [CrossRef]

66. Wamsley, E.J.; Stickgold, R. Dreaming of a Learning Task Is Associated with Enhanced Memory Consolidation: Replication in an Overnight Sleep Study. *J. Sleep Res.* **2019**, *28*, e12749. [CrossRef]

67. Klepel, F.; Schredl, M. Correlation of Task-Related Dream Content with Memory Performance of a Film Task—A Pilot Study. *Int. J. Dream Res.* **2019**, *12*, 112–118.

68. Powell, R.A.; Nielsen, T.A.; Cheung, J.S.; Cervenka, T.M. Temporal Delays in Incorporation of Events into Dreams. *Percept. Mot. Skills* **1995**, *81*, 95–104. [CrossRef] [PubMed]

69. Davidson, J.; Hart, K.; Haines, J. Contextualising Imagery in Dreams Following a September 11 Video from Television News. *Aust. Psychol.* **2005**, *40*, 202–206. [CrossRef]

70. Gackenbach, J.; Rosie, M.; Bown, J.; Sample, T. Dream Incorporation of Video-Game Play as a Function of Interactivity and Fidelity. *Dreaming* **2011**, *21*, 32–50. [CrossRef]

71. Davidson, J.; Lynch, S. Thematic, Literal and Associative Dream Imagery Following a High-Impact Event. *Dreaming* **2012**, *22*, 58–69. [CrossRef]

72. Solomonova, E.; Stenstrom, P.; Paquette, T.; Nielsen, T. Different Temporal Patterns of Memory Incorporations into Dreams for Laboratory and Virtual Reality Experiences: Relation to Dreamed Locus of Control. *Int. J. Dream Res.* **2015**, *8*, 10–26. [CrossRef]

73. Flockhart, C.; Gackenbach, J. The Nightmare Protection Hypothesis: An Experimental Inquiry. *Int. J. Dream Res.* **2017**, *10*, 1–9. [CrossRef]

74. Gott, J.; Bovy, L.; Peters, E.; Tzioridou, S.; Meo, S.; Demirel, Ç.; Esfahani, M.J.; Oliveira, P.R.; Houweling, T.; Orticoni, A.; et al. Virtual Reality Training of Lucid Dreaming. *Philos. Trans. R. Soc. Lond. B Biol. Sci.* **2021**, *376*, 20190697. [CrossRef]

75. Ribeiro, N.; Gounden, Y.; Quaglino, V. Enhancement of Spatial Memories at the Associative and Relational Levels after a Full Night of Sleep and Likelihood of Dream Incorporation. *Int. J. Dream Res.* **2021**, *14*, 67–79. [CrossRef]

76. Picard-Deland, C.; Pastor, M.; Solomonova, E.; Paquette, T.; Nielsen, T. Flying Dreams Stimulated by an Immersive Virtual Reality Task. *Conscious. Cogn. Int. J.* **2020**, *83*, 102958. [CrossRef]

77. Hartmann, E. Nightmare after Trauma as Paradigm for All Dreams: A New Approach to the Nature and Functions of Dreaming. *Psychiatry* **1998**, *61*, 223–238. [CrossRef]

78. Revonsuo, A. The Reinterpretation of Dreams: An Evolutionary Hypothesis of the Function of Dreaming. *Behav. Brain Sci.* **2000**, *23*, 877–901. [CrossRef]

79. Hudachek, L.; Wamsley, E.J. A Meta-Analysis of the Relation between Dream Content and Memory Consolidation. *Sleep* **2023**, *46*, zsad111. [CrossRef] [PubMed]

80. Hall, C.S.; Van de Castle, R.L. *The Content Analysis of Dreams*; Appleton-Century-Crofts: New York, NY, USA, 1966.

81. Gottschalk, L.A.; Gleser, G.C.; Springer, K.J. Three Hostility Scales Applicable to Verbal Samples. *Arch. Gen. Psychiatry* **1963**, *9*, 254–279. [CrossRef] [PubMed]

82. Hartmann, E.; Kunzendorf, R.; Rosen, R.; Grace, N.G. Contextualizing Images in Dreams and Daydreams. *Dreaming* **2001**, *11*, 97–104. [CrossRef]

83. Schredl, M. Dream Content Analysis: Basic Principles. *Int. J. Dream Res.* **2010**, *3*, 65–73. [CrossRef]

84. Revonsuo, A.; Valli, K. Dreaming and Consciousness: Testing the Threat Simulation Theory of the Function of Dreaming. *Psyche Interdiscip. J. Res. Conscious.* **2000**, *6*, 8.

85. Nielsen, T.; Saucier, S.; Stenstrom, P.; Lara-Carrasco, J.; Solomonova, E. Interactivity in a Virtual Maze Task Enhances Delayed Incorporations of Maze Features into Dream Content. *Sleep* **2007**, *30*, A376.

86. Schredl, M.; Hofmann, F. Continuity between Waking Activities and Dream Activities. *Conscious. Cogn.* **2003**, *12*, 298–308. [CrossRef]
87. Malinowski, J.; Horton, C.L. Evidence for the Preferential Incorporation of Emotional Waking-Life Experiences into Dreams. *Dreaming* **2014**, *24*, 18–31. [CrossRef]
88. Eichenlaub, J.-B.; Van Rijn, E.; Gaskell, M.G.; Lewis, P.A.; Maby, E.; Malinowski, J.E.; Walker, M.P.; Boy, F.; Blagrove, M. Incorporation of Recent Waking-Life Experiences in Dreams Correlates with Frontal Theta Activity in REM Sleep. *Soc. Cogn. Affect. Neurosci.* **2018**, *13*, 637–647. [CrossRef]
89. Schredl, M. Laboratory References in Dreams: Methodological Problem and/or Evidence for the Continuity Hypothesis of Dreaming? *Int. J. Dream Res.* **2008**, *1*, 3–6. [CrossRef]
90. Picard-Deland, C.; Nielsen, T.; Carr, M. Dreaming of the Sleep Lab. *PLoS ONE* **2021**, *16*, e0257738. [CrossRef] [PubMed]
91. Sikka, P.; Feilhauer, D.; Valli, K.; Revonsuo, A. How You Measure Is What You Get: Differences in Self- and External Ratings of Emotional Experiences in Home Dreams. *Am. J. Psychol.* **2017**, *130*, 367–384. [CrossRef]